RESTRICTION ENZYMES

ENZYMES

A History

RESTRICTION ENZYMES

A History

Wil A. M. Loenen

Leiden University Medical Center

www.restrictionenzymes.org

COLD SPRING HARBOR LABORATORY PRESS
Cold Spring Harbor, New York • www.cshlpress.org

Restriction Enzymes: A History

© 2019 by Cold Spring Harbor Laboratory Press, Cold Spring Harbor, New York
Printed in the United States of America
doi:10.1101/restrictionenzymes

Publisher and Acquisition Editor	John Inglis
Director of Editorial Development	Jan Argentine
Project Manager	Inez Sialiano
Permissions Coordinator	Carol Brown
Director of Publication Services	Linda Sussman
Production Editor	Kathleen Bubbeo
Production Manager	Denise Weiss
Cover Designer	Mike Albano

Front cover artwork: An example of an early REase catalog in the United Kingdom (1980), redrawn from Figure 1 in Chapter 5. Courtesy of Bayer.

Library of Congress Cataloging-in-Publication Data

Names: Loenen, Wil A. M., author.

Title: Restriction enzymes : a history / Wil A.M. Loenen, Leiden University Medical Center.
Description: Cold Spring Harbor, New York : Cold Spring Harbor Laboratory Press, [2019] | Includes bibliographical references and index.
Identifiers: LCCN 2018048589 (print) | LCCN 2018051960 (ebook) | ISBN 9781621822721 (ePub3) | ISBN 9781621822738 (Kindle-Mobi) | ISBN 9781621821052 (printed hardcover)
Subjects: LCSH: Restriction enzymes, DNA. | Gene amplification.
Classification: LCC QP609.R44 (ebook) | LCC QP609.R44 L64 2019 (print) | DDC 572/.76–dc23
LC record available at https://lccn.loc.gov/2018048589.

10 9 8 7 6 5 4 3 2 1

To Noreen Murray,
who passed away in 2011,
to the sadness of many

Contents

Preface

A HISTORIC MEETING ON RESTRICTION ENZYMES at Cold Spring Harbor Laboratory (CSHL) in 2013 was the impetus for this book. But who could have predicted back in 1953—when the first review on restriction enzymes appeared (Luria 1953) and the structure of DNA was published (Watson and Crick 1953)—that this baby girl would be born (Leiden, The Netherlands) and grow up to become a science writer? Imagine my delight to have been asked by Rich Roberts, who founded the REBASE website, to chronicle the history of these enzymes that revolutionized molecular science.

In many respects, 2018 is a memorable year: Sixty years ago Matt Meselson and Frank Stahl provided the first experimental proof for the semiconservative nature of DNA replication (Meselson and Stahl 1958); 50 years ago Matt Meselson and Bob Yuan published their classic paper on EcoKI (Meselson and Yuan 1968), the restriction enzyme of the workhorse of molecular biology, *E. coli* K12; and 40 years ago Werner Arber, Ham Smith, and Dan Nathans were awarded the Nobel Prize in Physiology or Medicine for their discovery of restriction enzymes. That same year, I went to Leicester for a PhD and succeeded, thanks to the efforts of Bill Brammar; and 30 years ago I started my investigation into the role of CD27 in lymphocyte development, making good use of my extensive knowledge of DNA cloning. Lack of funding, however, led me to return to the restriction enzyme field, which resulted in the Survey and Summary in *Nucleic Acids Research* that celebrated 50 years of research on EcoKI in 2003; this review was made possible with the extensive help of Noreen Murray and David Dryden.

This book is divided into chapters covering periods of roughly a decade, using the 2013 meeting and early reviews (about 60 years apart) as ending and starting points, the comprehensive *Restriction Endonucleases* (edited by Alfred Pingoud in 2004), and more recent reviews by experts who studied or study different aspects of restriction enzymes, including genetics, DNA cloning, biochemistry, biophysics, microscopy, X-ray crystallography, and nanotechnology.

The 2013 meeting that started this book was sponsored by Life Technologies (http://www.lifetechnologies.com/); New England Biolabs (http://www.neb.com/); ThermoFisher Scientific (http://www.thermofisher.com/); Promega (http://worldwide.promega.com/); Genentech (http://www.gene.com/); TAKARA/Clontech (http://www.clontech.com/); Nippon (http://nippongene.com/); and Molecular Biology Resources (http://molbiores.com/). This book was funded by the Genentech Center for the History of Molecular Biology and Biotechnology (GCHMBB).

I thank all the experts in the field for their willingness and generous help with letters, PDFs of papers, and figures (acknowledged in the legends), which made this book possible. A special thank you to the reading committee (Tom Bickle, Steve Halford, Stu Linn, and Rich Roberts), as well as Joe Bertani (who sadly passed away while preparing his comments on Chapter 2 of this book), and Werner Arber and Bill Brammar, who both commented on the chapters as they came along. Stu not only carefully read the contents, but he also helped with the difficult punctuation in English, which is so different from that in Dutch. I thank André Dussoix and Sandra Citi for information on Daisy Dussoix, one of the pioneers in the field, and Hiroshi Nikaido for the translation of the Japanese text about Tsutomu Watanabe, who discovered resistance transfer factors. Once the first complete draft of the book was ready nearly a year ago, Aneel Aggarwal, Herb Boyer, David Dryden, Steve Halford, Ken Horiuchi, Clyde Hutchison, Bill Linton, Matt Meselson, Ham Smith, Lise Raleigh, Giedrius Sasnauskas, Virgis Siksnys, Gintas Tamulaitis, Giedre Tamulaitiene, and Geoff Wilson were asked to comment on one, several, or all chapters. This greatly improved the final version—thank you all.

Many other people made this book possible: my parents, who helped me on the road to science, my brothers and sisters, and my colleagues and friends inside and outside of university. At Leiden University Medical Center, I am grateful to Micheline Giphart-Gassler and Leon Mullenders for hospitality at ToxGen, and Eduard Klasen, Ruud Kukenheim, and Jacqueline Ton for arranging current guest facilities at the Department of Directorate Research. Finally, at CSHL, I thank Mila Pollock, Executive Director, Library & Archives; the editorial and production staff of CSHL Press—John Inglis, Jan Argentine, Inez Sialiano, Denise Weiss, and Kathleen Bubbeo—for their help and patience in getting this book published; and Wayne Manos and Ted Roeder at CSHL Press for building the accompanying website.

Wil A. M. Loenen
November 29, 2018
Leiderdorp, The Netherlands
wloenen@xs4all.nl

References

Luria SE. 1953. Host-induced modifications of viruses. *Cold Spring Harb Symp Quant Biol* **18:** 237–244.

Meselson M, Stahl FW. 1958. The replication of DNA in *Escherichia coli. Proc Natl Acad Sci* **44:** 671–682.

Meselson M, Yuan R. 1968. DNA restriction enzyme from *E. coli. Nature* **217:** 1110–1114.

Pingoud A, ed. 2004. *Restriction endonucleases.* Springer, Berlin.

Watson JD, Crick FH. 1953. Molecular structure of nucleic acids: a structure for deoxyribose nucleic acid. *Nature* **171:** 737–738.

Abbreviations

aa   amino acid

In the context of protein:

A	Ala; alanine
C	Cys; cysteine
D	Asp; aspartic acid (aspartate)
E	Glu; glutamic acid (glutamate)
F	Phe; phenylalanine
G	Gly; glycine
H	His; histidine
I	Ile; isoleucine
K	Lys; lysine
L	Leu; leucine
M	Met; methionine
N	Asn; asparagine
P	Pro; proline
Q	Gln; glutamine
R	Arg; arginine
S	Ser; serine
T	Thr; threonine
V	Val; valine
W	Trp; tryptophan
X	any amino acid
Y	Tyr; tyrosine

In the context of DNA or RNA:

A	adenine
G	guanine
C	cytosine
T	thymine
U	uracil
Y	C or T (pyrimidine)
R	A or G (purine)
S	C or G (strong H-bonds)
W	A or T (weak H-bonds)

M	A or C (commonly modified bases)
K	G or T (not commonly modified)
H	A, C, or T (not G)
B	C, G, or T (not A)
V	A, C, or G (not T)
D	A, G, or T (not C)
N	A, C, G, or T (any base)
m	methyl
m5C	5-methylcytosine
m4C	*N*4-metyhylcytosine, cytosine with methyl group at 4th position
m6A	*N*6-methyladenine, adenine with methyl group at 6th position
hm5C	5-hydroxymethylcytosine, cytosine with methyl group at 5th position
hm5U	5-hydroxymethyluracil
ghm5C	glycosylated hm5C

Other:

SAM	*S*-adenosylmethionine (or AdoMet, but this name ignores the important S [sulfur]!), the universal methyl donor
PCR	polymerase chain reaction
PAGE	polyacrylamide gel electrophoresis
SMRT®	single-molecule, real-time (sequencing)
DNA	deoxyribonucleic acid
RNA	ribonucleic acid
oligo	oligonucleotide
bp	base pair(s)
nt	nucleotide(s)
kb	kilobase(s)
ds	double-stranded
ss	single-stranded
kDa	kilodalton(s)
MW	molecular weight

In the context of genetics and R-M systems:

E. coli	*Escherichia coli* bacteria; in recombinant DNA technology usually strain K12 (K-12)
EcoK	"K," old names for R-M system of *Escherichia coli* K12; currently called EcoKI
EcoB	"B," old names for R-M system of *Escherichia coli* B, currently called EcoBI

"P1"	old name for R-M system of phage P1; currently called EcoP1I
"P15"	old name for R-M system of P15 plasmid; currently called EcoP15I
HindII	*Haemophilus influenza* (or *influenzae*) strain D; initial isolate EndoR has two R-M systems
Hae III	*Haemophilus aegypticus,* initial isolate EndoZ has two R-M systems
Rgl	restricts glucose-less DNA; renamed Mcr (methylcytosine restriction)
Mrr	modified DNA rejection and restriction
Mcr	methylcytosine restriction
phage	bacterial virus or bacteriophage
e.o.p.	efficiency of plating
F	fertility factor
HCV	host-controlled variation
ts	temperature-sensitive
res	restriction gene of phage P1
mod	DNA recognition and modification gene of phage P1
pnk	T4 polynucleotide kinase
RTF	resistance transfer factor (plasmid or phage with antibiotic resistance gene[s])
R	restriction
M	modification
RM	combined restriction-modification protein
R-M	restriction-modification system with separate restriction and modification proteins
S	specificity subunit of Type I restriction enzymes
REase	restriction endonuclease or restriction enzyme
MTase	modification enzyme; prokaryotic DNA-methyltransferase
C	control protein of Type II system
CD	catalytic domain
DBD	DNA-binding domain
MBD	methyl-binding domain
TRD	target recognition domain
TET	ten-eleven translocase translocation
HR	homologous recombination
NHEJ	nonhomologous end joining
DSB	double-strand break
hsd S, M, R	host specificity determinant S (specificity), M (modification), R (restriction) of Type I system
ZF	zinc-finger protein

ZFN   engineered zinc-finger nuclease
TALE   transcription activator-like effector protein (from *Xanthomonas*)
TALEN   engineered TALE nuclease
CRISPR   clustered, regularly interspaced, short palindromic repeats
Cas9   CRISPR-associated nuclease (from *Streptococcus pyogenes*)

Introduction

This book describes the history of the development of the restriction enzymes and covers the major advances in the field over a period of more than 60 years. These bacterial endonucleases bind the DNA helix at specific base sequences and cut the DNA backbone. The bacteria that harbor such enzymes protect their own DNA against restriction via modification at the same recognition sites by their (usually adjacent) methyltransferase. A historical meeting at Cold Spring Harbor Laboratory in 2013 (see Appendix A: The History of Restriction Enzymes October 19–21, 2013 Program) brought together people from the origins of the field, as well as other scientists working on restriction enzymes, with this book as a result. The structure of the book in eight chapters is based on talks at this meeting, major reviews and research articles, and, in the first chapters, some useful older books listed in the References for further reading. Taken together, the book relates the history (from 1952 to early 2017) of this amazing, very large endonuclease family with four types (Types I, II, III, and IV).

The story starts in the early 1950s, an important period in the history of molecular science. The year of publication of the famous structure of the DNA double helix is 1953, but, less known, it is also the year of the first review that summarizes genetic experiments on "host-controlled variation." These experiments with bacterial viruses ("phages") in different bacterial strains were in fact providing evidence for the first Type I, III, and IV restriction enzyme systems and led to the slow but inevitable conclusion that the distinct picture of genotype and phenotype that had been built up with care during the first half of the twentieth century was breaking down (Chapter 1). During the 1960s, Werner Arber and coworkers showed host-controlled variation to be restriction and modification by methylation at the DNA level (Chapter 2). For this achievement, Arber was awarded the Nobel Prize in Physiology or Medicine in 1978, together with Hamilton O. Smith and Daniel Nathans, who discovered the Type II enzymes.

It should be noted that research into the phenomenon of restriction and modification relied heavily on major developments during the first half of the twentieth century, both technical progress (e.g., X-ray crystallography, ultracentrifuge, electron microscope, fractionation techniques) and in the field of microbial genetics (e.g., the discovery of both lytic and lysogenic bacterial viruses, conjugation, transduction, recombination, and use of isotopes for

Chapter doi:10.1101/restrictionenzymes_intro

labeling). The integration of genetic studies with biochemical and physical analyses of the synthesis, structure, and function of restriction enzymes led to an explosion of fundamental knowledge and details on their structure and function. Throughout the years improvements in these fields would go hand in hand with the progress in the restriction field, ranging, for example, from cloning vectors for use in bacteria, yeast, and other organisms ("shuttle vectors") and improved bacterial hosts for the analysis of DNA from all kingdoms to agarose gels, DNA sequencing, Southern blots, polymerase chain reaction, increased computer power, atomic force microscopy, single-molecule studies, and, in recent years, whole-genome sequencing and methylome analysis.

The discovery in 1970 of Type II restriction enzymes (such as EcoRI; Chapter 3) that produced "sticky" DNA ends that could be resealed resulted in worldwide interest in these enzymes (Chapter 4). It soon led to reagents for recombinant DNA, mapping and isolation of genes, DNA sequencing, and analysis of repeat sequences in DNA. In addition, some scientists saw these enzymes as a marvelous opportunity to study DNA–protein interactions. A decade later the first DNA sequences of genes encoding restriction enzymes appeared while research into the biochemistry and genetics of these systems continued (Chapter 5). The studies in the following years indicated a great variety in mechanisms and structures of an increasingly large number of restriction enzymes and led to attempts to build evolutionary trees (Chapter 6). In 2003 a subdivision of the Type II enzymes into 11 subtypes was proposed, and a year later a book was published totally dedicated to restriction enzymes (Chapter 7). It illustrated the progress made with many Type II enzymes in different species and the crystal structures that led to the definition of a common catalytic cleavage core with the PD...(D/E)XK motif, although some enzymes have other (HNH, GIY-YIG, or PLD) structural domains. During these years research on the Type I and III enzymes was basically limited to a few enzymes in *Escherichia coli*. But all this changed with the advent of whole-genome sequencing and methylome analysis. These and other improved detection methods, together with single-molecule studies, are providing novel insights into the structures, functions, and applications of Type I, II, and III enzymes, but also those of the modification-dependent Type IV enzymes (Chapter 8). The discovery of many restriction systems in pathogenic bacteria is starting to shed light on the way these organisms evade host immunity and persist in the host with or without periodic disease outbreaks. Also, restriction-modification systems are useful for other medical purposes: for example, for typing strains of pathogenic bacteria such as "nontypeable *Haemophilus influenzae* (NTHi)" and for following changes in the gut microbiome upon changes in diet or disease treatment regimes. Last but not least, another highlight in the recent past has been the elucidation of the long-awaited structures of the first Type I and III enzyme complexes, which shows that the Type I, II, and III enzymes may have more in common than initially surmised.

The emphasis of this book is on restriction enzymes, but occasionally the methyltransferases are mentioned: For example, in 1993 the first example of base flipping (by M·HhaI) became known, which proved to be limited not only to methyltransferases, as restriction enzymes, other endonucleases, and RNA enzymes can "do it." Methyltransferases cannot be avoided in the case of the Type I and III and several Type II subtypes, which are combined restriction-modification complexes or even single-chain polypeptides. The methyltransferases have been important in the identification of restriction enzymes in whole-genome sequencing projects. Although the restriction enzymes are highly diverse and not easy to identify by DNA sequence alone, they are usually located next to their accompanying ("cognate") methyltransferase. The latter show circular permutations (i.e., catalytic domains are put together with amino acids that may be adjacent in the coding region but also may be far apart), but the subdomain regions are recognizable. Hence many restriction enzymes have been (putatively) identified by searching for methyltransferase genes in whole genomes and by subsequent analysis of the flanking DNA to find accompanying restriction genes. As the modification-dependent Type IV enzymes are highly diverse and not accompanied by a methyltransferase, these enzymes are difficult to identify, but the importance of these enzymes for epigenetic studies will be clear.

Taken together, this book shows that the restriction enzymes are excellent tools to study the interactions of all sorts of proteins with DNA: How can we search the DNA for a DNA recognition sequence, whether a target site for nuclease activity, modification, or repair? How does the enzyme (complex) find that site via hopping, sliding, jumping, and/or looping? How do the enzymes bind and form a specific complex involving conformational changes in the protein and/or alterations in the DNA structure in order to position the catalytic domain in the right conformation to do its job, whether at that site or away from it? This latter process may involve more movement along the DNA and/or translocation in the case of enzymes with "molecular motor" subunits belonging to the SF2 superfamily. Many eukaryotic enzymes that are involved in, for example, DNA repair also belong to this family.

CHAPTER 1

Discovery of a Barrier to Infection and Host-Controlled Variation: 1952–1953

While I was focusing on P2 and the mechanism of lysogeny, some unexpected findings came up which deserved proper attention. One was the discovery of "host controlled variation," now more commonly called "restriction and modification," a phenomenon of great theoretical interest. I noticed it in P2 (using strain B as the restricting host, Shigella being the standard host) and did not know what to make of it. Jean Weigle noticed it in lambda (using strain C as the permissive host, K-12 being the standard host): being aware of my results, he immediately recognized the parallelism of the two "systems."

Thus Giuseppe (Joe) Bertani recalls events leading up to his joint publication with Jean Weigle of "Host controlled variation in bacterial viruses" (Bertani and Weigle 1953) in a letter to Noreen Murray (July 17, 2003).[1]

Joe Bertani obtained his doctor's degree in zoology at the University of Milan soon after World War II. Via Zürich, Naples, and Cold Spring Harbor, he started working on lysogeny in Bloomington in 1951 with Salvador Luria, a man of "brilliance, integrity, breadth of culture, and wicked sense of humor," according to Evelyn Witkin. Here Joe Bertani shared a bench for a while with James Watson, of later double helix fame.

Lysogeny is the ability of some viruses to be carried in a dormant "prophage" state in their bacterial host chromosome. The discovery by Esther Lederberg of prophage lambda in *Escherichia coli* K12, and that of P1 and P2 by Joe Bertani, opened research into the different aspects of genetic exchange in phage and host bacteria in the same genetic background. These phages would prove useful tools in molecular biology: P1 encodes its own restriction-modification (R-M) system, EcoP1I, later to be classified as Type III; it can package and transfer foreign DNA allowing genetic exchange with a new host (generalized transduction). This led to the development of the highly useful LoxP-Cre recombination system, a topic outside the scope of this book (see,

---

[1]See Appendix 1 (letter) and Appendix 2 (Joe Bertani's obituary).

Chapter doi:10.1101/restrictionenzymes_1

FIGURE 1. Joe Bertani and two other pioneers of phage and bacterial genetics (2001); from *left* to *right*: Abe Eisenstark, Joe Bertani, and Wacław Szybalski (taken at the meeting in Madison, Wisconsin). (Reprinted from Young 2002, with permission from the American Society for Microbiology.)

e.g., Yu and Bradley 2001). Phage P2 would prove useful in cloning schemes in the 1970s and 1980s because of the "spi⁻ phenotype" (susceptibility to P2 inhibition): P2 lysogens exclude growth of wild-type lambda (see, e.g., Hershey 1971, p. 146), allowing selection for recombinant phage in the generation of libraries in lambda gene bank vectors. In 2001, Joe Bertani was honored at the Molecular Genetics of Bacteria and Phages meeting (Fig. 1) that marked the 50th anniversary of the discovery of the three classic phages: lambda, P1, and P2 (Young 2002). The importance of lambda needs no further explanation (Hershey 1971).

Jean Weigle started his career at the University of Geneva in physics. After a heart attack he quit his professorship and joined Max Delbrück at the California Institute of Technology (Caltech), working on transduction and recombination until his death. He continued to spend summers at the Kellenberger laboratory in Geneva, which led Werner Arber to a postdoctoral year with Joe Bertani. Thus, Werner Arber recognized host-controlled variation (HCV) in his own experiments 7 years later. Renamed restriction and modification, he would be awarded the Nobel Prize in 1978, together with Hamilton (Ham) Smith and Daniel Nathans.

In his letter, Bertani recalls his interest in the mechanism of P2 lysogeny that led him to encounter the phenomenon of R-M. P2 was usually grown on *Shigella*, but this phage only rarely gave plaques on *E. coli* B, from which background P2 had been isolated. What about the opposite effect? Grow P2 on *E. coli* B and passage the outcoming phage on *Shigella*. In November

1951, Bertani did his crucial "one-step growth" experiment and cycled P2 on the two strains. The results were clear: Progeny from *E. coli* B (P2·B) plated with 100% efficiency (e.o.p.=1) on *E. coli* B and *Shigella*, but phage grown on *Shigella* (P2·Sh) only plated with 0.01% efficiency on *E. coli* B (e.o.p.=10^{-4}), compared to the titer on *Shigella*.

By this time, Joe Bertani was on good terms with Jean Weigle at Caltech, who noted a similar pattern cycling lambda on different *E. coli* strains. Lambda grown on *E. coli* C (lambda·C) grew with an e.o.p. of $\sim 2 \times 10^{-4}$ on a *E. coli* K12 derivative that had been cured of prophage lambda (called K12S; S in Fig. 2). Having passed this barrier to productive infection, the surviving phages were now fully capable again of growth on *E. coli* K12 but were "restricted" by *E. coli* B (lambda·B). Their preliminary results were published independently in the *Microbial Genetics Bulletin* (MGB6) in April 1952, a full year before the publication of the DNA helix structure. Both realized that this was probably a very general phenomenon. They published their unexpected results in a joint paper, although no satisfactory mechanistic explanation was in sight at the time (Bertani and Weigle 1953). At the end of their paper they present a picture to show the parallel between the barriers to lambda and P2 (Fig. 2).

It was clear from their experiments that lambda·K was not a genetic mutant of lambda·C, and it was concluded that the modifying property was "host-

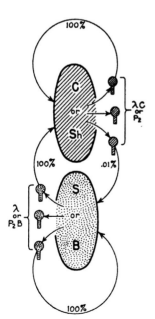

FIGURE 2. Homologies between the phage P2 and λ systems of host range variation. Lines and stipples in the phages represent phage structures whose specificity is completely or almost completely determined by the host cells in which the phage was produced and which are similarly lined or stippled. The percentages indicate the efficiency of plating of a phage on the type of host indicated by the arrow. (Reprinted from Bertani and Weigle 1953, with permission from the American Society for Microbiology.)

controlled." Apparently, *E. coli* C lacked such a barrier, allowing both lambda·K and lambda·C to grow with an e.o.p. of 1.0. The interpretation proposed assumed "the existence of a phage structure, the specificity of which is completely or almost completely under control of the host cell, and which is required for some step in the process of phage multiplication."

Around the same time, Salvador Luria and Mary Human published their paper on a barrier to infection by certain T phages (Luria and Human 1952). Their finding of T* phage active on *Shigella* and not on *E. coli* B/4o cells that produced it was also very striking. Later it would become clear that T* phage could no longer glycosylate its DNA, thus becoming sensitive to restriction by the *mcr* system present in *E. coli* B but absent in *Shigella* (later designated as a Type IV REase).

About their finding of phage growth on *Shigella*, but not on the cells that produced it, Joe Bertani wrote: "I don't seem to have thought of it at the time as more than a curiosity to be further investigated, and probably the same applies to some extent to Luria, in so far prior examples of transient phenotype changes were known.... Of course all this was before we had any idea of DNA structure and before we had fully digested the implications of the Hershey & Chase experiment. Besides, there was no certainty at the time that P2 and lambda would resemble the T phages in composition." It only slowly dawned upon him and Jean Weigle that they faced a breakdown of the distinct picture of genotype and phenotype that had been built up with great care during the first half of the century.

In 1953, Luria summarized all known examples of restriction and modification, called at the time "host-controlled variation," or "host-induced modifications of viruses" (Luria 1953). He combined and generalized the results of his own phage work, those of Bertani and Weigle, and other data on *E. coli*, *Salmonella*, and *Staphylococcus* phages (Table 1). He concluded that "its [host-controlled variation] outstanding characteristic is that it is strictly phenotypic, nonhereditary, and determined by the host cell, in which the virus has been produced." Furthermore, these modifications by successive hosts were not accumulative but mutually exclusive. His general scheme would apply to phages P2, lambda, T1, and P1. Evidence for the involvement of DNA in this phenomenon had to await experiments in the early 1960s.

Joe Bertani continued his research on lysogeny (Bertani 2004), and silence would reign on the topic of HCV until chance brought Werner Arber into the field 7 years later.

Some suggestions for further reading on these early days are Judson (1979), Luria (1984), Gribbin (1985), Fischer and Lipson (1988), and Lily (1993).

TABLE 1. *General scheme of adaptive host-induced modification*

Efficiency of plating on host

Phage	A	B	C
Phage·A	1	10^{-4}	10^{-6}
Phage·B	1	1	10^{-6}
Phage·C	1	10^{-4}	1
Phage·B,C (= P·C)	1	10^{-4}	1
Phage·C,B (= P·B)	1	1	10^{-6}
Phage·B,A (= P·A)	1	10^{-4}	10^{-6}
Phage·C,A (= P·A)	1	10^{-4}	10^{-6}

Adapted, with permission, from Luria 1953, © Cold Spring Harbor Laboratory Press.

This scheme would apply to phages P2, lambda, T1, and P1. Strain A is a nonrestricting host, allowing all phages to infect productively. Strains B and C have different restriction systems, and strain C poses a more effective barrier than strain B.

(Phage·A) Phage grown on strain A, etc., (Phage B,C) phage grown on strain B, then C; phage will have strain specificity of C, etc.

REFERENCES

Bertani G, ed. 2004. Lysogeny at mid-twentieth century: P1, P2, and other experimental systems. *J Bacteriol* **186:** 595–600.

Bertani G, Weigle JJ. 1953. Host controlled variation in bacterial viruses. *J Bacteriol* **65:** 113–121.

Fischer EP, Lipson C. 1988. *Thinking about science: Max Delbruck and the origins of molecular biology.* Norton, New York.

Gribbin J. 1985. *In search of the double helix. Quantum physics and life.* McGraw-Hill, New York.

Hershey AD. ed. 1971. *The bacteriophage lambda.* Cold Spring Harbor Laboratory, Cold Spring Harbor, NY.

Judson HF. 1979. *The eighth day of creation. The makers of the revolution in biology,* a Touchstone book. Simon and Schuster, New York.

Lily EK. 1993. *The molecular vision of life. Caltech, the Rockefeller Foundation and the rise of the new biology.* Oxford University Press, New York.

Luria SE. 1953. Host-induced modifications of viruses. *Cold Spring Harb Symp Quant Biol* **18:** 237–244.

Luria SE. 1984. *A slot machine, a broken test tube: an autobiography.* Harper & Row, New York.

Luria SE, Human ML. 1952. A nonhereditary, host-induced variation of bacterial viruses. *J Bacteriol* **64:** 557–569.

Young R. 2002. Molecular genetics of bacteria and phages, 2001. *J Bacteriol* **184:** 2572–2575.

Yu Y, Bradley A. 2001. Engineering chromosomal rearrangements in mice. *Nat Rev Genet* **2:** 780–790.

APPENDIX 1: LETTER FROM JOE BERTANI
TO NOREEN MURRAY, 2003

G. Bertani
Biology 156-29
Caltech
Pasadena, CA 91125
<gbertani@earthlink.net>

July 17, 2003

Dear Noreen,

More than a month ago I promised to write you within a week or two... My apologies!

The brief paper on my (and others') old work on lysogeny is not ready yet. Nevertheless, I am copying below what I'll say in it concerning restriction and modification, and follow with some comments.

"...While I was focusing on P2 and the mechanism of lysogeny, some unexpected findings came up which deserved proper attention. One was the discovery of "host controlled variation", now more commonly called "restriction and modification", a phenomenon of great theoretical interest. I noticed it in P2 (using strain B as the restricting host, *Shigella* being the standard host) and did not know what to make of it. Jean Weigle noticed it in *lambda* (using strain C as the permissive host, K-12 being the standard host): being aware of my results, he immediately recognized the parallelism of the two "systems". Shortly before that, a minor laboratory accident, as told by Luria (REF A), had led to the discovery of another, albeit more complex case of host controlled variation (REF B). Although no satisfactory mechanistic explanation was in sight at the time, Jean and I were encouraged by the parallelism between our two, totally independent "systems" and decided to publish together our findings (REF C). It rarely happens that a new phenomenon, observed in two different materials, in different labs, is described in the same paper, in a comparative manner. Of course, this strengthened the evidence, hinting at the generality of the phenomenon. It also scored a point for cooperation vs. competition in science and human affairs. A similar case, several years later, was that of a paper by René Thomas and Elizabeth Bertani (REF D), which reported parallel experiments with *lambda* and with P2 to more precisely define the mode of action of the immunity repressor."

Ref B=Luria & Human 1952 J.Bact. 64:557
Ref C=Bertani & Weigle 1953 J.Bact. 65:113
Ref D=Thomas & Bertani, L.E. 1964 Virology 24:241

REF A is Luria's autobiography ("A slot machine, a broken test tube", 1984). He describes the episode that clarified his problem with strains B/4o and B/4oo. I remember the episode a bit differently, but the essential facts are the same. This must have happened in late 1950 or early 1951, since Mary Human, who was doing the experiments, left our lab at the end of March 1951, and I remember having to convince her that using *Shigella* was not so risky. The finding of phage active on *Shigella* and not on the B/4o cells that produced it was very striking, but I don't seem to have thought of it at the time as more than a curiosity to be further investigated, and probably the same applies to some extent to Luria, in so far prior examples of transient phenotype changes were known. This is also the impression one gathers from reading the first description of the effect in Luria's abstract for the phage meeting at Cold Spring Harbor, August 20–22, 1951 (in PIS #6 / copy enclosed).

I was mostly interested in lysogeny and had isolated some P2 plaque type mutants to be used as markers in a variety of experiments. Having seen that P2 (which I usually grew on *Shigella* at that time) only rarely gave plaques on *coli* B, I presumed these were "host range mutants". Trying now to reconstruct from my lab notes, it seems that at first I did not see a clear cut effect of the passage from B to *Shigella*, probably because I was making mostly plate stocks and I was using relatively large phage inocula. I had however some suspicion because I also made, beginning in April 1951, a number of "single clone" experiments, to see the distribution of plaques formed by P2 (grown on *Shigella*) on *coli* B: there was no evidence of a clonal distribution. I don't know at which point I started worrying about the loss of the ability to plate on B by P2 grown on B and then passaged on *Shigella*: my first "neat" one-step growth experiment showing this is of November 1951.

I have not succeeded in reconstructing when I first met Jean Weigle, whether in 1950 or in 1951, but by the Fall of 1951 we were in very good terms, corresponding by letter and exchanging strains. He knew about my problems concerning P2 and *coli* B vs. *Shigella*. In late 1951 he wrote me a letter giving data on efficiencies of plating of *lambda* on various indicator strains, and showing that the "*lambda*/ K-12/122" (122 being what we later called strain C) pattern of plating could perfectly parallel the "P2/B/*Sh*" pattern, except that we were starting from opposite ends. That is when (I believe) we all realized that this was probably a very general phenomenon. I proposed to Jean that we publish a joint paper and he accepted "enthusiastically" (his word). I still have most of the correspondence from that point on. Very unfortunately, I cannot find his first letter about the common pattern of the two systems. It probably was misplaced or taken by someone in the lab. We briefly reported our findings in Microbial Genetics Bulletin #6, which appeared in April 1952 (I enclose copies). We agreed to do some more experiments to complete the comparison P2/*lambda*, and to write or finalize the joint paper together in Pasadena in March 1952. That is when Jean had his second heart attack, and my trip had to be postponed to mid May. As a consequence of this delay the Luria and Human paper came out ahead of ours, while we were hoping at first that the two papers would appear in the same issue of J.Bact. The discussions with Jean in the preparation of the paper were very interesting because, being a physicist only recently converted to biology, he viewed thing differently from me. Of course all this was before we had any idea of DNA structure and before we had fully digested the implications of the Hershey & Chase experiment. Besides, there was no certainty at the time that P2 and *lambda* would resemble the T phages in composition.

Looking back at the period when the first examples of restriction and modification were observed up to the first attempts at biochemical analysis, I'm struck by the following: (1) There were several examples of the phenomenon cropping up (for example the one on *Staphylococcus* phage by Ralston & Krueger, Proc. Soc. Exp. Biol. Med. 80:217, that was published as we were working on our manuscripts) and only a few happened to be seen as seriously problematic at the time and investigated further. [On reviewing my old protocols I discovered another example of this in my own work of February 1950 (!), a case which I had completely forgotten and failed to pursue at the time.] (2) The immediate reaction was to think in terms of known non-genetic effects, like "phenotypic mixing" or the heterogeneity of phage particles in respect to heat stability. (3) The realization that the distinction genotype/phenotype was breaking down (i.e. that one had reached the limits of standard formal genetics) came slowly. When Jean and I were working on our joint paper, we had several discussions on the meaning of "genetic". The question is also discussed at some length in Luria's Cold Spring Harbor paper (C.S.H.Symp. 18:237; 1953).

As intriguing as host modification was, I realized that I would not be able to continue on both it and lysogeny as my main research activities, so that I became a spectator as far as host controlled variation was concerned. An exception was when Allan Campbell got a bright idea for testing whether DNA was involved in the effect: he spent the summer of 1956 with us and we did together a few (very complicated) experiments with P2 in B. The experiments did not work well enough, though, and we abandoned them. Also, much later, in Stockholm, I believe in the summer of 1964, I organized a small informal meeting on nucleic acid methylation, which was attended by Arber, Campbell, Seymour Lederberg and a few others.

Jean Weigle came to Caltech in 1949, but he used to spend summers in Geneva in the biophysics lab that was part of his former physics department, so that he had quite a bit of influence on the direction of the work there and was for many years a direct connection between Geneva and Caltech. Werner Arber finished his doctorate in 1958 at Geneva working on Gal transduction by lambda, then spent a year in our lab at the University of Southern California, working on transduction by P1 of lambda prophages and of the F factor (Virology 11:250 & 11:273).

Sorry for taking so long to put together this letter.

With kind regards,

APPENDIX 2: BERTANI OBITUARY[2]

Giuseppe Bertani

Professor Giuseppe Bertani (Joe to friends) died on April 7, 2015 at the age of 91 in Pasadena, CA. As a pioneering microbial geneticist, his insights helped to develop both modern microbiology and the molecular biology of today. Born in Como, Italy, Joe was raised in Milan, where he earned his doctorate in zoology. After postgraduate studies in Naples and Zürich, he arrived at the Cold Spring Harbor Laboratory (CSHL) in October 1948 as a Carnegie Fellow working in Milislav Demerec's group. Here, he shifted his focus to bacterial genetics and was soon measuring reverse mutation rates in a streptomycin-dependent mutant strain of *Escherichia coli* after exposure to radiation and chemical agents; in fact, these experiments preceded what would later become the Ames test. Most importantly, it was here that Joe was shown phage plaques for the first time by his friend Gus Doermann, who was working on phage T4, and that he first encountered lysogeny.

Joe attended Max Delbrück's phage course at CSHL in 1949, after which he joined Salvador Luria at Indiana University in Bloomington. Here he began studying lysogeny, although at first Luria was somewhat reluctant. Using what he called a "modified single burst technique" Joe demonstrated that phage production by a lysogen was discontinuous, involving rare, large bursts of phage. He went on to characterize the establishment of lysogeny, the state of the prophage, and superinfection immunity. As it turned out, the Lisbonne strain Joe was using produced three different phages, which he named P1, P2, and P3. It was P2, the noninducible phage, which was to become his primary phage of study. During these studies Joe composed the now ubiquitous LB medium, which subsequently has been referred to as Luria broth, Lennox broth, or Luria-Bertani medium. For the historical record, Joe pointed out that the abbreviation LB was intended to stand for "lysogeny broth." In addition to his lysogeny work, Joe's discovery in

---

[2]Reprinted, with permission, from the American Society for Microbiology (*Microbe*, January 2006, pp. 20–24).

1953 of "host-controlled variation," together with Jean Weigle, ushered in our understanding of host restriction and modification, which influenced the discovery of restriction enzymes 15 years later.

Joe remained with Luria after the lab moved to the University of Illinois in 1950, where he met and married Betty, and then in 1954 Joe joined the laboratory of Max Delbrück at Caltech. In 1957 Joe took up a professorship in the medical school at University of Southern California in Los Angeles, where Werner Arber joined him as a research associate from 1958–59.

In the early 1960s Joe was appointed professor in microbial genetics at the Karolinska Institute and studies of phage P2 became the focus of the Bertani lab. In these years a steady stream of postdoctoral fellows filled his laboratory in addition to his students and many distinguished visitors. In addition to his obligations at the Karolinska Institute he was also responsible for the advanced teaching of microbiology at the University of Stockholm. His influence on the scientific community in Sweden was significant and his work was recognized by Uppsala University where he received an honorary doctorate in 1982. During this time he also participated in establishing the European Molecular Biology Organisation (EMBO). In 1981 he returned to California to take up a position at the Jet Propulsion Laboratory (JPL) in Pasadena, where he studied the genetics of methanogenic bacteria and described a curious phenomenon of transduction. After formally retiring from JPL in 1991, Joe continued as a voluntary scientist in the Division of Biology at Caltech.

Joe was highly critical but generous when it came to publishing. He rarely put his name on his students work when they were ready to publish their results. Joe Bertani was an outstanding scientist with a philosophical touch, belonging to that dwindling group of pioneers in microbial genetics with roots in the legendary Phage Group. We thank him for taking us on a marvelous journey in science, with him as our guide, and for his friendship. Our thoughts are with his wife Betty, their sons Christofer and Niklas, and their families.

RICHARD CALENDAR
University of California, Berkeley

ELISABETH HAGGÅRD-LJUNGQVIST
Stockholm University

BJÖRN H. LINDQVIST
University of Oslo

STEVEN E. FINKEL
University of Southern California

CHAPTER 2

Host-Controlled Variation Is Methylation and Restriction of DNA: The 1960s

INTRODUCTION

Werner Arber entered the restriction field by chance. His research into host-controlled variation (HCV) stands as one of the examples of "serendipity" in scientific discovery: the combination of a chance observation, an opportunity that favors the prepared mind, and being at the right place at the right time. Serendipity has been referred to by Salvador Luria in his autobiography entitled *A Slot Machine, a Broken Test Tube* (Luria 1984). The slot machine refers to his landmark paper on spontaneous mutation in bacteria with Max Delbrück from 1943 (Luria and Delbruck 1943), the broken test tube to the event that caused him to discover restriction of one of his mutant T* phages: He dropped his bacterial culture and asked Joe Bertani for some other cells. Bertani gave him some *Shigella*, which happily plated T*, in contrast to the *Escherichia coli* B/4o cells that had originally produced that phage. (Later it would become clear that T* had become sensitive to the *mcr* restriction system [see Chapter 8] present in the *E. coli* B strain, but absent in *Shigella*.) Arber's chance observation was that the generation of lambda^s derivatives of *E. coli* B/r resulted in strains that did not properly propagate lambda phage, as discussed below.

The Arber–Dussoix Papers (1962)

Werner Arber entered biophysics as a student of Edward Kellenberger in Geneva (Fig. 1) in 1953, the year of the double helix model, the Bertani and Weigle paper, and the first review by Salvador Luria on HCV (Chapter 1). During this period, Arber held a journal club seminar on the Watson and Crick model of the DNA double helix (Watson and Crick 1953). Sixty years later, he recalled that there was very little acceptance that genetic information was present in the linear sequence of base pairs (bp), because people had the

Chapter doi:10.1101/restrictionenzymes_2

FIGURE 1. (*Left* to *right*) Werner Arber, Edouard Kellenberger, and Jean Weigle after the defense of Arber's PhD thesis (1958). (Courtesy of Werner Arber.)

idea that genetic information was very complicated. They simply could not understand how nucleic acid, composed of only four different building blocks, the nucleotides (nt), was able to give that information rather than the proteins with 20 building blocks, the amino acids (aa), that would be much better carriers of highly complex information (http://library.cshl.edu/Meet ings/restriction-enzymes/v-Arber.php).

This is in contrast to Matthew (Matt) Meselson's recollection of this event (Meselson with Franklin Stahl provided the experimental evidence for the semiconservative nature of DNA replication in 1958 [Meselson and Stahl 1958]). Meselson recalls that "e.g., Erwin Chargaff (who determined that the ratios of adenine to thymine and of guanine to cytosine were always unity [see Judson 1979 for details]), upon reading the Avery et al. paper of 1944 [Avery et al. 1944] almost immediately stopped his research on lipids and began his well-known work on the nucleotide composition of DNA, from diverse organisms, showing wide variation in composition and thereby overthrowing Levene's tetranucleotide hypothesis. Further work in the late 1940s by Hotchkiss and others [for an overview see, e.g., Hotchkiss 1953, 1995] showed that highly purified DNA could transform specific bacterial genes. The 1952 finding by Al Hershey and Martha Chase [Hershey and Chase 1952] that upon infection most of the ^{32}P of phage T2 enters the bacterial cell while most of the ^{35}S stays outside, together with the earlier transformation experiments, was widely interpreted at the time to mean that the information resides in DNA, not protein" (M Meselson, pers. comm.; references added by the author). There is an interesting anecdote in this respect: Apparently, as late as 1956, Beadle asked an "eminent biochemist" whether he thought the double helix model was correct. The answer: "Yes, I am sure it is correct, but I do not think it has anything to do with replication!" (Berends 1977). But according to Meselson, this was an oddity by then.

This journal club activity and his training in a high-powered laboratory generated Arber's lifelong interest in research into DNA and phage genetics. For his PhD thesis he studied a defective phage called lambda d*gal*, obtained from Larry Morse in Joshua Lederberg's laboratory (for details of this phage, see http://www.sci.sdsu.edu/~smaloy/MicrobialGenetics/topics/phage/lambda-dgal.html). Afterward he spent a year with Joe and Betty Bertani to work on P1-mediated transduction of lambda prophage as well as the fertility (F) factor. Here he heard about the 1953 HCV paper. At that time people had started to realize the huge size of genomes of many organisms. How could you study the structure or function of a gene in such a large genome? Impossible! You had to take the gene out. But how? Could you perhaps incorporate genes of interest in a phage? In that case, one could harvest enough DNA for biochemical studies.

At the suggestion of Esther Lederberg, Arber went to work with the radiation[r] *E. coli* B/r strain after his return to Geneva. He first made a lambda[s] derivative of this strain, as *E. coli* B was not a normal host for lambda (it was used for T phages). However, the new strains did not efficiently propagate lambda (which had been prepared in its normal *E. coli* K12 host). Fortunately, he quickly realized that this was owing to the HCV phenomenon he had heard about before. Intriguing questions were to be answered: How could phage overcome this barrier? What exactly happened to the phage, when it was prevented to grow normally? How did phage become adapted to the new host? Was HCV caused by a host protein picked up by the phage during its replication that would allow it to return to the same host? Or, alternatively, rather than taking a protein along in the phage head, would the phage DNA get some kind of "imprinting" by the host? Using phage DNA labeled with ^{32}P, he could resolve this issue and test whether the new DNA of the progeny phage had no host modification. Not a protein in the phage head but something on the DNA caused modification.

In the meantime, Grete Kellenberger in the laboratory had found that irradiated phage after infection was rapidly destroyed. She suggested to Arber's PhD student Daisy Dussoix (Fig. 2) that Dussoix should test her restricting strains in the same way. The results of these experiments led to the two classical "Arber–Dussoix" papers on the inactivation of DNA by restriction, and modification of DNA owing to a host function (Arber and Dussoix 1962; Dussoix and Arber 1962). In paper I (Arber and Dussoix 1962) they analyzed the fate of lambda DNA in *E. coli* K12 carrying one of Bertani's original temperate phages,

FIGURE 2. Daisy Roulland-Dussoix (~1962). (Courtesy of André Dussoix.)

P1 (Table 1). This strain carries two restriction systems, that of *E. coli* K12 and phage P1 ("K" and "P1," currently called EcoKI and EcoP1I, respectively, later designated as members of the Type I and III REases [http://rebase.neb.com/rebase/rebase.html; Roberts et al. 2003, 2015]). Lambda cultured on *E. coli* C (lacking a restriction system), lambda·C, was restricted cumulatively by both K and P1, leading to a drop in efficiency of plating from 2×10^{-4} to 7×10^{-7}.

After adsorption of phage onto P1-restricting or nonrestricting cells, the DNA apparently entered both cell types within minutes, as judged from the presence of the label in the pellet after low-speed centrifugation. It became clear that the phage DNA was injected and degraded upon infection of different bacterial hosts (Dussoix and Arber 1962; Lederberg and Meselson 1964) unless it carried host-specific modification of that DNA. Similar DNA breakdown with other restriction systems supported these data.

Grete Kellenberger had already suggested that this degradation might be a two-step process. A highly specific "restriction enzyme" (REase) would be cleaving the DNA, followed by one or more less-specific nuclease(s) that would

TABLE 1. *Efficiency of plating of phage λ variants on different host strains*

Phage variant	Efficiency of plating on host strains			
	K12	K12(P1)	B251	C
λ · K	1	2×10^{-5}	10^{-4}	1
λ · K(P1)	1	1	10^{-4}	1
λ · B	4×10^{-4}	7×10^{-7}	1	1
λ · C	4×10^{-4}	4×10^{-7}	2×10^{-4}	1

Reprinted from Arber and Dussoix 1962, with permission from Elsevier.

degrade the cleavage products, a prediction later shown to be correct. Further evidence that restriction occurred at the level of DNA came from the observation that hybrid DNA molecules, with one strand modified and the other not, were not degraded and on transformation gave rise to new phage progeny (Arber and Dussoix 1962; Arber 1965; Meselson and Yuan 1968).

In paper II (Dussoix and Arber 1962), the "suicide" method was used to analyze the nature of HCV of newly synthesized DNA. This protocol was developed by Alfred Hershey in 1951 and extensively used by Gunther Stent: DNA heavily labeled with ^{32}P becomes noninfectious owing to radioactive damage, but the DNA will not be immediately degraded. Thus the modification status of new phage DNA, made in the cell after infection in medium free of label, could be easily analyzed. These experiments proved the modification to be physically linked to the DNA. In this paper, they also provided the first tentative evidence that modification might occur on nonreplicating DNA, and that modification was dominant—that is, the presence of modification on one of the two strands would protect the DNA, as was later shown definitively by the inability of the EcoKI enzyme to make even single-stranded (ss) breaks on lambda DNA having one modified chain and one nonmodified chain (Meselson and Yuan 1968).

The molecular aspects of restriction-modification (R-M) were becoming clear. Both R and M appeared to act somehow on the DNA, and apparently the modification was reversible upon replication, during which it did not lead to mutations. Phage DNA was modified upon its propagation and successfully infected the same host again. In contrast, not properly modified phage DNA was broken down to oligonucleotides in an r$^+$ host. Probably the R-M enzymes could detect minor changes in different DNA molecules. It also became clear that R and M acted on both phage and cellular DNA, a fact we now take for granted. Similar DNA breakdown was observed soon afterward for other restriction systems. In r$^-$ mutants the DNA remained intact as predicted.

The three main topics of subsequent study were the nature of the DNA modification, the localization of the host genes for R-M, and the purification of the enzymes responsible for these activities. The Geneva group headed by Arber continued their research resulting in a series of papers during the 1960s. The Americans Stuart (Stu) Linn and William (Bill) Wood came to help with the isolation of the enzymes and mutants, respectively. Stu Linn did his first degree at Caltech and his PhD at Stanford, where he shared a laboratory with Daisy Dussoix. He was a postdoc with Arber from 1966 to 1968 before moving to Berkeley. Bill Wood was a PhD student of Paul Berg, went to Geneva in 1963, and became assistant professor at Caltech afterward. Clearly, R and M were two separate events, and both topics generated widespread interest, inspiring others to embark on the genetics and enzymology of R-M systems.

THE ROLE OF METHIONINE IN DNA MODIFICATION

The first major problem to tackle was the nature of the modification. Although the modification was apparently attached to the DNA in a "phenol-insensitive way" (Dussoix and Arber 1965), there was no evidence about the nature of the actual change. Gunther Stent suggested that perhaps modification was methylation. When Arber visited Berkeley in 1963, he got the first inkling for the essential role of methionine in the modification process: lambda became poorly modified on propagation in an *E. coli* met⁻ host and withdrawal of this amino acid (aa) from the medium. The essential role for methionine led to the idea of nucleotide "alkylation" of special (presumably sequence-specific) sites on the DNA, via methyl transfer using *S*-adenosylmethionine (SAM) as donor.

Next, John Smith from Cambridge came to Geneva to look at the correlation of modification with DNA methylation. Unfortunately, basal levels of methyl groups on the lambda DNA were high. Hence, no significant differences in methylation could be found between phage grown on m⁺ or m⁻ bacteria. This high basal methylation level was not entirely surprising, as the bacterial chromosome was known to be considerably methylated, a property later shown to be due to the action of different bacterial methylases that modify adenine (to m6A) or cytosine (to m5C). The solution to this problem was provided by Hartmut Hoffmann-Berling, who suggested the use of phage fd, which has a single-stranded circular DNA chromosome and is a close relative of the well-known phage M13 of Sanger sequencing fame.

Stu Linn and Werner Arber established that the presence of the *E. coli* B REase (R·B) led to a drop in the phage fd titer. The phage DNA itself was degraded at a later stage, and in the electron microscope only limited double-stranded (ds) breaks were seen. Phage fd has two recognition sites for the B enzyme (Arber and Kuhnlein 1967), and on the ds-replicative form (RF) of its DNA two As per site were found to be methylated, suggesting one methylated A on each strand (Kuhnlein and Arber 1972). Attempts to determine the nt sequence around this A were initially not successful (van Ormondt et al. 1973; Linn et al. 1974). Phage mutants that lacked these restriction sites were no longer modified. These data proved (1) a link between the number of methylated sites and the restriction targets; (2) the B enzyme methylated an A, presumably within the specificity site; (3) restriction resulted in scission of the DNA, presumably at the same site (which would later prove untrue!); and

(4) both activities acted on a substrate lacking this modification. This protective effect to DNA cleavage was the first evidence for a biological function of DNA methylation. In today's terminology this is clearly one of the first described epigenetic effects. In his 1965 review, Arber suggested to name the phenomenon host-controlled *modification*, as the term *variation* could indicate some permanent change in the genetic message, which was clearly not the case (Arber 1965). He anticipated that R and M enzymes would be very helpful for functional and structural studies of genetic information.

THE LOCALIZATION AND ALLELISM OF THE GENES ENCODING THE *E. coli* K12 AND B GENES

The second issue to solve was the location of the R-M genes. Throughout the 1960s, research concentrated mainly on the systems of *E. coli* K12 and B and phage P1, although restriction activity by so-called "resistance transfer factors" (R or RTF factors) was reported by Tsutomu Watanabe and coworkers in Tokyo during this time. This laboratory was the first to show restriction of infectious lambda DNA in vitro using a bacterial extract containing an R factor, which conferred antibiotic[r] to the host (Takano et al. 1966). Watanabe published groundbreaking papers on this topic (see, e.g., Watanabe 1963; Watanabe et al. 1962, 1964, 1966; Takano et al. 1966, 1968; and Appendix 1 with translation of Japanese Wikipedia entry for Watanabe). This would inspire Herbert (Herb) Boyer to ask PhD student Robert Yoshimori to screen clinical isolates for REases, which led to the discovery of EcoRI (and EcoRII [Yoshimori 1971]; see Chapter 3).

Because of the experimental and genetic tools and knowledge available, the hunt for the genes responsible for R-M was dependent on gene transfer by conjugation and transduction (see Chapters 9 and 13 in Stent and Calendar 1978), before the breakthrough by Mandel and Higa, who managed to transform *E. coli* with the $CaCl_2$ plus heat shock method in 1970 (Mandel and Higa 1970). Survival and modification of incoming phage or bacterial DNA in the recipient was used to answer questions: Where were the R-M genes on the *E. coli* chromosome? How many genes were involved? What was the exact nature of the modification? Where was it located? How and when and where was the DNA cut and degraded? Was it the same for the *E. coli* K12, B, and phage P1 systems? More and more evidence would emerge that the phage P1 system was different from that of *E. coli* K12 and B, and also from the restriction system that attacked Luria's mutant T* phages, as reported in 1952 (Luria and Human 1952; Chapter 1). These facts slowly emerged and would eventually lead to the current classification in the four types of restriction systems and subdivisions (http://rebase.neb.com/rebase/rebase.html; Roberts et al. 2003, 2015; http://www.library.cshl.edu/Meetings/restriction-enzymes/v-Rob erts.php).

The genetic mapping experiments were giving a reasonably coherent picture of the genomic locus encoding the R-M enzymes of *E. coli* K12 and B. Experiments with partial dipoids ("merozygotes") provided convincing evidence that (1) K and B mutants (either r⁻m⁺ or r⁻m⁻) complement each other; (2) the systems were functionally allelic: recombinants with *E. coli* K12 and B/r restricting and modifying properties never superimposed on one another but were always mutually exclusive (in contrast to the K and P1 activities in the Arber and Dussoix experiments); (3) the R-M genes in both strains had to be linked, as it was easy to transfer them together; and (4) there were three genes located near *thr* on the *E. coli* chromosome. The most-conclusive evidence for the three-gene model was obtained with temperature-sensitive (ts) *hsm* mutants made by Josef Hubáček and Stuart Glover (Hubáček and Glover 1970). All three genes were probably needed for restriction, although the involvement of additional genes could not be formally excluded at the time. The K and B genes were called *hss*, *hsm*, and *hsr* (currently host specificity determinant *hsdS*, *hsdM*, and *hsdR*), and it was tentatively concluded that *hsr* was not needed for modification. The P1 system would prove to be a two-gene system, named *res* and *mod*, respectively.

The picture emerging was quite clear: Modification acted on the DNA itself at a limited number of sites in the shape of methylation, leading to host specificity, and restriction occurred only if proper host specificity was absent. The implications were obvious, and the logical assumption was that both R-M enzymes recognized the same particular base sequences. The enzymes for R and M would thus share that part that recognized these specific sites, so that mutation arising in this part would cause the loss of both functions in question. The same explanation could apply independently of whether only one gene product exerted the functions of sequence recognition, modification, and restriction or whether different gene products were assembled as subunits to the specific enzymes.

The major transition about to occur at the turn of the decade can be easily seen in the reviews published just before and after 1970 (Arber and Linn 1969; Boyer 1971; Meselson et al. 1972; also see Chapter 3). The 1969 review by Arber and Linn discusses three possible models for hyphenated and/or palendromic recognition sites, which are familiar to scientists today. In model I, the specificity would be conferred by one strand only, and the recognition site would be a sequence of nucleotides on that strand with one (perhaps more) modifiable base(s). In model II, both strands would be involved. In this case, a sequence of nucleotide pairs would dictate specificity, with each strand carrying at least one modifiable base. Model III was an extension of the second model in which the specificity site would possess internal symmetry (a palindrome). Such sites could be either contiguous (i.e., all individual bases were essential for recognition to occur) or hyphenated, which would allow one or more ambiguities during recognition. The authors also worked out various estimates of the frequency of the recognition sequence as a function of the length of the specificity site. Based on the number of sites found in phage, this site would be 6- to 8-nt long (later shown to be correct for the enzymes analyzed). Expectations were high that such a putative mechanism of base-sequence specific recognition might provide a tool for the sequence-specific cleavage of DNA. The ability to use different enzymes should allow the sequence determination of DNA molecules. A similar hope (i.e., to use these enzymes to sequence DNA) would lead Richard (Rich) Roberts into the restriction field a few years later. In 1969, the exact sequence of such recognition sites had to await biochemical experiments involving end labeling of broken ends with ^{32}P, and the arrival of DNA sequencing.

In conclusion, during the 1960s, much of the groundbreaking work was performed and many facts became clear that we now so easily take for granted: (1) R and M could take place on nonreplicating DNA; (2) R levels depended on the number of specificity sites per DNA molecule and varied from enzyme to enzyme; and (3) R-M was a general phenomenon: Host, plasmid, and phage DNA were all sensitive to different systems, depending on the bacterial host. Restriction emerged as the bacterial defense system against foreign DNA.

PURIFICATION OF THE RESTRICTION ENZYMES OF *E. coli* K12 AND B

The groups of Werner Arber in Geneva and Matt Meselson at Harvard University set out to purify the REases from *E. coli* K12 (EcoKI) and B (EcoBI). Meselson had first detected restriction activity as breakage of unmodified lambda DNA assayed in a sucrose gradient, and started experiments to detect and purify EcoK1. Although he found ATP-dependent EcoKI restriction activity in crude extracts, he could not detect the activity in DEAE column eluates. At that point, Robert (Bob) Yuan came into his laboratory. They used a combination of column chromatography, glycerol gradients, and preparative gels to isolate EcoKI (Meselson and Yuan 1968). They knew that Bill Wood had found evidence that restriction is impaired in methionine auxotrophs if methionine is withheld, so they added methionine and ATP, which restored the activity. But upon further purification the activity again vanished, until they realized that the enzyme needed to be replenished with ATP and that the enzyme also needed SAM for activity (Meselson and Yuan 1968; see Appendix 2 with emails [July 2, 2004] between Mattt Meselson and Noreen Murray for an historical account). Once this absolute requirement for SAM became evident, EcoKI could be purified ~5000-fold to homogeneity, as determined by gel electrophoresis. During these purification attempts they used breakage of lambda DNA in the presence of ATP, SAM, and Mg^{2+} as an assay (Meselson and Yuan 1968). For later studies of enzyme binding to DNA and ATP hydrolysis a rather simple and convenient assay was used (Yuan and Meselson 1970; Yuan et al. 1972); filter retention of unmodified DNA. If the EcoKI enzyme was incubated with unmodified and modified lambda DNA and the mixture passed through a nitrocellulose filter, only the unmodified DNA was retained on the filter. Maximum retention required ATP, SAM, and Mg^{2+}, and EcoKI mutant enzymes failed to cause retention.

The enzyme did not break at specific sites and broke the strands sequentially (Meselson and Yuan 1968). The molecular weight (MW) of the native enzyme was ~400 kDa, based on sedimentation and gel filtration rates relative to proteins of known MW. The complex could be dissociated with SDS, revealing three subunits with MWs of 135, 62, and 52 kDa, respectively. The relative amounts indicated that the complex contained two of each of the larger subunits and only one of the smallest (Meselson et al. 1972). The complex had both REase and MTase activities.

In Geneva, and then in Berkeley, Stu Linn started out to purify EcoBI (Linn and Arber 1968; Eskin and Linn 1972a,b). The protocol was a little different, and eventually yielded a ~1000-fold purification. Around the same time, Daisy Roulland-Dussoix also purified EcoBI ~1000-fold in the laboratory of Herb Boyer (Roulland-Dussoix and Boyer 1969). The enzyme proved to be very similar to EcoKI: a large complex with a MW of ~400 kDa and three types of subunits of MW 135, 60, and 55 kDa (Arber and Linn 1969; Boyer 1971; Eskin and Linn 1972b). It also had an absolute requirement for SAM, needed Mg^{2+} and ATP as cofactors, and degraded ATP during the reaction. However, in contrast with the stable pentameric EcoKI enzyme, EcoBI purified as several active oligomeric species with a MW ranging from 450 to 750 kDa. Was this difference due to differences in the purification procedures? The predominant form had a proposed subunit composition of two larger, four medium, and two of the smallest subunits (Eskin and Linn 1972b). Two active forms, which were enzymatically indistinguishable, were also isolated by native gel electrophoresis. What was the role of SAM in the restriction process? Technical reasons (the small number of enzyme molecules in the cell and the instability of SAM) made it hard to address this issue (Linn et al. 1977). And why did the enzymes remain attached to the DNA after restriction? Was this a signal to another enzyme molecule (Rosamond et al. 1979)?

The EcoBI MTase, M·EcoBI, could be purified by essentially the same procedure used to prepare the EcoBI REase except that the MTase was separated from the REase during column chromatography (Kuhnlein et al. 1969) and further purified on DNA cellulose (Lautenberger and Linn 1972). M·EcoBI contained the same small subunits as the EcoBI REase (Lautenberger and Linn 1972). Depending on the time after purification, the enzyme had various subunit compositions upon storage, all enzymatically active (Linn et al. 1974). An attempt to match the enzyme subunits to the respective genes proved difficult. Bacteria with mutations in the three *hsd* genes could complement in vivo, as mentioned earlier, but also in vitro (Linn and Arber 1968). Also, purified MTase could supply the two subunits to generate restriction in *hsdS⁻* and *hsdM⁻* extracts (Linn 1974). This indicated that the M and S subunits of the MTase had an active role in restriction. In this way, it could be established that the large subunit of 130 kDa was the subunit encoded by the *hsdR* gene. At the time, the two smaller subunits could not be assigned to either the *hsdM* or *hsdS* genes (Linn 1974).

For further details on this topic, see Arber et al. (1975) and Endlich and Linn (1981); for a biography of Daisy Roulland-Dussoix, see Appendix 3; and for more background reading, see Hershey (1971) and Hayes (1968), Stahl (1969), Watson (1970), Portugal and Cohen (1977), Gribbin (1985), Cairns et al. (1992), and Holmes (2001).

REFERENCES

Arber W. 1965. Host-controlled modification of bacteriophage. *Ann Rev Microbiol* **19:** 365–378.

Arber W, Dussoix D. 1962. Host specificity of DNA produced by *Escherichia coli*. I. Host controlled modification of bacteriophage lambda. *J Mol Biol* **5:** 18–36.

Arber W, Kuhnlein U. 1967. [Mutational loss of the B-specific restriction in bacteriophage fd]. *Pathol Microbiol* **30:** 946–952.

Arber W, Linn S. 1969. DNA modification and restriction. *Ann Rev Biochem* **38:** 467–500.

Arber W, Yuan R, Bickle TA. 1975. Strain-specific modification and restriction of DNA in bacteria. In *Proceedings of the FEBS Ninth Meeting, Budapest 1974. Post-synthetic modification of macromolecules* (ed. Antoni F, Farago A), Vol. 34, pp. 3–22. FEBS, Oxford.

Avery OT, Macleod CM, McCarty M. 1944. Studies on the chemical nature of the substance inducing transformation of pneumococcal types: induction of transformation by a desoxyribonucleic acid fraction isolated from pneumococcus type Iii. *J Exp Med* **79:** 137–158.

Berends W. 1977. 'Biochemistry Lecture Notes,' Leiden University, The Netherlands. Translation by the author: p. 323.

Boyer HW. 1971. DNA restriction and modification mechanisms in bacteria. *Ann Rev Microbiol* **25:** 153–176.

Cairns J, Stent GS, Watson JD. 1992. *Phage and the origins of molecular biology*, expanded ed. Cold Spring Harbor Laboratory Press, Cold Spring Harbor, NY.

Dussoix D, Arber W. 1962. Host specificity of DNA produced by *Escherichia coli*. II. Control over acceptance of DNA from infecting phage lambda. *J Mol Biol* **5:** 37–49.

Dussoix D, Arber W. 1965. Host specificity of DNA produced by *Escherichia coli*. IV. Host specificity of infectious DNA from bacteriophage lambda. *J Mol Biol* **11:** 238–246.

Endlich B, Linn S. 1981. Type I restriction enzymes. In *The enzymes,* 3rd ed. (ed. Boyer PD), Vol XIV, Part A, pp. 137–156. Academic, New York.

Eskin B, Linn S. 1972a. The deoxyribonucleic acid modification and restriction enzymes of *Escherichia coli* B. *J Biol Chem* **247:** 6192–6196.

Eskin B, Linn S. 1972b. The deoxyribonucleic acid modification and restriction enzymes of *Escherichia coli* B. II. Purification, subunit structure, and catalytic properties of the restriction endonuclease. *J Biol Chem* **247:** 6183–6191.

Gribbin J. 1985. *In search of the double helix. Quantum physics and life.* McGraw-Hill, New York.

Hayes W. 1968. *The genetics of bacteria and their viruses. Studies in basic genetics and molecular biology,* 2nd ed. Blackwell Scientific, Oxford.

Hershey AD. ed. 1971. *The bacteriophage lambda.* Cold Spring Harbor Laboratory, Cold Spring Harbor, NY.

Hershey AD, Chase M. 1952. Independent functions of viral protein and nucleic acid in growth of bacteriophage. *J Gen Physiol* **36:** 39–56.

Holmes FL. 2001. *Meselson, Stahl, and the replication of DNA. A history of the most beautiful experiment in biology.* Yale University Press, New Haven, CT.

Hotchkiss RD. 1953. The genetic chemistry of the pneumococcal transformations. *Harvey Lect* **49:** 124–144.

Hotchkiss RD. 1995. DNA in the decade before the double helix. *Ann NY Acad Sci* **758:** 55–73.

Hubáček J, Glover SW. 1970. Complementation analysis of temperature-sensitive host specificity mutations in *Escherichia coli. J Mol Biol* **50:** 111–127.

Judson HF. 1979. *The eighth day of creation. The makers of the revolution in biology,* a Touchstone book. Simon and Schuster, New York.

Kellenberger G, Zichini ML, Weigle JJ. 1961. Exchange of DNA in the recombination of bacteriopage λ. *Proc Natl Acad Sci* **47:** 869–878.

Kuhnlein U, Arber W. 1972. Host specificity of DNA produced by *Escherichia coli*. XV. The role of nucleotide methylation in in vitro B-specific modification. *J Mol Biol* **63:** 9–19.

Kuhnlein U, Linn S, Arber W. 1969. Host specificity of DNA produced by *Escherichia coli*. XI. In vitro modification of phage fd replicative form. *Proc Natl Acad Sci* **63:** 556–562.

Lautenberger JA, Linn S. 1972. The deoxyribonucleic acid modification and restriction enzymes of *Escherichia coli* B. I. Purification, subunit structure, and catalytic properties of the modification methylase. *J Biol Chem* **247:** 6176–6182.

Lederberg S, Meselson M. 1964. Degradation of non-replicating bacteriophage DNA in non-accepting cells. *J Mol Biol* **8:** 623–628.

Linn S, Arber W. 1968. Host specificity of DNA produced by *Escherichia coli*, X. In vitro restriction of phage fd replicative form. *Proc Natl Acad Sci* **59:** 1300–1306.

Linn S, Eskin B, Lautenberger JA, Lackey D, Kimball M. 1977. Host-controlled modification and restriction enzymes of *Escherichia coli* B and the role of adenosylmethionine. In *The biochemistry of adenosylmethionine* (ed. Salvatore F, et al.), pp. 521–535. Columbia University Press, New York.

Linn S, Lautenberger B, Eskin B, Lackey D. 1974. The host-controlled restriction and modification enzymes of *E. coli* B. *Fed Proc* **33:** 1128–1134.

Luria SE. 1984. *A slot machine, a broken test tube: an autobiography.* Harper & Row, New York.

Luria SE, Delbruck M. 1943. Mutations of bacteria from virus sensitivity to virus resistance. *Genetics* **28:** 491–511.

Luria SE, Human ML. 1952. A nonhereditary, host-induced variation of bacterial viruses. *J Bacteriol* **64:** 557–569.

Mandel M, Higa A. 1970. Calcium-dependent bacteriophage DNA infection. *J Mol Biol* **53:** 159–162.

Meselson M, Stahl FW. 1958. The replication of DNA in *Escherichia coli*. *Proc Natl Acad Sci* **44:** 671–682.

Meselson M, Yuan R. 1968. DNA restriction enzyme from *E. coli*. *Nature* **217:** 1110–1114.

Meselson M, Yuan R, Heywood J. 1972. Restriction and modification of DNA. *Ann Rev Biochem* **41:** 447–466.

Portugal FH, Cohen JS. 1977. *A history of the discovery of the structure and function of the genetic substance*. MIT (to the Lighthouse Press), Boston.

Roberts RJ, Belfort M, Bestor T, Bhagwat AS, Bickle TA, Bitinaite J, Blumenthal RM, Degtyarev S, Dryden DT, Dybvig K, et al. 2003. A nomenclature for restriction enzymes, DNA methyltransferases, homing endonucleases and their genes. *Nucl Acids Res* **31:** 1805–1812.

Roberts RJ, Vincze T, Posfai J, Macelis D. 2015. REBASE—a database for DNA restriction and modification: enzymes, genes and genomes. *Nucl Acids Res* **43:** D298–D299.

Rosamond J, Endlich B, Telander KM, Linn S. 1979. Mechanisms of action of the type-I restriction endonuclease, ecoB, and the recBC DNase from *Escherichia coli*. *Cold Spring Harb Symp Quant Biol* **43**(Pt 2): 1049–1057.

Roulland-Dussoix D, Boyer HW. 1969. The *Escherichia coli* B restriction endonuclease. *Biochim Biophys Acta* **195:** 219–229.

Spector DH, Smith K, Padgett T, McCombe P, Roulland-Dussoix D, Moscovici C, Varmus HE, Bishop JM. 1978. Unaffected avian cells contain RNA related to the transforming gene of avian sarcoma viruses. *Cell* **13:** 371–379.

Stahl FW. 1969. *The mechanisms of inheritance*, 2nd ed. *Foundations of modern genetics series* (ed. Suskind SS, Hartman PE). Prentice-Hall, Englewood Cliffs, NJ.

Stent GS, Calendar R. 1978. *Molecular genetics. An introductory narrative*, 2nd ed. W.H. Freeman, San Francisco.

Takano T, Watanabe T, Fukasawa T. 1966. Specific inactivation of infectious lambda DNA by sonicates of restrictive bacteria with R factors. *Biochem Biophys Res Commun* **25:** 192–198.

Takano T, Watanabe T, Fukasawa T. 1968. Mechanism of host-controlled restriction of bacteriophage lambda by R factors in *Escherichia coli* K12. *Virology* **34:** 290–302.

van Ormondt H, Lautenberger JA, Linn S, de Waard A. 1973. Methylated oligonucleotides derived from bacteriophage fd RF-DNA modified in vitro by *E. coli* B modification methylase. *FEBS Lett* **33:** 177–180.

Watanabe T. 1963. Infective heredity of multiple drug resistance in bacteria. *Bacteriol Rev* **27:** 87–115.

Watanabe T, Fukasawa T, Takano T. 1962. Conversion of male bacteria of *Escherichia coli* K12 to resistance to f phages by infection with the episome "resistance transfer factor." *Virology* **17:** 217–219.

Watanabe T, Nishida H, Ogata C, Arai T, Sato S. 1964. Episome-mediated transfer of drug resistance in enterobacteriaceae. VII. Two types of naturally occurring R factors. *J Bacteriol* **88:** 716–726.

Watanabe T, Takano T, Arai T, Nishida H, Sato S. 1966. Episome-mediated transfer of drug resistance in *Enterobacteriaceae*. X. Restriction and modification of phages by fi^- R factors. *J Bacteriol* **92:** 477–486.

Watson JD. 1970. *Molecular biology of the gene*, 2nd ed. W.A. Benjamin, New York.

Watson JD, Crick FH. 1953. Molecular structure of nucleic acids: a structure for deoxyribose nucleic acid. *Nature* **171:** 737–738.

Yoshimori R. 1971. "A genetic and biochemical analysis of the restriction and modification of DNA by resistance transfer factors." PhD thesis, University of California, San Francisco.

Yoshimori R, Roulland-Dussoix D, Boyer HW. 1972. R factor–controlled restriction and modification of deoxyribonucleic acid: restriction mutants. *J Bacteriol* **112:** 1275–1279.

Yuan R, Heywood J, Meselson M. 1972. ATP hydrolysis by restriction endonuclease from *E. coli* K. *Nature: New Biol* **240:** 42–43.

Yuan R, Meselson M. 1970. A specific complex between a restriction endonuclease and its DNA substrate. *Proc Natl Acad Sci* **65:** 357–362.

WWW RESOURCES

https://en.wikipedia.org/wiki/Daisy_Roulland-Dussoix A biography of Daisy Roulland-Dussoix.

https://en.wikipedia.org/wiki/Grete_Kellenberger-Gujer A biography of Grete Kellenberger-Gujer.

http://www.estherlederberg.com/home.html The Esther M. Zimmer Lederberg memorial website.

http://library.cshl.edu/Meetings/restriction-enzymes/v-Arber.php History of restriction enzymes meeting archive.

http://www.library.cshl.edu/Meetings/restriction-enzymes/v-Roberts.php History of restriction enzymes archive.

http://rebase.neb.com/rebase/rebase.html A restriction enzyme database.

http://www.sci.sdsu.edu/~smaloy/MicrobialGenetics/topics/phage/lambda-dgal.html Formation of lambda LFT and HFT lysates.

https://en.wikipedia.org/wiki/Esther_Lederberg A biography of Esther Lederberg.

APPENDIX 1: TSUTOMU WATANABE[1]

Born in Gifu city, Gifu prefecture. Graduated from Keio Gijuku University School of Medicine in 1948 and studied microbial genetics. He spent his entire life in the study of the mechanism involved in the acquisition of drug resistance by bacteria until his death by stomach cancer on Nov. 4, 1972, and achieved major contributions in not only modern molecular biology but also therapeutic medicine and public health. In the early decades of his scientific life, he carefully examined the genetic mechanism by which bacteria become resistant to streptomycin, a drug that produced much benefit in the treatment of tuberculosis in those days, and proved experimentally the "spontaneous mutation and selection" mechanism. This achievement brought revision to the then-prevailing hypothesis that resistance occurred by "induction" by drugs. The latter half of his scientific life began with the discovery that the simultaneous

---

[1]Translation from the Japanese Wikipedia, courtesy Hiroshi Nikaido at Berkeley.

acquisition of resistance to many drugs by various pathogenic bacteria occurs by the transmission of extranuclear genetic material; he created the name resistance transfer factor (RTF) for this material. Furthermore, he found that RTF belongs to the category of episomes of François Jacob, through examination from various angles. These studies represent a major achievement in modern molecular biology, especially molecular genetics, as attested to by his invitation to a number of international meetings and his giving of memorable lectures on these occasions. For this achievement he received the award from the Japanese Society of Bacteriology and the Purkinje prize from the Czechoslovak Academy of Sciences. It should be emphasized that these studies gave theoretical foundations on the rational use of antibiotics in the treatment of infectious disease patients. That is, if we make a mistake in the proper use of antibiotics, the most powerful weapon for us in the continuing battle between humans and microorganisms, it will cause the extensive spread of resistant microorganisms in nature. He appealed to the scientists, doctors, and public about the dangers of the spreading multidrug-resistant plasmids and warned about the use of excessive amounts of antibiotics for farm animals as well as in aquaculture. In his last years, he tried to examine this situation himself by collaborating with the Egusa laboratory of Tokyo University School of Agriculture, Department of Fishery. He spent most of his scientific life in the Department of Microbiology, Keio University School of Medicine, taught numerous medical students, and trained many research scientists. He also taught molecular biology as a part-time lecturer at Tokyo University School of Agriculture and at Ochanomizu University School of Science.

APPENDIX 2: E-MAILS BETWEEN MATT MESELSON AND NOREEN MURRAY

Re: Your question about EcoKI

Matthew S. Meselson <msm@wjh.harvard.edu>   Mon, Jul 5, 2004 at 9:09 PM
To: Noreen Murray <Noreen.Murray@ed.ac.uk>
Cc: 'PTASHNE, Mark -- Mark Ptashne' <m-ptashne@mskmail.mskcc.org>,
Mark Ptashne <m-ptashne@ski.mskcc.org>, 'Matthew S. Meselson' <msm@wjh.harvard.edu>

Dear Noreen,

We always wondered about Jean and Joe's UV effect and had only hand-waving explanations for it. Astounding that, already knowing not to attack semi-methylated DNA, the minuscule bacteria know to inactivate restriction if their own DNA fails to get methylated. Congratulations on your elucidation of the mechanism.

Last October Mark Ptashne asked me about the isolation of EcoKI. So I have taken the easy way out by pasting my reply to him below. After e-mailing it, I found or maybe Mark told me that I had confused John Lis with Stuart Linn, for which I am embarrassed.

You are right about Bill Wood and methionine. Seymour Lederberg and I had earlier done some things with restriction degradation of DNA that made us look into it ourselves. As you are delving into history, you may like to have the list of our restriction-modification publications below, including the one with Seymour.

At the end of what I sent to Mark, I have added an account of the wrong hypothesis about lambda injection that caused me to try ATP at the start of my attempt to isolate a K restriction enzyme. Again taking the easy way out, I am copying this e-mail to Mark.

Matthew Meselson
Department of Molecular and Cellular Biology
Harvard University
7 Divinity Avenue
Cambridge, MA 02138 USA
email: <msm@wjh.harvard.edu>
telephone: (617) 495-2264
telefax: (617) 496-244

******

'Luria had our MS for PNAS for many weeks before telling me that I should shorten it. So I sent it to Nature. As was the practice then, I sent a pre-publication copy of the MS to colleagues. That was at about the time we sent it to Salva for PNAS. I sent a copy to John Lis [read Stuart Linn] who was with Arber and maybe a copy to Bill Wood.

The whole trick to isolating the type I restriction enzymes was to know the right co-factors. I had discovered the ATP requirement in a somewhat hilarious way because of a wrong hypothesis about phage lambda injection that you can get me to tell you about over a drink. So long as the lysate was fairly crude, ATP was all that was needed for endonucleolytic action. Adding ATP to crude lysate of restricting bacteria, I got good cleavage of tritium-labeled unmodified lambda DNA with P-32 labeled modified DNA (or the reverse) as control. I measured cleavage by sedimentation in sucrose gradients and dripping and distinguished the radioactive isotopes in Ed Lenhof's scintillation counters on the fifth floor of the Biolabs.

At that point, Bob Yuan came and I invited him to join me in attempting purification of the presumed enzyme. During initial steps of purification the activity went down. But from some experiments of Bill Wood on the effect of methionine on restriction we got the idea that methionine might also be required. With ATP and methionine added we continued to purify but again the activity went down as purification continued. At that point Bob Yuan, knowing biochemistry which he learned in part from Bernie Horecker, and which as you know I do not know very well, realized that S-adenosylmethyl-transferase in the still impure preparation might be making SAM from methionine and ATP.

At first, we thought that only SAM would be needed. But it soon turned out that SAM and ATP are both needed, allowing us to purify the endonuclease to "homogeneity". We discovered to our surprise that even in the limit digest, not all lambda molecules are broken in the same places. The argument followed from the sedimentation distribution of the digested lambda pieces. For example, although there were pieces of size in the range of 40 percent of a lambda chromosome, they accounted for less than 40 mass percent of the digest. (The DNA moves with respect to the enzyme for variable distances from its recognition sites.) This, alas, made the type I endos not useful for genetic engineering. In the same Nature paper we showed that the enzyme nicks before it makes a double-strand break and that hybrid lambda DNA made by annealing modified and unmodified strands is neither nicked nor cleaved, meaning that the enzyme looks at both chains before deciding what to do.

Whether Linn and Arber independently discovered the rather complex co-factor requirements of the endonuclease or instead first learned of the requirements from our MS I do not know.

Looking at Werner's Nobel account, I gather that they got the co-factor requirements for the B restriction endonuclease by assuming they were the same as those we had found for the K enzyme.

******

ABOUT LAMBDA AND ATP: I had been trying entirely without success to find an endonuclease activity in Rec+ coli not present in Rec−. I had isolated some Rec− mutants even before I knew of John Clark's work at Berkeley. For an assay, I looked for reduction in sedimentation velocity of P-32 labeled lambda DNA in sucrose gradients. Finally, dispirited, disgusted and feeling that I had to prove to myself that I was not simply incompetent in biochemistry, I decided to abandon recombination enzymes and look for some other kind of endonuclease--a restriction endonuclease. The assay I had been using for Rec+ endo could be made better by mixing modified and non-modified lambda DNA, one labeled with 32-P and the other with 3-H. Now, there were six tubes in the swinging bucket rotor I had for the experiment. One tube for the mix by itself, one for the mix with r- bacterial extract, one for r+ extract but what to do with the other three tubes? I decided to use up two of the remaining tubes by adding ATP and can't remember what I did with the sixth tube. Why ATP? I knew that cyanide and azide prevent phage T4 injection. Phage workers used azide to synchronize T4 injections. So I thought maybe lambda too needed ATP for injection. The next step in this faulty argument came from asking where would you fight an invader. At the point of entry -- the city gates-- of course. So that is where cells should position their restriction apparatus to defend against hostile DNA. Then it could be that the apparatus that participated in injection and needed ATP to do so would also need ATP to do restriction. A real non-sequitur. Still, it "smelled" right and even though it wasn't right, it worked. From then on, one could try to be a biochemist, as related above.

Of course another reason for adding ATP is that biochemists always add ATP!

******

Ihler, G. and M. Meselson. 1963. Genetic Recombination in Bacteriophage lambda by Breakage and Joining of DNA Molecules. Virology, 21: 7–10.

Lederberg, S. and M. Meselson. 1964. The Degradation of Non-Replicating Bacteriophage DNA in Non-accepting Cells. Journal of Molecular Biology, 8: 623–628.

Menninger, J.R., M. Wright, L. Menninger, and M. Meselson. 1968. Attachment and Detachment of Bacteriophage lambda DNA in Lysogenization and Induction. Journal of Molecular Biology, 32: 631–637.

Meselson, M. and R. Yuan. 1968. An Endonuclease of Host-Controlled Restriction in *E. coli*. Federation Proceedings, 27: 395.

Meselson, M. and R. Yuan. 1968. DNA Restriction Enzyme from *E. coli*. Nature, 217: 1110–1114.

Yuan, R. and M. Meselson. 1969. Binding of lambda DNA by a Restriction Endonuclease. Federation Proceedings, 28: 465.

Yuan, R. and M. Meselson. 1970. A Specific Complex Between a Restriction Endonuclease and its DNA Substrate. PNAS, 65: 357–362.

Meselson, M. and R. Yuan. 1971. DNA Restriction Enzyme from *E. coli*. *Procedures in Nucleic Acid Research*, eds. G. Cantoni and R. Davies, Harper and Row, New York, 2: 889–895.

Yuan, R. and M. Meselson. 1971. DNA Restriction Enzyme from *E. coli. Methods in Enzymolgy*, eds. Grossman and Moldave, Academic Press, 21: 269.

Meselson, M., R. Yuan, and J. Heywood. 1972. Restriction and Modification of DNA. Annual Review of Biochemistry, 41: 447–466.

Haberman, A., J. Heywood, and M. Meselson. 1972. DNA Modification Methylase Activity of *E. coli* Restriction Endonuclease K and P. PNAS, 69: 3138–3141.

Yuan, R., J. Heywood, and M. Meselson 1972. ATP Hydroysis by Restriction Endonuclease from *E. coli*. K. Nature New Biology, 240: 42–43.

******

>From Arber's Nobel Address:

"This work would not have been possible without a very fruitful help by a large number of collaborators in my own laboratory and of colleagues working on related topics in their own laboratories. I was extremely lucky to receive in my laboratory in the basement of the Physics Institute of the University of Geneva a number of first class graduate students, postdoctoral fellows and senior scientists. It is virtually impossible to list them all in this context, but my warmest collective thanks go to all of them. In 1964 Bill Wood laid out a solid basis for the genetics of the restriction and modification systems EcoK and EcoB. Later, Stuart Linn, profiting from his fruitful contacts with Bob Yuan and Matt Meselson, who worked in the USA on the enzymology of EcoK restriction, set the basis for in vitro studies with EcoB restriction and modification activities. These studies culminated in the final proof that modification in *E. coli* B and K is brought about by nucleotide methylation. This concept had found its first experimental evidence during my two months' visit in 1963 with Gunther Stent at the University of California in Berkeley. Several years later Urs Kühnlein, a Ph.D. student, and John Smith, working for various lengths of time with us, succeeded in careful in vivo and in vitro measurements on methylation to validate and extend the earlier conclusions. Their experiments also brought important conclusions with regard to the concept of the sites of recognition on the DNA for the restriction and modification enzymes."

******

On Fri, 2 Jul 2004, Noreen Murray wrote:
Date: Fri, 2 Jul 2004 14:05:42 +0100
From: Noreen Murray <Noreen.Murray@ed.ac.uk>
To: Professor Matthew S Meselson <msm@wjh.harvard.edu>
Subject: History of the purification of EcoKI

Dear Matt,

I write to ask if you would enlighten me of some early facts re: the history of the purification of EcoKI. I have often wondered how you managed to find the appropriate co-factors to permit the purification of this complex enzyme. It is difficult to appreciate how in 1967 it would have been obvious that either ATP or AdoMet (SAM) would be cofactors, although I am aware that the choice of "SAM" was stimulated by the unpublished observation by Bill Wood that restriction in *E. coli* K12 was impaired when the cells were deprived of methionine. I'm also aware of your concept expanded in your review in Ann. Revs. 1972, that SAM would serve to control restriction activity. This ties in with our recent work in which our information on the control of restriction activity has become even more complex.

I'm to give a plenary lecture at an NEB Symposium on R/M systems later this summer, and I'd like to cover some of our more recent work on control of restriction activity together with the pioneering experiments of Bertani and Weigle and Meselson and Yuan.

I can't expect that you will have kept up with the R-M field, but you may be interested to learn that for some families of Type I R/M systems ClpXP degrades the R polypeptide if there are unmodified targets in the bacterial chromosome. A mutation in hsdM that blocks methyltransferase activity, but not the binding of AdoMet leaves an enzyme with endonuclease activity, but the R polypeptide gets degraded if the enzyme translocates the DNA of the bacterial chromosome (Makovets et al. 1999, PNAS 96, 9757, and Doronina and Murray, 2001, Mol. Microb. 39, 416).

My apologies for troubling you, but I would very much like to have accurate historical facts, and I doubt that many are aware of the problems facing purification of the first restriction endonuclease.

With best wishes.

Yours sincerely,

Noreen Murray

Emeritus Professor of Molecular Genetics
Institute of Cell and Molecular Biology
University of Edinburgh
Darwin Building, Mayfield Road
Edinburgh EH9 3JR
Scotland, UK

Tel. ++ 44 131 650 5374

APPENDIX 3: MORE ABOUT DAISY ROULLAND-DUSSOIX

After completion of the last chapter of this book, several people pointed out that Daisy Roulland-Dussoix's contribution to the early restriction field was not limited to the two hallmark papers published with Werner Arber in *Journal of Molecular Biology* in 1962 (Arber and Dussoix 1962; Dussoix and Arber 1962), as she also played an important role in scientific discoveries later on. Hence, this appendix gives a short resume of her scientific career, based on https://en.wikipedia.org/wiki/Daisy_Roulland-Dussoix, which was written by a scientist at the University of Geneva, with the help of Daisy's brother André Dussoix.

PhD work

Daisy Dussoix (1936–2014) joined the Biophysics group of the University of Geneva in 1959. She obtained her PhD in 1963 with Edouard Kellenberger (former PhD student of Jean Weigle) and Werner Arber as advisors. She studied the nature of the barrier to infection and HCV described in the first chapter of this book. This outstanding work showed both phenomena to occur at the DNA level (Arber and Dussoix 1962; Dussoix and Arber 1962), which would lead to the Nobel Prize for Arber in 1978 (together with Hamilton Smith and Daniel Nathans, whose contributions are described in Chapter 3).

After Weigle moved to Caltech in 1948 (see the dedication to Weigle in the first lambda book; Hershey 1971), the lambda work in Geneva heavily relied on Grete Kellenberger-Gujer (1919–2011), who gave "Werner Arber the conceptual basis and practices for his future studies in the genetics of bacteriophages" and was a pioneer in the "early development of molecular biology," in particular, the genetic analysis of phages (https://en.wikipedia.org/wiki/Grete_Kellenberger-Gujer).

Grete's major scientific contribution was the discovery that recombination was due to a physical exchange of DNA (and not to selective replication (Kellenberger et al. 1961, p. 869). Grete also developed novel methods to prepare and analyze biological samples for the EM. Finally, it is worth mentioning that these experiments were made possible by earlier discoveries by Esther Zimmer Lederberg (1922–2006), the American microbiologist and pioneer of bacterial genetics, who discovered phage lambda, specialized transduction, and the bacterial fertility factor (the F plasmid) and also developed the method of replica plating (https://en.wikipedia.org/wiki/Esther_Lederberg). See also the Esther M. Zimmer Lederberg Memorial website (http://www.estherlederberg.com/home.html) for more information on these early days.

Postdoctoral Work

In 1964, Dussoix went to Stanford University as a postdoc and married Daniel Roulland in San Francisco. She became an assistant professor at UCSF in 1968, where she published five papers with Herb Boyer, including the famous 1972 paper on the discovery and analysis of EcoRI, in which her experience and knowledge were invaluable to their PhD student Robert Yoshimori (Yoshimori et al. 1972). She later lectured at UCSF and contributed to the *Cell* paper on avian sarcoma proto-oncogenes (src; Spector et al. 1978) with Harold E. Varmus, who would receive the Nobel Prize for the discovery of the cellular origin of retroviral oncogenes in 1989 (together with J. Michael Bishop).

In 1980 she moved to the Pasteur Institute in Paris, where she worked on PCR-based mycoplasma methods. In 1987 she became Group Head of the Mycoplasma laboratory, part of the Viral Oncology Unit of Luc Montagnier, which resulted in eight significant publications between 1985 and 1998.

Malaria

Sadly, during one of her many travels she contracted malaria and went into a coma. She never fully recovered from this illness. After the death of Daniel Roulland in Paris in 2005, André Dussoix moved his sister back to Switzerland, where she passed away in 2014.

CHAPTER 3

The Discovery of Type II Restriction Enzymes: The 1970s

THE DISCOVERY OF HindII IN THE LABORATORY OF HAMILTON SMITH[1]

As described in the previous chapters, early work on R-M enzymes had focused on the enzymes of *Escherichia coli* K12 and B and phage P1. The work by the Geneva group of Werner Arber supported the notion that the enzymes would recognize a specific sequence, but the determination of this sequence was no easy matter. As chance brought Werner Arber into the restriction field, it was also by chance that Hamilton (Ham) Smith at Johns Hopkins identified the first recognition site of a REase that belongs to a family of enzymes that would change the landscape of molecular biology. Trained as a clinician, Ham Smith's vocation in life had changed some years earlier when reading the Watson and Crick paper on the model of the DNA double helix (Watson and Crick 1953). In a journal club he discussed another interesting paper, that of the purification and in vitro study of the EcoKI enzyme (Chapter 2; Meselson and Yuan 1968). Kent Wilcox from his laboratory was present at this seminar. When a little later he encountered rapid degradation of phage ^{32}P-labeled P22 DNA after transformation in competent cells of *Haemophilus influenza*, Kent wondered: Was the degradation of P22 DNA perhaps also due to restriction? Although skeptical at first, that night at home Ham Smith realized that this could be tested in the Ostwald viscometer (Fig. 1) (http://library.cshl.edu/Meet ings/restriction-enzymes/v-Smith.php. 2013). Changes in the viscosity of the DNA would relate to changes in the size of that DNA!

And indeed, in the presence of Mg^{2+} the viscosity of the P22 DNA dropped rapidly within 2 minutes after addition of the cell extract, whereas that of the control *H. influenzae* DNA remained constant. The rate of this degradation of the "foreign" P22 DNA was proportional to the time (Smith and Wilcox

---

[1]http://library.cshl.edu/Meetings/restriction-enzymes/v-Smith.php. 2013.
Chapter doi:10.1101/restrictionenzymes_3

Viscometric assay for restriction enzymes.

The reaction begins when extract is added to the DNA plus buffer solution in the Ostwald viscometer.

Reaction mixture is sucked up to level A.

Time in seconds for the level to drop from A to B gets shorter as the DNA is broken down.

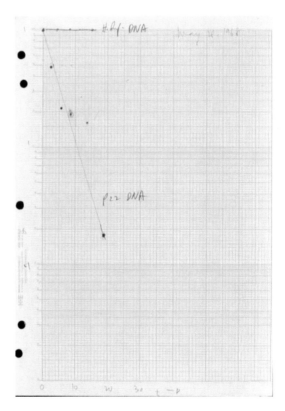

FIGURE 1. The viscosity meter experiment, May 1968, to measure degradation of foreign P22 DNA in *H. influenza*. (Courtesy of Ham Smith.)

1970). Purification of this activity confirmed the existence in the extract of a nuclease that recognized the foreign DNA but not the bacterial DNA. This enzyme, called EndoR, later renamed HindII, not only cleaved DNA of P22, but also that of phage T7 (±70 times). Alkaline gradients (that separate the DNA strands) showed the cuts to be double-strand breaks. Did this enzyme cut a specific sequence, and if so, could one sequence the actual end(s)?

Ham Smith recalls their luck in having Bernard (Bernie) Weiss on the same floor of their building (Fig. 2, left). Bernie Weiss had come from Charles (Charlie) Richardson's laboratory at Harvard (where he had discovered T4 DNA ligase). Bernie Weiss supplied them with T4 polynucleotide kinase (pnk) and homemade γ-^{32}P ATP "that was so hot that it turned brown from radiation" (http://library.cshl.edu/Meetings/restriction-enzymes/v-Smith.php. 2013). Labeling the 5' termini of the cleavage products with pnk allowed digestion of the DNA with various nucleases to obtain labeled oligonucleotides, which

FIGURE 2. (*Left*) Members of the Department of Microbiology at Johns Hopkins in the mid-1970s. (From *left* to *right*) Ken Berns, Tom Kelly, Dan Nathans, Ham Smith, and Bernie Weiss. (*Right*) Kathleen Danna and Dan Nathans at the CSH Tumor Virus Meeting, summer 1970. (Courtesy of Ham Smith.)

were analyzed by chromatography and electrophoresis. These procedures revealed the dinucleotide to be AA and GA. Ham Smith gave a talk about this at the Federation meeting in Atlantic City in April 1969 and recalls that there was little general interest in their results, although Stu Linn recalls that he was very much interested indeed.

But what was the nature of the structure? When Tom Kelly arrived as a postdoc, he could use newly commercially available ^{33}P in combination with $5'$ ^{32}P label and thin-layer electrophoresis. This allowed him to prove the specific recognition sequence to consist of six bases. There were three possibilities: the breaks could be opposite each other (Fig. 3A) or away from each other with either a $5'$ or $3'$ overhang (Fig. 3B or 3C).

Subsequent experiments showed the cut to be blunt end with no intervening bases; hence, structure A in Figure 3 was correct. The recognition sequence of EndoR/HindII $5'$-GTPy-PuAC-$3'$ was published in 1970, back-to-back with a paper on the purification of the enzyme (Kelly and Smith 1970; Smith and Wilcox 1970). In contrast to the lack of interest the year before at the Federation meeting, this time the impact was enormous and Salva Luria and Werner Arber sent their congratulations.

Around that time, Dan Nathans, who also worked on the same floor as Ham Smith, was about to return from sabbatical leave in Israel, where he had learned to handle the oncogenic simian virus SV40. He felt that the cancer field was about to explode and he wanted to look at transformation by this eukaryotic virus. Could he test EndoR on SV40 DNA? Initially he and PhD student Kathleen Danna (Fig. 2, right) used sucrose gradients and polyacrylamide tube gels, which failed to give clean and clear results. When they resorted to radiography of dried-down gels, however, resolution was excellent. In the

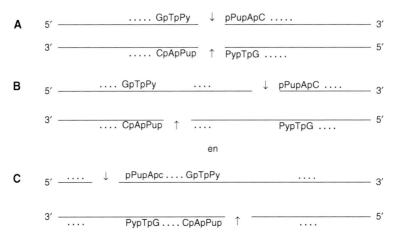

FIGURE 3. Possible structures of the EndoR/HindII recognition site. The cuts on the two DNA strands could be blunt end (*A*) or staggered with either a 3′ or 5′ overhang (*B* and *C*).

absence of DNA markers, Kathleen Danna mapped the sites on the SV40 DNA using hundreds of gels with partial digests: partials had to be isolated, digested again with the same or another enzyme, ordered, and the sizes determined with the help of the label, a tedious procedure (Danna and Nathans 1971). In 1973 they published the first map of SV40 (Danna et al. 1973). Dan Nathans then proposed a nomenclature for R-M systems (Smith and Nathans 1973): a three-letter abbreviation for the host strain plus a subtype designation, where necessary (e.g., EcoK for *E. coli* K12 or Hind for *H. influenza* strain d). Different R-M systems in a strain would be indicated by roman numerals (e.g., EcoKI).

Together with Werner Arber and Ham Smith, Dan Nathans was awarded the Nobel Prize in Physiology or Medicine in 1978. Sadly, he passed away in 1999 (Danna 2010).

THE DISCOVERY OF EcoRI

One of the best-known enzymes discovered in the early 1970s was identified in the laboratory of Herb Boyer at University of California, San Francisco (UCSF). Herb Boyer developed a strong interest in genetics and chemistry fueled by his science and math teacher Pat Bucci and was "completely blown away by the structure and heuristic value of the structure" of the model of the DNA double helix (http://library.cshl.edu/Meetings/restriction-enzymes/v-Boyer.php. 2013). How he entered the restriction field will sound familiar: For his PhD thesis he had to analyze the regulation of the arabinose operon using conjugation experiments between *E. coli* K12 and *E. coli* B strains. Thus he encountered host-controlled variation. After his move to UCSF, he asked Daisy Roulland-Dussoix to purify the *E. coli* B enzyme, EcoBI, which she did (Chapter 2). But when Stu Linn told them that EcoBI cut randomly, the project was abandoned.

In the meantime, Herb Boyer had come across the pioneering papers by Tsutomu Watanabe and coworkers in Tokyo on multidrug resistance in clinical strains, a major health hazard in Japan in the early 1950s (Watanabe et al. 1964, 1966; Takano et al. 1966, 1968). Apparently, drug resistance was located on extrachromosomal DNA, and, in addition, some of these episomes also showed nuclease activity. They called these resistance transfer factors (RTFs). During the analysis of these RTFs, they were the first to provide evidence for restriction of lambda DNA in vitro, which required Mg^{2+} for activity (Watanabe et al. 1964; Takano et al. 1966). Herb Boyer wondered: Would clinical strains collected in California also carry such restricting RTFs? He asked PhD student Bob Yoshimori to identify such restricting RTF factors in the hospital collection. He should try to test this by transfer of the RTF plasmids from the clinical backgrounds into *E. coli* K12. This search resulted in two different restriction specificities: one was the same as one of Watanabe's RTFs, named EcoRII, but the other one had a novel specificity, called EcoRI. Like HindII, and in contrast to EcoKI and EcoBI, both required only Mg^{2+} for activity. Moreover, the recognition site of EcoRI (GAATTC) did not have an ambiguity like the HindII site, and to their surprise and delight, the enzyme cut after the first G on both strands producing "AATT" sticky ends (Yoshimori et al. 1972).

Could the symmetrical sticky ends make it possible to mix fragments of different DNAs and transform them into *E. coli*? The answer to that was positive: Genetic engineering was born. EcoRI would become one of the most important enzymes of the decade, as it made a limited number of cuts on lambda DNA and various plasmids and a single one on SV40 DNA (Hedgpeth et al. 1972; Jackson et al. 1972; Cohen et al. 1973). This led to the construction of cloning vectors based on lambda, plasmids, and SV40. In these constructions the new protocol for the analysis of DNA fragments using agarose gels combined with staining with ethidium bromide greatly simplified the analysis of both size and number of fragments made by the enzymes (Sharp et al. 1973).

THE CLASSIFICATION OF THE RESTRICTION-MODIFICATION ENZYMES INTO FOUR MAIN TYPES

The differences between the enzymes encoded by *E. coli*, *H. influenza*, and the RTF plasmids led to the classification in *S*-adenosylmethionine (SAM)-

dependent Type I and SAM-independent Type II enzymes (Boyer 1971). A little later, new evidence warranted a further division of the SAM-dependent enzymes: Methionine analogs (such as ethionine, which in contrast to methionine does allow protein synthesis to proceed) revealed two classes of enzymes: those truly dependent on SAM for restriction (Type I enzymes: no REase activity in the absence of SAM → stable DNA in the presence of analogs → survival of the cells), and those that were less dependent on SAM (REase active → cells self-destruct in the presence of analogs). The latter type, which includes the phage P1 enzyme, was renamed Type III in 1978 (Kauc and Piekarowicz 1978).

The story does not end there, as Luria's T* mutant phages (Chapter 1) led to the discovery of another class of restriction enzymes. T* mutants are unable to add glucose residues to the hydroxymethylcytosine (hm5C) bases in their DNA. hm5C blocks a great many restriction enzymes, but not the *E. coli* Rgl1 and Rgl2 (restriction of glucose-less DNA) enzymes. Thomas (Tom) Bickle recalls that "these enzymes were considered to be more or less an anomaly by the late 1970s, and certainly, at that time, not at the forefront of research" (http://library.cshl.edu/Meetings/restriction-enzymes/v-Bickle.php 2013). He had started as a "diploma" student of Werner Arber, and in 1973 joined Robert (Bob) Yuan at the new Basel Biozentrum, where Werner Arber had moved to in 1971. Tom Bickle started his own group in Basel in 1977 and would work on restriction enzymes for the rest of his career.

The methylation- and/or other modification-dependent enzymes caused a lot of cloning trouble when people tried to clone often considerably methylated, eukaryotic DNA in *E. coli* strains (Raleigh et al. 1988). This led to the discovery that Rgl restricts both m5C and hm5C, which also meant cloning trouble when trying to (over)express some modification enzymes in *E. coli*. The enzymes were renamed Mcr (methyl cytosine restriction) and classified as Type IV in a multi-authored review in 2003 (Roberts et al. 2003). This review details the characteristics of different subclasses of the Type I and Type II enzymes (see http://rebase.neb.com/rebase/rebase.html for current details).

SUBDIVISIONS

The original definition of Type II enzymes was that they recognize specific DNA sequences and cleave at a constant position at or close to that sequence. As a rule they require only Mg^{2+} as cofactor, though more and more exceptions to this rule were and are being identified. The Type II enzymes are highly heterogeneous though the subtypes sometimes overlap, and an enzyme may belong to more than one subtype (Roberts et al. 2003). For example, BcgI is both a Type IIB enzyme (cuts on both sites of its recognition sequence) and Type IIC enzyme (DNA cleavage and modification are carried out by one and the same polypeptide). EcoRI is an enzyme that fits the original definition, and this subtype is called Type IIP.

Tom Bickle compared the heterogeneity with "Cuvier's pachyderms" (http://library.cshl.edu/Meetings/restriction-enzymes/v-Bickle.php). Cuvier was a French zoologist who tried to classify animals above species level and grouped animals with thick skins such as rhinos and hippos (called pachyderms in Greek). In this way, organisms (and by extension proteins) that are not genetically and/or evolutionarily linked are classified based on a single particular trait. In the case of REases, Type II enzymes do not (usually) require ATP and/or SAM, which sets them apart from the Type I and Type III enzymes. Similarly, their general inability to cut the recognition site when modified sets them apart from the Type IV enzymes, although some Type II enzymes (such as Type IIM DpnI) will only cut methylated DNA. In evolutionary terms, the Type II enzymes are very highly heterogeneous, both in protein sequences and reaction mechanisms.

SUBSEQUENT ENZYME DISCOVERIES

After the discovery of HindII, EcoRI, EcoRII, and their recognition sequences, several other enzymes soon followed. By 1974, recognition sequences of five additional enzymes (HaeIII, HindIII, HpaI, HpaII, and AvaI) were known, as well as nine other enzymes that recognize the same sequence, so-called "isoschizomers," and 16 less well-characterized REases (Nathans and Smith 1975). All known sites were 4- to 6-bp-long with twofold rotational symmetry. This led to the idea that the REase would also have twofold symmetry, which would be in line with EcoRI having two identical subunits. Cleavage could be blunt end, like HindII, or staggered, like EcoRI. The enzymes could have a 5' overhang with 2, 3, 4, or 5 nucleotides, or a 3' overhang of 2 or 4 nucleotides, posing an interesting mechanistic problem to protein chemists for years to come. In addition to these enzymes cutting at the site, the first example of a Type II enzyme that cuts away from the site was identified with the help of known lambda DNA sequences (Kleid et al. 1976).

How fast the field was expanding becomes clear from the next major survey of the field a little later (Roberts 1976). By 1976, 86 REases were known with 45 different specificities, 22 of these identified in *Haemophilus* species. The finding of an enzyme in the Gram-positive *Bacillus subtilis* raised expectations that REases might not be limited to Gram-negative bacteria but might be widespread throughout the bacterial kingdom (Bron et al. 1975).

Several facts were puzzling: Why would one strain make a lot of enzyme, whereas another one such a tiny amount that it was impossible to characterize properly? What was the pattern or origin of the REases? They were found on plasmids, phages, (defective) prophages, and on the chromosome. Richard (Rich) Roberts wondered whether perhaps REases were coded by these mobile DNA elements, which would use them "to shuffle genetic information and create new recombinant genomes in the same way that has recently been accomplished in vitro" (Roberts 1976). Did REases exist in higher organisms such as the lower eukaryotes yeast, fungi, or ciliates? Nothing emerged, but such enzymes might be missed because of low amounts, instability, or peculiar cofactors.

EARLY USES OF TYPE II REases IN RECOMBINANT DNA

This topic has been extensively covered in reviews and books, and here a short overview is given of the birth of genetic engineering.

Mapping

One of the first applications of REases was physical mapping of chromosomes (see Nathans and Smith 1975 for details). Specific restriction sites could be correlated with the genetic map and serve as physical reference points. In this way, genes, as well as start and stop signs for transcription and translation, could be mapped. Direction of transcription, origin of DNA replication, and direction of replication could be resolved (Nathans and Smith 1975). Binding sites for enzymes could be located on specific DNA fragments, and the map could also be used to find deletions, rearrangements, insertions, or substitutions that previously relied on EM visualization. Such research could reveal differences between strains of viruses and their evolutionary changes or show recombination events (Nathans and Smith 1975). One interesting early example was the discovery of maternal inheritance of mitochondria: The mule and the hinny never have the same mitochondrial DNA, as this is always inherited from the mother horse or mother donkey and never from the father (Hutchison et al. 1974).

Sequencing

Another application of REases was to simplify nucleotide sequencing (Salser 1974) by providing small DNA restriction fragments from <20 to ±1000 bp. These would be generated by sequential enzyme digestion, starting with enzymes that cleave once or only a few times in a given genome (such as that of SV40, adenovirus, or phage lambda), followed by other enzymes producing overlapping smaller fragments amenable to sequencing. Around this time, novel DNA sequencing protocols (Sanger and Coulson 1975; Maxam and Gilbert 1977) and the useful "Southern blot" technique were published (Southern 1975a).

Isolation of Genes and Analysis of Repeat Sequences

In addition, one could isolate genes (Brown and Stern 1974) or analyze repetitive DNA sequences that gave discrete bands superimposed on the large and long smear of a million or more bands, generated when cutting a mammalian genome (Mowbray and Landy 1974). For example, EcoRII cleaves purified mouse satellite DNA into some 20-odd regularly spaced bands, which are multiples of a 240-bp repeat (Southern 1975b). Methods were not available yet to undertake "detailed analysis of the unique fraction of eukaryotic DNA [which] will require amplification of these sequences artificially, e.g. by molecular cloning. Such DNA segments might then be amenable to the same type of analyses applied to viral chromosomes and satellite DNA" (Nathans and Smith 1975). With the then available techniques, it became easier to analyze repetitive DNA in eukaryotes—satellite repeats, as well as the multiple copies of histone genes and rRNA genes.

Recombinant DNA

One of the most far-reaching applications of REases was the in vitro construction of DNA molecules with novel biological activities. Insertion or deletion mutants or artificial recombinants could be made by joining DNA fragments via the sticky ends (made at the cut site by the REase) or by homopolymer tailing with oligo(dT). Excision, addition, and rearrangement of fragments in a given genome or combination of DNA from various sources into recombinant molecules became possible (Jackson et al. 1972; Cohen and Chang 1973; Cohen et al. 1973; Lobban and Kaiser 1973; Hershfield et al. 1974; Lai and Nathans 1974; Murray and Murray 1974). Such recombined molecules could be cloned in suitable cells and propagated to yield large quantities of the new DNA, and, in some cases, specific gene products (e.g., *X. laevis* rDNA in pSC101 [Morrow et al. 1974] or *trp* in lambda [Murray and Murray 1974]). In this way, one also could produce specific transducing viruses for animal cells (Brockman and Nathans 1974; Nathans et al. 1974).

All these exciting new uses of REases in molecular biology, to analyze the DNA of many different organisms, made many forget the biological properties

of the REases. A few laboratories did not, and currently the world of REases
and their MTases is at least as exciting as the field of genetic engineering and
DNA sequencing, as detailed later.

See, for some further background reading, Judson (1979) and Watson and
Tooze (1981).

REFERENCES

Boyer HW. 1971. DNA restriction and modification mechanisms in bacteria. *Ann Rev Microbiol*
  25: 153–176.
Brockman WW, Nathans D. 1974. The isolation of simian virus 40 variants with specifically
  altered genomes. *Proc Natl Acad Sci* **71:** 942–946.
Bron S, Murray K, Trautner TA. 1975. Restriction and modification in *B. subtilis*. Purification
  and general properties of a restriction endonuclease from strain R. *Mol Gen Genet* **143:**
  13–23.
Brown DD, Stern R. 1974. Methods of gene isolation. *Ann Rev Biochem* **43:** 667–693.
Cohen SN, Chang AC. 1973. Recircularization and autonomous replication of a sheared R-factor
  DNA segment in *Escherichia coli* transformants. *Proc Natl Acad Sci* **70:** 1293–1297.
Cohen SN, Chang AC, Boyer HW, Helling RB. 1973. Construction of biologically functional
  bacterial plasmids in vitro. *Proc Natl Acad Sci* **70:** 3240–3244.
Danna D. 2010. Biographical memoir. *Proc Am Philos Soc* **154:** 338–354.
Danna K, Nathans D. 1971. Specific cleavage of simian virus 40 DNA by restriction endonu-
  clease of *Hemophilus influenzae*. *Proc Natl Acad Sci* **68:** 2913–2917.
Danna KJ, Sack GH Jr, Nathans D. 1973. Studies of simian virus 40 DNA. VII. A cleavage
  map of the SV40 genome. *J Mol Biol* **78:** 363–376.
Hedgpeth J, Goodman HM, Boyer HW. 1972. DNA nucleotide sequence restricted by the RI
  endonuclease. *Proc Natl Acad Sci* **69:** 3448–3452.
Hershfield V, Boyer HW, Yanofsky C, Lovett MA, Helinski DR. 1974. Plasmid ColEl as a
  molecular vehicle for cloning and amplification of DNA. *Proc Natl Acad Sci* **71:** 3455–3459.
Hutchison CA III, Newbold JE, Potter SS, Edgell MH. 1974. Maternal inheritance of mammalian mitochondrial DNA. *Nature* **251:** 536–538.
Jackson DA, Symons RH, Berg P. 1972. Biochemical method for inserting new genetic informa-
  tion into DNA of Simian Virus 40: Circular SV40 DNA molecules containing lambda phage
  genes and the galactose operon of *Escherichia coli*. *Proc Natl Acad Sci* **69:** 2904–2909.
Judson HF. 1979. *The eighth day of creation. The makers of the revolution in biology*, a Touchstone
  book. Simon and Schuster, New York.
Kauc L, Piekarowicz A. 1978. Purification and properties of a new restriction endonuclease from
  Haemophilus influenzae Rf. *Eur J Biochem/FEBS* **92:** 417–426.
Kelly TJ Jr, Smith HO. 1970. A restriction enzyme from *Hemophilus influenzae*. II. *J Mol Biol*
  51: 393–409.
Kleid D, Humayun Z, Jeffrey A, Ptashne M. 1976. Novel properties of a restriction endonuclease
  isolated from *Haemophilus parahaemolyticus*. *Proc Natl Acad Sci* **73:** 293–297.
Lai CJ, Nathans D. 1974. Deletion mutants of simian virus 40 generated by enzymatic excision
  of DNA segments from the viral genome. *J Mol Biol* **89:** 179–193.
Lobban PE, Kaiser AD. 1973. Enzymatic end-to end joining of DNA molecules. *J Mol Biol* **78:**
  453–471.

Maxam AM, Gilbert W. 1977. A new method for sequencing DNA. *Proc Natl Acad Sci* **74:** 560–564.

Meselson M, Yuan R. 1968. DNA restriction enzyme from *E. coli. Nature* **217:** 1110–1114.

Morrow JF, Cohen SN, Chang AC, Boyer HW, Goodman HM, Helling RB. 1974. Replication and transcription of eukaryotic DNA in *Escherichia coli. Proc Natl Acad Sci* **71:** 1743–1747.

Mowbray SL, Landy A. 1974. Generation of specific repeated fragments of eukaryote DNA. *Proc Natl Acad Sci* **71:** 1920–1924.

Murray NE, Murray K. 1974. Manipulation of restriction targets in phage lambda to form receptor chromosomes for DNA fragments. *Nature* **251:** 476–481.

Nathans D, Smith HO. 1975. Restriction endonucleases in the analysis and restructuring of DNA molecules. *Ann Rev Biochem* **44:** 273–293.

Nathans D, Adler SP, Brockman WW, Danna KJ, Lee TN, Sack GH Jr. 1974. Use of restriction endonucleases in analyzing the genome of simian virus 40. *Fed Proc* **33:** 1135–1138.

Raleigh EA, Murray NE, Revel H, Blumenthal RM, Westaway D, Reith AD, Rigby PW, Elhai J, Hanahan D. 1988. McrA and McrB restriction phenotypes of some *E. coli* strains and implications for gene cloning. *Nucleic Acids Res* **16:** 1563–1575.

Roberts RJ. 1976. Restriction endonucleases. *CRC Crit Rev Biochem* **4:** 123–164.

Roberts RJ, Belfort M, Bestor T, Bhagwat AS, Bickle TA, Bitinaite J, Blumenthal RM, Degtyarev S, Dryden DT, Dybvig K, et al. 2003. A nomenclature for restriction enzymes, DNA methyltransferases, homing endonucleases and their genes. *Nucleic Acids Res* **31:** 1805–1812.

Salser WA. 1974. DNA sequencing techniques. *Ann Rev Biochem* **43:** 923–965.

Sanger F, Coulson AR. 1975. A rapid method for determining sequences in DNA by primed synthesis with DNA polymerase. *J Mol Biol* **94:** 441–448.

Sharp PA, Sugden B, Sambrook J. 1973. Detection of two restriction endonuclease activities in *Haemophilus parainfluenzae* using analytical agarose–ethidium bromide electrophoresis. *Biochemistry* **12:** 3055–3063.

Smith HO, Nathans D. 1973. Letter: A suggested nomenclature for bacterial host modification and restriction systems and their enzymes. *J Mol Biol* **81:** 419–423.

Smith HO, Wilcox KW. 1970. A restriction enzyme from *Hemophilus influenzae*. I. Purification and general properties. *J Mol Biol* **51:** 379–391.

Southern EM. 1975a. Detection of specific sequences among DNA fragments separated by gel electrophoresis. *J Mol Biol* **98:** 503–517.

Southern EM. 1975b. Long range periodicities in mouse satellite DNA. *J Mol Biol* **94:** 51–69.

Takano T, Watanabe T, Fukasawa T. 1966. Specific inactivation of infectious lambda DNA by sonicates of restrictive bacteria with R factors. *Biochem Biophys Res Commun* **25:** 192–198.

Takano T, Watanabe T, Fukasawa T. 1968. Mechanism of host-controlled restriction of bacteriophage lambda by R factors in *Escherichia coli* K12. *Virology* **34:** 290–302.

Watanabe T, Nishida H, Ogata C, Arai T, Sato S. 1964. Episome-mediated transfer of drug resistance in enterobacteriaceae. VII. Two types of naturally occurring R factors. *J Bacteriol* **88:** 716–726.

Watanabe T, Takano T, Arai T, Nishida H, Sato S. 1966. Episome-mediated transfer of drug resistance in enterobacteriaceae X. Restriction and modification of phages by fi R factors. *J Bacteriol* **92:** 477–486.

Watson JD, Crick FH. 1953. Molecular structure of nucleic acids: a structure for deoxyribose nucleic acid. *Nature* **171:** 737–738.

Watson JD, Tooze J. 1981. *The DNA story: a documentary history of gene cloning*. WH Freeman, San Francisco.

Yoshimori R, Roulland-Dussoix D, Boyer HW. 1972. R factor-controlled restriction and modification of deoxyribonucleic acid: Restriction mutants. *J Bacteriol* **112:** 1275–1279.

WWW RESOURCES

http://library.cshl.edu/Meetings/restriction-enzymes/v-Bickle.php   Bickle, T. 2013. Variations on a theme: the families of restriction/modification enzymes.

http://library.cshl.edu/Meetings/restriction-enzymes/v-Boyer.php   Boyer, H. 2013. The discovery of EcoRI and its uses in recombinant DNA.

http://library.cshl.edu/Meetings/restriction-enzymes/v-Smith.php   Smith, H. 2013. Discovery of the first Type II restriction enzyme and its aftermath.

http://rebase.neb.com/rebase/rebase.html   The Restriction Enzyme Database.

CHAPTER 4

Expansion and Cloning Restriction Enzymes: The 1970s and Early 1980s

EXPANSION AND IDENTIFICATION OF REases IN DIFFERENT BACTERIAL STRAINS

Like others experiencing their "Eureka" or "Aha" moment, Rich Roberts remembers his own "incredible moment" at Harvard in 1972, when he heard a talk by Dan Nathans about specific restriction of SV40 DNA by endonuclease R (HindII; Chapter 3). He immediately realized that REases were the answer to the burning question of how one could generate small DNA fragments that would be ideal for the development of DNA sequencing methods. (Fred Sanger had already developed methods for sequencing RNA, because small molecules such as 5S RNA upon which new approaches could be tested were available. However, no comparably small DNA molecules were available.) After moving to Cold Spring Harbor Laboratory (CSHL) in the autumn of 1972, Rich Roberts entered the R-M field and moved to the forefront right away. In the process, he discovered split genes in adenovirus in 1977 (Chow et al. 1977), for which he would share the Nobel Prize in Physiology or Medicine with Phillip (Phil) Sharp in 1993.

At CSHL, he started by collecting the half dozen known REases: EcoRI and endonuclease R from Ham Smith (Chapter 3), HpaI and HpaII from Phil Sharp (Sharp et al. 1973), and endonuclease Z from Clyde Hutchison III (Middleton et al. 1972). During the purification of endonuclease R, it turned out that the preparation actually contained a second enzyme, and the two enzymes became known as HindII and HindIII following the nomenclature rules first proposed by Smith and Nathans (Smith and Nathans 1973). Likewise, during purification of endonuclease Z it turned out that a second enzyme was present that was called HaeII, and endonuclease Z became renamed to HaeIII. By screening a wide variety of strains from the collection of Jack Strominger that Rich had brought with him to CSHL, 17 new REases

Chapter doi:10.1101/restrictionenzymes_4

were soon identified (including AluI, HhaI, MboI, MboII, and SalI). For the memorable EMBO Workshop (Fig. 1) on "Restriction Enzymes and DNA Sequencing" (organized by Walter Fiers, Jozef [Jeff] Schell, and Marc Van Montagu in Ghent, Belgium in 1974), Rich (Fig. 1A) prepared the first complete REase list, and came up with the name "isoschizomer" to indicate enzymes that recognize the same DNA sequence (Fig. 1B). He expanded and formally published the first comprehensive lists (Roberts 1976a; Roberts 1978), which he would update regularly for many years to come under the name of "The Restriction Enzyme List" (Fig. 2). In the early 1990s, this was stored as a relational database, and in 1993, it was set up as the REBASE database and accompanying website with Dana Macelis (Fig. 3; Roberts and

A

B

Richard J. Roberts, Phyllis A. Myers and John R. Arrand

New Specific Endonucleases

New type II restriction-like endonucleases have been isolated from Arthrobacter luteus (Alu I), Haemophilus aphrophilus (Hap I), Haemophilus haemolyticus (Hha I), Moraxella bovis (Mbo I and Mbo II), Streptomyces achromogenes (Sac I), Streptomyces albus G (Sal I) and Xanthomonas amaranthicola (Xam I). In addition further examples of known restriction endonucleases have been found in Bordetella bronchiseptica (Bbr I = Hind III), Haemophilus haemoglobinophilus [Hhg I = Hae III (endonuclease)[and Moraxella nonliquefaciens (Mno I = Hpa II). The overall distribution of these enzymes and some interesting correlations between isoschizomers will be discussed.

FIGURE 1. EMBO Workshop "Restriction Enzymes and DNA sequencing" in Ghent, 1974. (*A*) (*Left* to *right*) Ulf Pettersson, Werner Arber, and Rich Roberts at this meeting. (*B*) Abstract that introduces the word "isoschizomer" to describe enzymes that recognize the same DNA sequence. (Courtesy of Rich Roberts.)

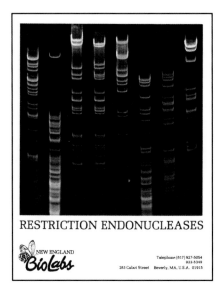

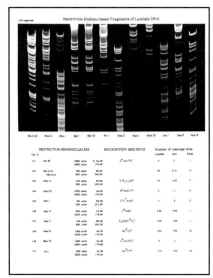

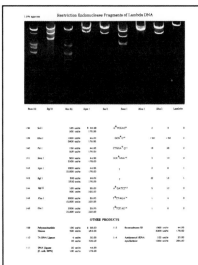

SalI AluI HhaI PstI SmaI KpnI BglI BglII XbaI XhoI
EcoRI HincII HindII HpaI HpaII HaeII HaeIII BamHI

FIGURE 2. Growth of the number of restriction enzymes. From the first NEB catalog. (Courtesy of Rich Roberts.)

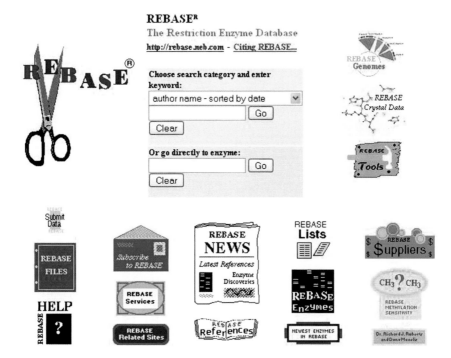

FIGURE 3. Homepage of REBASE (http://rebase.neb.com/rebase/rebase.html).

Macelis 1994), which became a valuable resource for the field (Roberts et al. 2015). Luckily, Fred Sanger was also present at the Ghent meeting. He spoke about his novel plus and minus sequencing method (Sanger et al. 1973) and quite happily shared his protocols with Rich, obtaining REases in return.

After the Ghent meeting the total number of REases identified rapidly increased to 75 and then >100. The analysis of these was greatly aided by the use of agarose slab gel electrophoresis (Sugden et al. 1975), which rapidly replaced the laborious and slippery vertical tube gels. Slab gels were also used to separate digests of large DNA fragments (e.g., the arms of phage lambda vectors), which had to be run on 0.7% agarose gels for 16–24 h at low voltage to prevent the gel from heating and melting in the middle. Hence, in the laboratory of William (Bill) Brammar in Leicester, the arrival of the submerged slab gel, lovingly called "Concorde" gel after the supersonic jet, was met with great enthusiasm.

The vast majority of the 100-odd REases identified were Type II enzymes, for the simple reason that they gave clear bands on gels and, therefore, analysis was easy. In 1976, Rich published a major overview of all known REases (Roberts 1976b). He tried to get Jim Watson to start a company at CSHL with the profits being used to support research at CSHL, but Jim was not interested because it was considered "unseemly" to be involved with commercial ventures and he also did not think that money could be made from it. Rich then joined forces with Don Comb, who had started a company to sell reagents for research and return the profits back to support research at the company (Biolabs, currently New England Biolabs [NEB]). The first catalog listed 18 purified enzymes and brought in $200,000 within the first year, quite a good return in those days.

Purification of REases was one thing, but getting to know the recognition site was quite another. One had to use 2D electrophoresis and painstakingly analyze the spots. In 1977, the year in which RNA splicing was discovered (Chow et al. 1977), the recognition sequence of BamHI was published in *Nature*, emphasizing both the difficulty of determining recognition sequences and also illustrating the value of knowing those recognition sequences (Fig. 4; Roberts et al. 1977).

In the following years, computer programs were developed to aid scientists in finding restriction sites and prepare restriction maps, which were the precursors of the programs currently available, such as REBpredictor (Gingeras et al. 1978), CUTTER, MUTATE, and REMAP (Blumenthal et al. 1982), currently NEBcutter (Vincze et al. 2003).

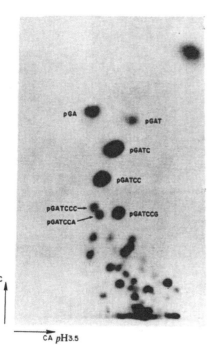

FIGURE 4. Determination of the recognition sequence of BamHI by analysis of radioactive spots after 2D electrophoresis. (Reprinted from Roberts et al. 1977, courtesy of Rich Roberts.)

CLONING OF THE GENES ENCODING REases:
THE FIRST R-M SYSTEMS

After the publication of the EcoRI and HindII/III R-M systems, many laboratories tried to clone the genes encoding REases in order to make their purification easier. Others wanted to sequence the systems to see if they could find common motifs. Could one use the corresponding protein sequence to find a common framework for a common DNA specificity? If so, one might be able to generate mutants that would recognize brand-new recognition sequences and save the labor of screening hundreds of new organisms! Despite numerous efforts, this proved impossible because REases are nearly always very different from one another (this is in contrast to the MTases [Pósfai et al. 1989]). Whether this incredible variety and lack of similarity means that they are diverging rapidly from a common ancestor or whether they have evolved independently, remains a matter of debate. It is only relatively recently that a family of restriction enzymes, the Type IIG enzymes, has been shown to be amenable to such engineering, mainly because they have restriction, methylation, and specificity functions in a single polypeptide chain (Morgan and Luyten 2009).

The first R-M system to be cloned was that of *Escherichia coli* K12 in the laboratories of Noreen Murray and Bill Brammar in Edinburgh in 1976 (Borck et al. 1976). The authors generated an *E. coli* K12 gene library in phage lambda and selected for recombinant phages that were resistant to EcoKI restriction because of self-modification by M·EcoKI. Some of the other early R-M systems to be identified occurred naturally on small plasmids, making it relatively simple to locate, clone, and sequence the genes and to overproduce the proteins. In addition to EcoRI and EcoRII (Chapter 3) this included EcoRV, PaeR7I, and PvuII (Kosykh et al. 1980; Greene et al. 1981; Newman et al. 1981; Theriault and Roy 1982; Gingeras and Brooks 1983; Bougueleret et al. 1984). The EcoRI and EcoRV REases would be among the first to be crystallized and studied in great detail (Rosenberg et al. 1978; Winkler et al. 1993). In 1978, Ham Smith's group tried to clone the HhaI R-M system from *Haemophilus haemolyticus* by selection for resistance to phage infection but instead ended up cloning a previously unknown system, which they called HhaII (Mann et al. 1978).

CLONING OF THE GENES ENCODING REases
AT NEW ENGLAND BIOLABS, USA[1]

Around that time Geoffrey (Geoff) Wilson left Noreen Murray's laboratory, where he had cloned the phage T4 gene encoding DNA ligase (Wilson and

---

[1]http://library.cshl.edu/Meetings/restriction-enzymes/v-GeoffWilson.php; http://library.cshl
.edu/Meetings/restriction-enzymes/v-Roberts.php.

Murray 1979). He moved to the United States and in 1980 joined NEB, where he started cloning R-M systems with the aim of making overproducing strains for commercial purposes, as well as for fundamental studies. He had his first success by building on the previous work of Walder and colleagues (Walder et al. 1981), successfully making an *E. coli* strain overexpressing PstI by duplicating the REase gene and expressing it from the strong early lambda promoter p_L. The result was a 100-fold increase in the yield of enzyme—and an immediate 10-fold drop in the price NEB charged for it, a win–win result establishing a precedent that continues to this day! In the process of cloning PstI he identified another R-M specificity in *Providencia stuartii*, named PstII, later shown to be a Type III R-M system (Sears et al. 2005).

Although phage selection led to the successful cloning of the EcoKI (Borck et al. 1976), HhaII (Mann et al. 1978), and PstI (Walder et al. 1981; Loenen 1982) R-M genes, attempts to clone other R-M systems by this means generally failed (e.g., HhaI, HaeII, HaeIII, HpaI, and HpaII). Finding reasons for and solutions to these failures was slow, as new problems were encountered time and time again. Could one perhaps clone enzymes using an in vitro variant of the "methylase-selection" method (Mann et al. 1978)? This procedure, sometimes called the "Hungarian trick" of Pál Venetianer, involved selection for self-modified clones by in vitro DNA restriction and the recovery of survivors by transformation into a nonrestricting host (Szomolanyi et al. 1980). Although this method yielded several cloned MTase genes by 1983, none of these appeared to express the corresponding restriction enzyme or to restrict phage lambda. Did this indicate that some R-M systems might produce too much REase too soon? If so, that would kill their host cell before all of the recognition sites in its DNA had been protectively modified by the incoming MTase. The strategy to solve this "establishment problem" was to clone the MTase gene first, allow full modification of the host DNA to occur, and in a second step acquire the adjacent REase gene. This strategy indeed worked for Joan Brooks at NEB, who used it to clone the DdeI, BamHI, and BglII systems (Howard et al. 1986; Brooks et al. 1989; Anton et al. 1997). However, several R-M systems continued to make little or no restriction enzyme or to restrict phage (e.g., MspI, TaqI, HpaII, and BglI [Walder et al. 1983; Slatko et al. 1987; Nwankwo and Wilson 1988; Lin et al. 1989; Card et al. 1990; Kulakauskas et al. 1994; Newman et al. 1998]). Much later, evidence would emerge for transcriptional regulation of the R genes of a number of Type II enzymes by control (C) proteins and antisense promoters (Tao et al. 1991; Tao and Blumenthal 1992; Liu et al. 2007; Liu and Kobayashi 2007; Mruk et al. 2011).

Quite a different type of cloning problem emerged: Certain *E. coli* strains do not tolerate m5C methylation and hence prohibit the cloning of this type of MTase genes (Noyer-Weidner et al. 1986; Raleigh and Wilson 1986; Raleigh et al. 1988). The genes involved, named modified cytosine restriction (*mcr*), would prove to be linked to Luria's paper on restriction of T* mutant phages (by Rgl enzymes: restriction of glucoseless DNA [Luria and Human 1952; Revel and Luria 1970; Noyer-Weidner et al. 1986; Raleigh and Wilson 1986]). And there was more trouble to come: Joe Heitman and Peter Model found another REase activity in *E. coli* that threw a spanner in the works of cloning people. They discovered that some m6A MTases induced an SOS response in the cell because of DNA damage, which activated yet another enzyme, named Mrr (modified DNA rejection and restriction) (Heitman and Model 1987; see also Halford 2009).

By 1988, Keith Lunnen and others in Geoff Wilson's group had cloned part or all of 38 different R-M systems, and three other laboratories had been set up to do the same by Joan Brooks, Elizabeth (Lise) Raleigh, and Jack Benner (Lunnen et al. 1988; Wilson 1988). That year the first NEB meeting on R-M systems was held in Gloucester, Massachusetts and the proceedings were published in a special issue of *Gene*, Vol. 74 (Issue 1, 1988). By the time of the CSHL meeting in 2013, NEB scientists had published more than 825 papers, the majority on R-M systems.

CLONING OF THE GENES ENCODING REases IN VILNIUS, LITHUANIA[2]

NEB was not the only company to look for REases for commercial purposes. At the Institute of Biotechnology (Vilnius, Lithuania), Arvydas Janulaitis started his research and development of REases in 1975 under Soviet rule. He would slowly build an "Enzyme Empire" against all odds (Dickman 1992) and created one of the largest REase collections in the world. From 1985 onward, Arvydas Lubys was involved in this laboratory with the production of optimized tools for molecular biology, including suitable plasmid vectors. Like at NEB, initially they also used phage screening (Mann et al. 1978). The advantage of this procedure was that no purified enzyme was needed for in vitro selection, but unfortunately false positives abounded. They switched to biochemical selection (Szomolanyi et al. 1980), which had the advantage that the library could be screened for multiple MTases in the gene pool.

In 1991, Lithuania became an independent state and the company Fermentas emerged as a spin-off from the Institute of Applied Enzymology. With the aid of EU Structural Fund money, they screened a large number of

---

[2]http://library.cshl.edu/Meetings/restriction-enzymes/v-Janulaitis.php; http://library.cshl.edu/Meetings/restriction-enzymes/v-Lubys.php.

bacteria for isoschizomers and enzymes of new specificity, purified and evaluated biochemical properties and yields of the REases, cloned R-M systems and sequenced the genes, and developed overexpressing strains and optimal purification protocols. By 1994 they had screened approximately 19,000 bacterial strains; by 2008 this figure had risen to more than 80,000. They implemented ISO standards, producing enzymes in so-called clean rooms, and designed a universal "FastDigest" buffer that enables any combination of restriction enzymes to work simultaneously in one reaction tube allowing full digestion by two (or more) REases at the same temperature in 5 min (Janulaitis 2013), rather than the often laborious and time-consuming traditional way of sequential double digests in different buffers. The combination of biochemical selection and two-step cloning enabled them to clone nearly any R-M system in *E. coli* devoid of the Mcr and Mrr REases. By the time of the CSHL meeting in 2013, they had cloned and expressed 80 Type IIP, 30 Type IIS, and four Type IIB R-M systems.

In 2003, Rich Roberts and Lise Raleigh visited Fermentas International, Inc. to negotiate product licensing (Fig. 5). In May 2010, all shares of Fermentas International, Inc. were sold to ThermoFisher Scientific.

FIGURE 5. The delegation from NEB in Vilnius to negotiate licensing with Fermentas International, Inc. (*Left* to *right*) Arvydas Lubys, Egle Radzeviciene, Algimantas Markauskas (then follow NEB people Elizabeth Raleigh, Rich Roberts, Jurate Bitinaite), Viktoras Butkus, and Arvydas Janulaitis. (Courtesy of Arvydas Janulaitis.)

METEORIC RISE IN THE DISCOVERY OF NEW REases
AND THEIR RECOGNITION SEQUENCES

As mentioned earlier, in contrast to the REases, the MTases share identifiable, well-defined common amino acid sequence motifs (Fig. 6). This is independent of whether the MTases flip adenine or cytosine out of the helix for methylation to m6A, m4C, or m5C (Klimasauskas et al. 1989, 1994). These motifs made them highly useful for the discovery of new RM systems in DNA sequences. In 1992, a program called SeqWare, developed by Janos Pósfai, became available at NEB based on this knowledge: SeqWare uses the MTase motifs as a starting point to identify potential R-M systems in GenBank and find the accompanying REases, which are invariably located next, or close, to the MTase genes.

Sanger sequencing simplified, accelerated, and facilitated DNA sequencing for three decades, but in recent years next-generation sequencing from several companies has transformed our ability to determine bacterial and archaeal DNA sequences. Most recently, single-molecule sequencing has become possible (Eid et al. 2009) and importantly can be used to find sites of DNA methylation (Flusberg et al. 2010). This so-called SMRT sequencing method from Pacific Biosciences basically measures how long it takes to add a nucleotide from one position to the next. If on the template a modified base is present, it takes the polymerase a little longer to incorporate across from a methylated base. This shows up as a lag in the profile such that by simply measuring the exact time of the lag, one can work out where the methylated bases are. This works really well for unknown MTases that methylate m6A and m4C but not so well for m5C (Clark et al. 2012). The procedure applies not only to the Type II enzymes but also and most importantly to the Type I and Type III systems that had always

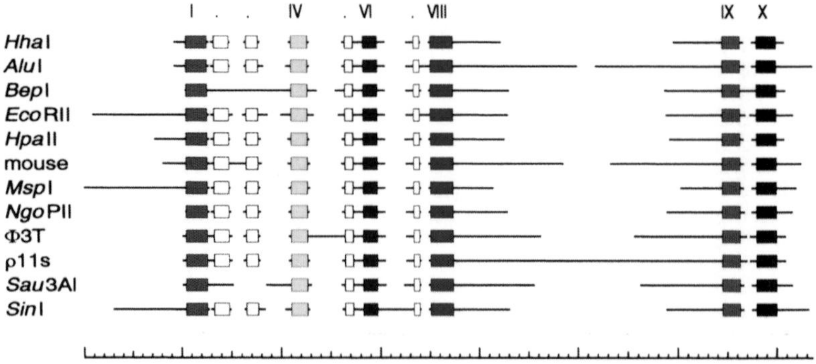

FIGURE 6. Motifs in m5C MTases associated with catalytic methylation (Pósfai et al. 1988, 1989). (Adapted from Kumar et al. 1994; originally adapted from Pósfai et al. 1988, with permission from Elsevier.)

been difficult to work with. Type I enzymes cut randomly and give no discrete bands on gels. Type III enzymes rarely give complete cleavage; hence, they give no clean and useful fragmentation patterns. Their specificity can only be deduced by monitoring their methylation patterns. This is the reason that toward the end of the twentieth century, only 20 Type I and 5 Type III enzymes had been identified versus approximately 2400 Type II REases (with more than 200 specificities) (Bickle and Kruger 1993; Roberts and Halford 1993). For the Type I and Type III R-M systems, the motifs and overall sequences of the MTases enable unambiguous identification of their types and the specificity and restriction subunits are usually located close by. It is becoming clear that many bacterial strains have multiple R-M systems—either complete, partial, or mutated. Currently, for every Type I there are roughly two Type II enzymes, although it varies a lot from one strain to the other. How easy is it to find all of them in a whole genome? Can you get that sort of information out of a whole genome? The answer is yes (Clark et al. 2012). In one study, six genomes were analyzed in this way and 27 MTases were found, 13 of which proved to have novel specificities (Murray et al. 2012). By the time of the CSHL meeting in October 2013, data on no less than 360 genomes had been obtained in 18 months.

The era of a brand-new discovery phase had arrived. Although outside the scope of this book, many of these MTases have no corresponding REases. Does this indicate extensive epigenetics in bacteria? If so, is this because in the "wild" bacteria have to adapt rapidly to environmental changes and DNA methylation helps them do so?

REFERENCES

Anton BP, Heiter DF, Benner JS, Hess EJ, Greenough L, Moran LS, Slatko BE, Brooks JE. 1997. Cloning and characterization of the BglII restriction-modification system reveals a possible evolutionary footprint. *Gene* **187:** 19–27.

Bickle TA, Kruger DH. 1993. Biology of DNA restriction. *Microbiol Rev* **57:** 434–450.

Blumenthal RM, Rice PJ, Roberts RJ. 1982. Computer programs for nucleic acid sequence manipulation. *Nucleic Acids Res* **10:** 91–101.

Borck K, Beggs JD, Brammar WJ, Hopkins AS, Murray NE. 1976. The construction in vitro of transducing derivatives of phage lambda. *Mol Gen Genet* **146:** 199–207.

Bougueleret L, Schwarzstein M, Tsugita A, Zabeau M. 1984. Characterization of the genes coding for the Eco RV restriction and modification system of *Escherichia coli. Nucleic Acids Res* **12:** 3659–3676.

Brooks JE, Benner JS, Heiter DF, Silber KR, Sznyter LA, Jager-Quinton T, Moran LS, Slatko BE, Wilson GG, Nwankwo DO. 1989. Cloning the BamHI restriction modification system. *Nucleic Acids Res* **17:** 979–997.

Card CO, Wilson GG, Weule K, Hasapes J, Kiss A, Roberts RJ. 1990. Cloning and characterization of the HpaII methylase gene. *Nucleic Acids Res* **18:** 1377–1383.

Chow LT, Gelinas RE, Broker TR, Roberts RJ. 1977. An amazing sequence arrangement at the 5′ ends of adenovirus 2 messenger RNA. *Cell* **12**: 1–8.

Clark TA, Murray IA, Morgan RD, Kislyuk AO, Spittle KE, Boitano M, Fomenkov A, Roberts RJ, Korlach J. 2012. Characterization of DNA methyltransferase specificities using single-molecule, real-time DNA sequencing. *Nucleic Acids Res* **40**: e29.

Dickman S. 1992. Lithuanian biochemist builds enzyme empire. *Science* **257**: 1473–1474.

Eid J, Fehr A, Gray J, Luong K, Lyle J, Otto G, Peluso P, Rank D, Baybayan P, Bettman B, et al. 2009. Real-time DNA sequencing from single polymerase molecules. *Science* **323**: 133–138.

Flusberg BA, Webster DR, Lee JH, Travers KJ, Olivares EC, Clark TA, Korlach J, Turner SW. 2010. Direct detection of DNA methylation during single-molecule, real-time sequencing. *Nat Meth* **7**: 461–465.

Gingeras TR, Brooks JE. 1983. Cloned restriction/modification system from *Pseudomonas aeruginosa*. *Proc Natl Acad Sci* **80**: 402–406.

Gingeras TR, Milazzo JP, Roberts RJ. 1978. A computer assisted method for the determination of restriction enzyme recognition sites. *Nucleic Acids Res* **5**: 4105–4127.

Greene PJ, Gupta M, Boyer HW, Brown WE, Rosenberg JM. 1981. Sequence analysis of the DNA encoding the Eco RI endonuclease and methylase. *J Biol Chem* **256**: 2143–2153.

Halford SE. 2009. The (billion dollar) consequences of studying why certain isolates of phage lambda infect only certain strains of *E. coli*: restriction enzymes. *Biochemist* **31**: 10–13.

Heitman J, Model P. 1987. Site-specific methylases induce the SOS DNA repair response in *Escherichia coli*. *J Bacteriol* **169**: 3243–3250.

Howard KA, Card C, Benner JS, Callahan HL, Maunus R, Silber K, Wilson G, Brooks JE. 1986. Cloning the DdeI restriction-modification system using a two-step method. *Nucleic Acids Res* **14**: 7939–7951.

Janulaitis A. 2013. Cut and go—FastDigest with all restriction enzymes at same temperature and buffer: a new paradigm in DNA digestion. In *Modern biopharmaceuticals: recent success stories* (ed. Knäblein J), pp. 135–146. Wiley-VCH, Weinheim.

Klimasauskas S, Timinskas A, Menkevicius S, Butkiene D, Butkus V, Janulaitis A. 1989. Sequence motifs characteristic of DNA[cytosine-N4]methyltransferases: similarity to adenine and cytosine-C5 DNA-methylases. *Nucleic Acids Res* **17**: 9823–9832.

Klimasauskas S, Kumar S, Roberts RJ, Cheng X. 1994. HhaI methyltransferase flips its target base out of the DNA helix. *Cell* **76**: 357–369.

Kosykh VG, Buryanov YI, Bayev AA. 1980. Molecular cloning of *Eco*RII endonuclease and methylase genes. *Mol Gen Genet* **178**: 717–718.

Kulakauskas S, Barsomian JM, Lubys A, Roberts RJ, Wilson GG. 1994. Organization and sequence of the *Hpa*I restriction-modification system and adjacent genes. *Gene* **142**: 9–15.

Kumar S, Cheng X, Klimasauskas S, Mi S, Pósfai J, Roberts RJ, Wilson GG. 1994. The DNA (cytosine-5) methyltransferases. *Nucleic Acids Res* **22**: 1–10.

Lin PM, Lee CH, Roberts RJ. 1989. Cloning and characterization of the genes encoding the *Msp*I restriction modification system. *Nucleic Acids Res* **17**: 3001–3011.

Liu Y, Kobayashi I. 2007. Negative regulation of the EcoRI restriction enzyme gene is associated with intragenic reverse promoters. *J Bacteriol* **189**: 6928–6935.

Liu Y, Ichige A, Kobayashi I. 2007. Regulation of the EcoRI restriction-modification system: identification of *ecoRIM* gene promoters and their upstream negative regulators in the ecoRIR gene. *Gene* **400**: 140–149.

Loenen WAM. 1982. "The construction and use of lambda vectors in molecular cloning." PhD thesis, University of Leicester.

Lunnen KD, Barsomian JM, Camp RR, Card CO, Chen SZ, Croft R, Looney MC, Meda MM, Moran LS, Nwankwo DO, et al. 1988. Cloning type-II restriction and modification genes. *Gene* **74:** 25–32.

Luria SE, Human ML. 1952. A nonhereditary, host-induced variation of bacterial viruses. *J Bacteriol* **64:** 557–569.

Mann MB, Rao RN, Smith HO. 1978. Cloning of restriction and modification genes in *E. coli*: the HhaII system from *Haemophilus haemolyticus. Gene* **3:** 97–112.

Middleton JH, Edgell MH, Hutchison CA 3rd. 1972. Specific fragments of φX174 deoxyribonucleic acid produced by a restriction enzyme from *Haemophilus aegyptius*, endonuclease Z. *J Virol* **10:** 42–50.

Morgan RD, Luyten YA. 2009. Rational engineering of type II restriction endonuclease DNA binding and cleavage specificity. *Nucleic Acids Res* **37:** 5222–5233.

Mruk I, Liu Y, Ge L, Kobayashi I. 2011. Antisense RNA associated with biological regulation of a restriction-modification system. *Nucleic Acids Res* **39:** 5622–5632.

Murray IA, Clark TA, Morgan RD, Boitano M, Anton BP, Luong K, Fomenkov A, Turner SW, Korlach J, Roberts RJ. 2012. The methylomes of six bacteria. *Nucleic Acids Res* **40:** 11450–11462.

Newman AK, Rubin RA, Kim SH, Modrich P. 1981. DNA sequences of structural genes for *Eco* RI DNA restriction and modification enzymes. *J Biol Chem* **256:** 2131–2139.

Newman M, Lunnen K, Wilson G, Greci J, Schildkraut I, Phillips SE. 1998. Crystal structure of restriction endonuclease BglI bound to its interrupted DNA recognition sequence. *EMBO J* **17:** 5466–5476.

Noyer-Weidner M, Diaz R, Reiners L. 1986. Cytosine-specific DNA modification interferes with plasmid establishment in *Escherichia coli* K12: involvement of rglB. *Mol Gen Genet* **205:** 469–475.

Nwankwo DO, Wilson GG. 1988. Cloning and expression of the *Msp*I restriction and modification genes. *Gene* **64:** 1–8.

Pósfai J, Bhagwat AS, Roberts RJ. 1988. Sequence motifs specific for cytosine methyltransferases. *Gene* **74:** 261–265.

Pósfai J, Bhagwat AS, Pósfai G, Roberts RJ. 1989. Predictive motifs derived from cytosine methyltransferases. *Nucleic Acids Res* **17:** 2421–2435.

Raleigh EA, Wilson G. 1986. *Escherichia coli* K-12 restricts DNA containing 5-methylcytosine. *Proc Natl Acad Sci* **83:** 9070–9074.

Raleigh EA, Murray NE, Revel H, Blumenthal RM, Westaway D, Reith AD, Rigby PW, Elhai J, Hanahan D. 1988. McrA and McrB restriction phenotypes of some *E. coli* strains and implications for gene cloning. *Nucleic Acids Res* **16:** 1563–1575.

Revel HR, Luria SE. 1970. DNA-glucosylation in T-even phage: genetic determination and role in phage-host interaction. *Ann Rev Genet* **4:** 177–192.

Roberts RJ. 1976a. Restriction and modification enzymes and their recognition sequences. In *CRC handbook of biochemistry and molecular biology*, 3rd ed., Nucleic Acids Vol II, pp. 532–535. CRC Press, Boca Raton, FL.

Roberts RJ. 1976b. Restriction endonucleases. *CRC Crit Rev Biochem* **4:** 123–164.

Roberts RJ. 1978. Restriction and modification enzymes and their recognition sequences. *Gene* **4:** 183–194.

Roberts RJ, Halford SE. 1993. Type II restriction enzymes. In *Nucleases*, 2nd ed. (ed. Linn SM, et al.), pp. 35–88, Cold Spring Harbor Laboratory Press, Cold Spring Harbor, NY.

Roberts RJ, Macelis D. 1994. REBASE—restriction enzymes and methylases. *Nucleic Acids Res* **22:** 3628–3639.

Roberts RJ, Wilson GA, Young FE. 1977. Recognition sequence of specific endonuclease BamH.I from *Bacillus amyloliquefaciens* H. *Nature* **265:** 82–84.

Roberts RJ, Vincze T, Pósfai J, Macelis D. 2015. REBASE—a database for DNA restriction and modification: enzymes, genes and genomes. *Nucleic Acids Res* **43:** D298–D299.

Rosenberg JM, Dickerson RE, Greene PJ, Boyer HW. 1978. Preliminary X-ray diffraction analysis of crystalline EcoRI endonuclease. *J Mol Biol* **122:** 241–245.

Sanger F, Donelson JE, Coulson AR, Kossel H, Fischer D. 1973. Use of DNA polymerase I primed by a synthetic oligonucleotide to determine a nucleotide sequence in phage fl DNA. *Proc Natl Acad Sci* **70:** 1209–1213.

Sears A, Peakman LJ, Wilson GG, Szczelkun MD. 2005. Characterization of the Type III restriction endonuclease PstII from *Providencia stuartii*. *Nucleic Acids Res* **33:** 4775–4787.

Sharp PA, Sugden B, Sambrook J. 1973. Detection of two restriction endonuclease activities in *Haemophilus parainfluenzae* using analytical agarose–ethidium bromide electrophoresis. *Biochemistry* **12:** 3055–3063.

Slatko BE, Benner JS, Jager-Quinton T, Moran LS, Simcox TG, Van Cott EM, Wilson GG. 1987. Cloning, sequencing and expression of the Taq I restriction-modification system. *Nucleic Acids Res* **15:** 9781–9796.

Smith HO, Nathans D. 1973. Letter: a suggested nomenclature for bacterial host modification and restriction systems and their enzymes. *J Mol Biol* **81:** 419–423.

Sugden B, De Troy B, Roberts RJ, Sambrook J. 1975. Agarose slab-gel electrophoresis equipment. *Anal Biochem* **68:** 36–46.

Szomolanyi E, Kiss A, Venetianer P. 1980. Cloning the modification methylase gene of *Bacillus sphaericus* R in *Escherichia coli*. *Gene* **10:** 219–225.

Tao T, Blumenthal RM. 1992. Sequence and characterization of *pvuIIR*, the *PvuII* endonuclease gene, and of *pvuIIC*, its regulatory gene. *J Bacteriol* **174:** 3395–3398.

Tao T, Bourne JC, Blumenthal RM. 1991. A family of regulatory genes associated with type II restriction-modification systems. *J Bacteriol* **173:** 1367–1375.

Theriault G, Roy PH. 1982. Cloning of *Pseudomonas* plasmid pMG7 and its restriction-modification system in *Escherichia coli*. *Gene* **19:** 355–359.

Vincze T, Pósfai J, Roberts RJ. 2003. NEBcutter: a program to cleave DNA with restriction enzymes. *Nucleic Acids Res* **31:** 3688–3691.

Walder RY, Hartley JL, Donelson JE, Walder JA. 1981. Cloning and expression of the Pst I restriction-modification system in *Escherichia coli*. *Proc Natl Acad Sci* **78:** 1503–1507.

Walder RY, Langtimm CJ, Chatterjee R, Walder JA. 1983. Cloning of the *Msp*I modification enzyme. The site of modification and its effects on cleavage by *Msp*I and *Hpa*II. *J Biol Chem* **258:** 1235–1241.

Wilson GG. 1988. Cloned restriction-modification systems—a review. *Gene* **74:** 281–289.

Wilson GG, Murray NE. 1979. Molecular cloning of the DNA ligase gene from bacteriophage T4. I. Characterisation of the recombinants. *J Mol Biol* **132:** 471–491.

Winkler FK, Banner DW, Oefner C, Tsernoglou D, Brown RS, Heathman SP, Bryan RK, Martin PD, Petratos K, Wilson KS. 1993. The crystal structure of EcoRV endonuclease and of its complexes with cognate and non-cognate DNA fragments. *EMBO J* **12:** 1781–1795.

WWW RESOURCES

http://library.cshl.edu/Meetings/restriction-enzymes/v-GeoffWilson.php   Wilson G. 2013. The cloning efforts at NEB.

http://library.cshl.edu/Meetings/restriction-enzymes/v-Janulaitis.php Janulaitis A. 2013. Science and politics: three phases of commercialization at Fermentas.

http://library.cshl.edu/Meetings/restriction-enzymes/v-Lubys.php Lubys A. 2013. The cloning efforts at Fermentas.

http://library.cshl.edu/Meetings/restriction-enzymes/v-Roberts.php Roberts R. 2013. Many more REs at CSHL, the start of REBASE and more recent work.

http://rebase.neb.com/rebase/rebase.html The Restriction Enzyme Database.

CHAPTER 5

The First Decade after the Discovery of EcoRI: Biochemistry and Sequence Analysis during the 1970s and Early 1980s

INTRODUCTION

The discovery of EcoRI was one of the most important scientific events of the 1970s (Yoshimori 1971; Yoshimori et al. 1972). It opened the way to molecular cloning as well as fundamental research into a large family of endonucleases with an extraordinary high specificity and fidelity. Interest in the Type II REases focused worldwide on their role as reagents, although initially the number of commercially available REases was small (Fig. 1). In several laboratories, hopes were high that REases might be good proteins to study in order to understand their enzymatic properties in DNA–protein interactions. Would this lead to new insights into DNA sequence specificity of REases (and other proteins)? Was there a single mechanism for the discrimination of base pairs? And how did REases distinguish their recognition sequences from the high background of nonspecific sequences?

The period between 1972 and 1982 marked the discovery of many new Type II enzymes, but few Type I and Type III enzymes. EcoRV is currently one of the best-characterized Type II enzymes, but the first important publications on EcoRV date to 1984–1985 (Bougueleret et al. 1984; Bougueleret 1985; Bougueleret et al. 1985; Chapter 6). After Hind II and EcoRI, EcoRII was among the very first REases for which the DNA substrate site was identified (5′C/C[A/T]GG) (Bigger et al. 1973; Boyer et al. 1973). A decade later, EcoRII would prove to be an interesting enzyme with rather different properties from EcoRI (Chapter 6).

During the 1970s, research focused mainly on the biochemistry and sequence recognition of EcoRI and a handful of other enzymes using plasmid,

Chapter doi:10.1101/restrictionenzymes_5

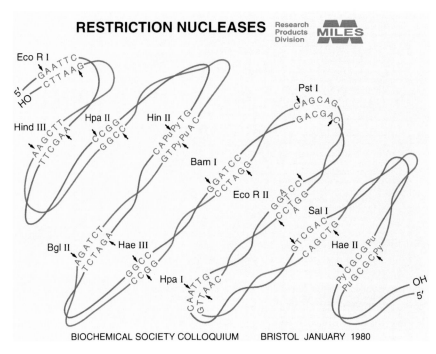

FIGURE 1. Miles Labs' restriction enzymes catalog, circa 1980. (Courtesy of Bayer.)

phage, or artificial DNA as substrate. Degradation was initially measured as the loss of biological activity or change in viscosity (Smith and Wilcox 1970; Gromkova and Goodgal 1972; Middleton et al. 1972; Takanami 1973; Chapters 1–4), but agarose gels soon simplified measurements greatly (Aaij and Borst 1972; Sharp et al. 1973). In particular, the arrival of submerged slab gels was met with great enthusiasm, as mentioned in Chapter 4. Note that although officially the EcoRI proteins should be called R·EcoRI and M·EcoRI to denote the REase and MTase, respectively, by convention the R is dropped in the case of the REase.

TYPE II ENZYMES

New Specificities and Isoschizomers of Type II Enzymes

The major feature that distinguishes one REase from another is its specificity: Where and how does the enzyme cut? The initial approach used by Kelly and Smith in 1970 (Kelly and Smith 1970) could reveal blunt end and/or 5′ or 3′ extensions at, or near, the recognition site (Roychoudhury and Kossel 1971; Murray 1973; Roychoudhury et al. 1976). In the 1970s, thanks to the computer (Brown and Smith 1977) (and agarose gels), banding patterns of new enzymes could be compared with those obtained with REases with known recognition sites. Was it a novel sequence or an isoschizomer of a known enzyme (Fuchs et al. 1978; Gingeras et al. 1978)? Additional experiments using primed synthesis reactions would then be used to confirm those predictions (Brown and Smith 1977; Brown et al. 1980). Isoschizomers were REases from different bacterial species that recognized the same sequence (Roberts 1976). This did not necessarily mean the same cleavage site (e.g., SmaI cleaves CCC/GGG and XmaI C/CCGGG; Endow and Roberts 1977). Such enzymes were later named neoschizomers (Hamablet et al. 1989).

By 1982, 360 Type II enzymes had been recognized, with more than 20 different sequence patterns of 85 specificities (Modrich and Roberts 1982). This was only the beginning of a meteoric rise: By 1993 this figure was nearly 2400 REases with 188 specificities (Roberts and Halford 1993; Roberts and Macelis 1993).

Recognition sites could be palindromic (like EcoRI), but also asymmetric, with ambiguities, or with unspecified bases in between the specific bases. Isoschizomers proved useful for various reasons: Amounts of REases were often found to vary substantially from strain to strain. This made one enzyme more commercially attractive than another or, alternatively, more feasible for biochemical or structural analysis. Also, isoschizomers could be differentially sensitive to methylation—for example, MboI could not cleave Gm6ATC, whereas Sau3A cleaved Gm6ATC (but not GATm5C [Sussenbach et al. 1976; Gelinas et al. 1977]). As the *Escherichia coli* Dam MTase (EcoDam, usually called Dam) methylates adenine in GATC sequences, Sau3A (/GATC) was extensively used to generate overlapping partial digests of DNA isolated from *E. coli* for gene libraries using the BamHI (G/GATTC) site in lambda (Loenen and Brammar 1980). Similarly, HpaII and MspI both recognized CCGG (Garfin and Goodman 1974), but only MspI could cleave Cm5CGG. This allowed screening eukaryotic genomes for the presence or absence of m5C sequences that might be involved in gene control (Waalwijk and Flavell 1978; Bird et al. 1979). Of course, such isoschizomers were also of particular scientific interest: Would these enzymes share common folds or have a common evolutionary origin?

Biochemistry of Type II Systems

By 1982, the MWs of more than 30 REases and MTases were known (Modrich and Roberts 1982). The few purified REases apparently contained a single polypeptide that existed in solution often as one or more oligomers, with some exceptions (Modrich and Roberts 1982). Did the aggregation state of the protein relate to the reaction mechanism (Lee and Chirikjian 1979)? In the case of

EcoRI, the answer appeared to be positive: In solution, EcoRI was a mixture of dimers and tetramers, the former being stable at catalytic concentrations (10^{-10} M) (Modrich and Zabel 1976). This and other data supported the notion that EcoRI interacted with the DNA recognition site as a homodimer.

Although the REases varied considerably in MW (range of ~22–70 kDa), the MTases were all more similar in length (range of ~30–40 kDa). Another striking, and surprising, difference between the REases and MTases was that the five MTases analyzed (M·EcoRI, M·HpaI, M·HpaII, M·RsuI, and Dam) appeared to be monomers (Marinus and Morris 1973; Rubin and Modrich 1977; Yoo and Agarwal 1980b; Gunthert et al. 1981). Did this mean that the REase and MTase interacted with the shared ("cognate") DNA recognition site in a significantly different way (Rubin and Modrich 1977)? And thus did not share a common evolutionary origin, despite recognizing the same DNA sequence?

The DNA Sequence of the EcoRI Genes

In 1981, the first sequence of a Type II R-M system was published—that of EcoRI. This allowed an answer to the burning question about the relationship between the REase and MTase (Greene et al. 1981; Jack et al. 1981; Newman et al. 1981). The REase gene encoded a protein of 277 amino acids (MW = 31 kDa) and the MTase gene a protein of 326 amino acids (MW = 38 kDa) (Newman et al. 1981; Rubin et al. 1981). Homology at the DNA or protein level was minimal; further experiments and computer predictions also made the existence of common features at a higher structural level unlikely (Chou and Fasman 1978a,b; Greene et al. 1981; Newman et al. 1981). This suggested that the two proteins would interact with DNA in different ways, which fit in with EcoRI being dimeric and M·EcoRI monomeric. The crystal structure of EcoRI in complex with DNA was eagerly awaited (McClarin et al. 1986; Kim et al. 1990).

Purification of EcoRI

The mechanism of DNA cleavage by EcoRI was examined in detail. The enzyme required only Mg^{2+} and unmodified DNA. Purification and early studies on the mechanism of cleavage were published by the groups of Paul Modrich in Durham (USA) (Modrich and Zabel 1976; Modrich and Rubin 1977; Rubin and Modrich 1978; Modrich 1979, 1982; Jack et al. 1980, 1981, 1982; Lu et al. 1981; Newman et al. 1981; Rubin et al. 1981; Young et al. 1981; Modrich and Roberts 1982; Cheng et al. 1984; Kim et al. 1984), Stephen (Steve) Halford in Bristol (UK) (Halford et al. 1979; Halford 1980; Halford and Johnson 1980, 1983), and John Rosenberg in Pittsburgh (USA), where he was joined later by Patricia (Pat) Greene (Rosenberg et al. 1978, 1980, 1981; Greene et al. 1981; Rosenberg and Greene 1982). Paul Modrich was awarded a PhD on *E. coli* DNA ligase with Robert Lehman at Stanford and was a postdoc with Charles Richardson at Harvard. At Duke University Medical Center he published on EcoRI from 1976 onward and became the leading expert in the field of strand-directed DNA mismatch repair, for which work he was awarded the Nobel Prize in Chemistry in 2015 (with Tomas Lindahl and Aziz Sancar). Steve Halford had done a PhD and postdoc with enzymologist Herbert Frederick (Freddie) Gutfreund on classical enzymes (e.g., lysozyme) at Bristol University. He switched to REases because he was intrigued by the extraordinary specificity of these enzymes and would study this in great detail. He was lucky to receive help from Nigel Brown, one of the first few people who had actually done a restriction digest! Nigel had mapped φX174 in Fred Sanger's laboratory, a technique he had learned from Rich Roberts' technician Phyllis Myers (Halford 2013). John Rosenberg had obtained a PhD with Alexander Rich at Massachusetts Institute of Technology and a postdoc at Caltech with Richard Dickerson, before moving to Pittsburgh to elucidate the crystal structure of the EcoRI–DNA complex.

Both Halford and Rosenberg recalled their struggle to obtain enough EcoRI enzyme for their studies (Halford 2013; Rosenberg 2013). Halford's protocol for the purification of EcoRI in 1978 was a herculean task: "An enzymologist starts with about 100 mg enzyme!" (Halford 2013). He needed 800 L of bacteria with the EcoRI plasmid (Yoshimori et al. 1972), a 400-L fermentor at Porton Down, a bathtub, a rowing oar for stirring in a sackful of DEAE-cellulose (to absorb the EcoRI enzyme), and an end product of 10 mL of enzyme at 30,000,000 units/mL.

Rosenberg was even more desperate: He needed grams of EcoRI! Fortunately, Marc Zabeau made a strain overproducing EcoRI (Botterman and Zabeau 1985). This strain was useless for biochemical studies, as it produced insoluble protein, but useful for work on the crystal structure (Rosenberg et al. 1978; McClarin et al. 1986). Initially Rosenberg used the "Dickerson" dodecamer as DNA substrate (Drew et al. 1981). Sadly, this EcoRI–dodecamer complex gave poor crystals, but technical reasons ("Serendipity or how 13 can be a lucky number") led to the addition of an extra T at the 5′ side of the 12-mer to give 5′-*T*CGCGAATTCGCG-3′, and those crystals were much better (Rosenberg 2013).

For his restriction analysis, Halford used a set of mutant lambda phages, each with only one of the five EcoRI recognition sites (Halford 2013). These were made by Noreen Murray from Edinburgh University, one of the "architects of the recombinant DNA revolution" (Gann and Beggs 2014), who would study the biology of EcoK and relatives for decades to come (Murray 2000,

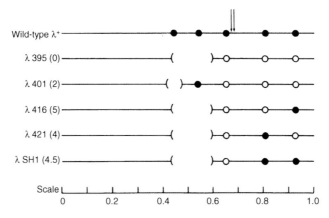

FIGURE 2. Bacteriophage lambda DNA contains five EcoRI sites. Lambda mutants lacking one or more EcoRI sites were useful to assay REase activity both in vivo and in vitro at individual sites. (Closed circle) EcoRI site, (open circle) mutant site, ( ) deletion removing one or two EcoRI sites. (Reprinted from Halford et al. 1980, with permission from Portland Press.)

2002). Stuart (Stu) Linn met Noreen and her husband Kenneth (Ken) Murray in Stanford in the 1960s, where they started their work that would result in many important contributions to the field of molecular biology and cloning DNA (see, e.g., Brockes et al. 1972; Bigger et al. 1973; Murray et al. 1973, 1975; Murray and Murray 1974; Old et al. 1975). In these papers (and in two complete books [Hershey 1971; Hendrix et al. 1983]) phage lambda played a prominent role and led the author of this book into the field.

Noreen had made these lambda mutants by cycling phages alternatively on an *E. coli* host with and without the EcoRI plasmid (Yoshimori et al. 1972) (Fig. 2). This work not only led to the first lambda vector for gene libraries (Murray and Murray 1974), but it also allowed Steve to study the kinetics of the reaction at each individual site in the same DNA context. Although the titer of the phage would drop ~10-fold per site, the efficiency with which each of these sites was cut varied considerably (Halford and Johnson 1980), in line with similar observations by Thomas and Davis (1975). Did this support the notion of involvement of sequences external to the recognition site?

DNA Cleavage by Type II Enzymes

The study of DNA cleavage reactions was facilitated by a convenient biochemical manipulation involving the retention of site-specific complexes of EcoRI with DNA on nitrocellulose filters in the absence of Mg^{2+} (which blocks DNA cleavage) (Modrich and Zabel 1976; Modrich 1979; Halford and Johnson 1980; Jack et al. 1980; Rosenberg et al. 1980). These complexes underwent rapid cleavage upon addition of Mg^{2+}. Depending on enzyme concentration and substrate, the enzyme dimer cut either one or both strands at the same time with a k_{cat} value of 1–4 double-strand cuts/min/dimer at 37°C (Modrich and Roberts 1982). The results suggested a symmetrical complex between the EcoRI homodimer and the palindromic recognition site.

Further support for the influence of DNA outside the recognition site (Greene et al. 1975; Thomas and Davis 1975) came from limited data on seven other Type II enzymes—BamHI, HindIII, SalI, ClaI, BspI, HpaI, and HpaII—and studies with short synthetic DNA duplexes (Modrich and Roberts 1982). For example, different isoschizomers revealed differential strand preference for DNA duplexes with the sequence 5'-GAACCGGAA-3': HpaII cut the complementary bottom (pyrimidine-rich) strand of the duplex three times faster than the upper strand; MspI cleaved the upper strand two times as fast; and a third isoschizomer, MnoI, attacked both strands at an equal rate (Baumstark et al. 1979; Yoo and Agarwal 1980a).

Methyl Transfer by Type II MTases

The kinetics of five purified MTases were simple: A monomer would transfer one methyl group at a time to the recognition site (Modrich and Roberts 1982). In contrast to Dam, EcoK, and EcoB (Vovis et al. 1974), M·EcoRI showed no preference for hemimethylated DNA (Modrich and Roberts 1982). These MTases were a novelty at the time: They were the first examples of monomeric proteins that interacted with twofold symmetrical DNA sequences.

Fidelity of Type II Enzymes

By the early 1970s, a variety of sequence-specific proteins were known to be able to bind DNA molecules lacking recognition sites, albeit with reduced affinity (von Hippel and McGhee 1972). Were Type II enzymes also capable of binding DNA nonspecifically? If so, was this sequence-independent or limited to sites that differed from the true ("canonical") site by only one base pair? In vivo such errors could generate dangerous double-strand breaks: These had to be infrequent in order for cells to survive. Yet another question: Was infidelity a common feature of both REase and MTase, and, if so, would they recognize the same additional site(s)?

On a different track, various experiments suggested that EcoRI could bind sequences other than its cognate GAATTC. Was this an artifact? This question was raised by the fact that some, but not all, batches of EcoRI enzyme nicked DNA sites that differed by only one base pair from the canonical site

(Tikchonenko et al. 1978; Bishop 1979; Bishop and Davies 1980; Maxam and Gilbert 1980).

In addition to this suspected purification artifact in some enzyme batches, EcoRI possessed solution-dependent relaxed (*) activity (much feared by early makers of gene libraries). EcoRI* appeared to preferentially cut G/AATTC but also /AATT sites (Polisky et al. 1975), but we now know (from Linda Jen-Jacobson and others) that that it is almost certain to be because of cutting at sites 1 and not 2 bp different from the recognition sequence. It is just that, under "odd" conditions such as low salt+high pH or Mn^{2+} in place of Mg^{2+} or with dimethylsulfoxide (DMSO) added, the difference in rate between cognate sites and sites 1 bp different is smaller than the more than million-fold difference under ideal conditions. The rates at sites 2 bp different are usually too slow to measure under any conditions. Various other enzymes, including MTases, showed relaxed specificity because of altered assay conditions: high enzyme concentrations, variations in pH and ionic strength, replacement of Mg^{2+} by Mn^{2+}, or addition of glycerol or DMSO (Modrich and Roberts 1982). von Hippel and colleagues (Woodbury et al. 1980) made a valiant effort to examine the sequences cleaved under EcoRI* conditions by exploiting the availability of the total DNA sequence of φX174, which lacks EcoRI sites (Sanger et al. 1978). Although this analysis suggested preference for some sequences, results were not clear-cut and the origin and nature of EcoRI* activity remained unresolved.

Recognition of RNA–DNA Hybrids and/or ssDNA by Type II Enzymes

Both REases and MTases recognized dsDNA: Could they also recognize RNA–DNA hybrids and/or single-strand DNA (ssDNA)? The answer appeared to be a tentative yes, maybe: EcoRI, HaeIII, HhaI, HindII, and MspI cleaved RNA–DNA hybrids (prepared by reverse transcription) in a sequence-specific way (Molloy and Symons 1980). Experiments with denatured salmon sperm or T7 DNA indicated methylation by M·HaeII, M·HaeIII, and Dam (but not M·EcoRI or M·HpaI), albeit at reduced rates (Modrich and Roberts 1982). Likewise, some REases opened up single-stranded circular DNA of phages fl, M13, and φX174, again at reduced rates (Blakesley and Wells 1975; Horiuchi and Zinder 1975; Godson and Roberts 1976). Did this restriction or modification indeed occur at ssDNA, or did this happen within regions of (transient) secondary structure? Studies on HaeII, HpaI (Blakesley et al. 1977), and MspI (Yoo and Agarwal 1980a) led to two different models: In the case of the former (Blakesley et al. 1977), REase recognition was purported to occur only within regions of preformed secondary structure, whereas in the latter case, MspI would be able to form the duplex itself (Yoo and Agarwal 1980a). In contrast to these enzymes, EcoRI and HpaII clearly cut only the duplex form of their recognition sites (Greene et al. 1975; Baumstark et al. 1979). As the oligonucleotides used for HpaII and MspI were identical, apparently these enzymes differed in their interaction with their common recognition sequence (CCGG).

Concluding, neither in the experiments with RNA–DNA hybrids nor in those with ssDNA was evidence sufficient to exclude methylation or restriction occurring in (transient) duplex regions. Three decades later it was shown that a handful of REases (6/223 analyzed) were able to cut RNA–DNA hybrids—for example, Sau3AI can cut 80% of the DNA strand of an RNA–DNA hybrid (http://rebase.neb.com/rebase/rebase.html; Murray et al. 2010), and MspI can cut 5% of the DNA strand of a DNA–RNA duplex (http://rebase.neb.com/rebase/rebase.html). Such data suggest additional roles for some REases in vivo.

Specific versus Nonspecific Enzyme–DNA Interactions by Type II Enzymes

The above experiments suggested differential interaction of different REases and MTases with their recognition sequences. But at what stage did sequence specificity occur? Did the enzyme find the target site and bind? Or was the interaction weak and unstable until the moment of restriction or modification itself (i.e., at the level of catalyzing the phosphodiester bond by the REase) or by transferring the methyl group from SAM to DNA by the MTase? What were the contacts between protein and DNA: the bases and/or the backbone? Did the enzyme bind the major groove or the minor groove of the DNA?

The filter-binding experiments demonstrated that the EcoRI dimer interacted with the GAATTC sequence, with an apparent equilibrium dissociation constant for specific complexes between EcoRI and DNA with a single EcoRI site of $\sim 10^{-9}$–10^{-11} M (DNA of pBR322 or one of the lambda mutants used by Halford) (Modrich and Roberts 1982). In contrast, the nonspecific binding constant was in the micromolar range (Woodhead and Malcolm 1980). Like the *lac* repressor (Jovin 1976; von Hippel 1979), EcoRI also bound nonspecifically to a variety of polynucleotides in the presence of Mg^{2+} (Halford and Johnson 1980; Langowski et al. 1980), the nature of which was not understood (Modrich and Roberts 1982).

Specific complexes between EcoRI and DNA formed in the absence of Mg^{2+} were very stable with $t_{1/2}$ ranging, surprisingly enough, from 16 to 140 min dependent on the chain length (range 6200 to 34 bp; Jack et al. 1980, 1982). This again implicated DNA sequences outside the recognition site (Jack et al. 1982). The rate of specific complex formation was also enhanced by DNA chain length, in line with preferential cleavage of longer DNAs in the presence of Mg^{2+} (Jack et al. 1982). The results were consistent with one-dimensional facilitated diffusion (Richter and Eigen 1974; Schranner and Richter 1978; Berg et al. 1981): that is, a sequence-specific protein would bind nonspecific sites and then diffuse randomly along the DNA helix until the recognition site was found (over a maximum distance of ~1000 bp). Similar effects of DNA chain length on the kinetics of the *lac* repressor were noted (Winter et al. 1981).

Despite the ability of EcoRI to form site-specific complexes in the absence of Mg^{2+}, attempts to detect specific binding by M·EcoRI in the absence of SAM were not successful (Modrich 1979; Jack et al. 1980; Woodhead and Malcolm 1980; Modrich and Roberts 1982).

Interactions of the EcoRI Enzymes with DNA

The above sequences of events indicated that EcoRI would bind the DNA and then slide along in order to find its cognate site. But how did it make specific contacts with the DNA: the major groove, the minor groove, and/or the backbone? To approach this question, base analogs were employed as well as partial chemical alkylation and alkylation interference. These provided the first tentative evidence that DNA contacts in the major groove were important in sequence recognition, and that the DNA–EcoRI complex possessed elements of twofold symmetry (Kaplan and Nierlich 1975; Berkner and Folk 1977; Modrich and Rubin 1977; Berkner and Folk 1979; Lu et al. 1981; Modrich and Roberts 1982). With respect to backbone contacts, ethylation interference studies suggested that four phosphates in each DNA strand were important in specific DNA–enzyme complexes, indicative of interactions with backbone phosphates outside the recognition site (Lu et al. 1981).

Attempts to elucidate the DNA contacts made by M·EcoRI and DNA involved analysis of analog effects on the kinetics of methyl transfer. The results of these studies were surprising and in marked contrast to those obtained with the REase—for example, T4 DNA has a bulky ghm5C base in the major groove, which makes the DNA resistant to EcoRI cleavage but an excellent substrate for M·EcoRI (Berkner and Folk 1977; Modrich and Rubin 1977; Modrich and Roberts 1982). These results were striking and lend support to the conclusion that the two proteins not only interacted with their cognate sites in different ways but also used different mechanisms to discriminate a given base pair.

Other Type II R-M Systems

The only other Type II system examined by 1982 in some detail was HpaI (Mann and Smith 1979). The results resembled those of EcoRI (Mann and Smith 1979). In an interesting experiment, a synthetic duplex oligonucleotide ($5'$-N_xCCGGCCN$_x$) with overlapping HpaII (CCGG) and HaeIII (GGCC) sites was used to examine the effects of m5C methylation on cleavage by the restriction enzymes (Mann and Smith 1979). Did M·HpaII modification affect HaeIII restriction and/or vice versa? Apparently they did: HaeIII cleaved 5'-GGCm5C as opposed to the normally modified site 5'-GGm5CC, which is resistant. Similarly, HpaII cleaved 5'-CCGG with m5C opposite the second G, but not 5'-CCGG.

Preliminary studies with other systems also supported the notion that different enzymes used different mechanisms to recognize a given base pair (Marchionni and Roufa 1978; Berkner and Folk 1979). This was further supported by the use of a random copolymer d(G,C) and analogs of this polynucleotide (Mann and Smith 1979), although other copolymer studies were inconclusive (Mann et al. 1978; Modrich and Roberts 1982).

TYPE I SYSTEMS
Genes and Proteins of Type I Systems

Although the hunt was on for new Type II enzymes, no extensive survey of bacteria was made for the presence of ATP-dependent REases. No techniques were available in the 1980s for the rapid identification of Type I sites but at that time sites of cleavage by Type II REases could easily be identified. Besides, unlike Type II, there was no commercial/technological driving force to find new Type I systems. During this decade, Noreen Murray in Edinburgh and Bob Yuan and Tom Bickle in Basel instigated much of the genetics and biochemistry. By 1982, all known Type I REases had been found in Enterobacteriaceae, with the exception of one enzyme in *Haemophilus* (HindI) (Gromkova and Goodgal 1976; Bickle 1982). These systems were carried by different strains of *E. coli* and *Salmonella typhimurium*, were mapped to the same location on the chromosome as EcoK and EcoB, and were allelic (Chapter 2; Boyer 1964; Arber and Wauters-Willems 1970; Glover 1970; Bullas and Colson 1975; Bachmann and Low 1980; Bullas et al. 1980). Two genes, *hsdM* and *hsdS*, were essential for modification, and a third gene, *hsdR*, was essential for restriction.

The *hsdK* locus of *E. coli* K12 was the first Type I system to be cloned (Borck et al. 1976). This allowed DNA sequence analysis of the smallest gene, *hsdS* (Sain and Murray 1980). By deletion analysis the gene order was shown to be *hsdR*, *hsdM*, *hsdS*, with the *hsdM* and *hsdS* genes transcribed from one promoter and the *hsdR* gene from a separate promoter in the same direction (Sain and Murray 1980).

What was the physical relationship between these systems? Using the novel Southern blot technique (Southern 1975), strong DNA cross-hybridization was detected with DNA probes derived from *hsdM* and *hsdR* of *E. coli* K12 and B; DNA derived from *E. coli* C did not have DNA sequences that hybridized to these probes, in line with the lack of R-M activity (Sain and Murray 1980; Murray et al. 1982). This killed the argument, often made at the time, that the Type I enzymes were "so complicated" that they must perform an additional, vital role in the cell. As *E. coli* C was perfectly viable, this was clearly not the case.

Antibodies raised against the purified HsdM and HsdR subunits of EcoK cross-reacted with purified HsdR and HsdM proteins of EcoB as expected (Murray et al. 1982). These antibodies also cross-reacted with HsdR and HsdM from three *S. typhimurium* strains (SB, SP, and SQ), but not with another *E. coli* strain, *E. coli* A (Arber and Wauters-Willems 1970). This would lead later to a subdivision into Type IA and IB.

Recognition Sequences of Type I Systems

As the Type I enzymes did not cut at their recognition sites, they were difficult to find. The analysis of mutant sequences was the solution to this problem. By the late 1970s the recognition sequences of EcoK and EcoB finally became known and were thought rather unusual: two specific sequences, a trimer and a tetramer, separated by a nonspecific sequence, and without any symmetry. The EcoB recognition sequence was 5'TGA(N8)TGCT (Lautenberger et al. 1978, 1979; Ravetch et al. 1978; Sommer and Schaller 1979), that of EcoK 5'-AAC(N6)GTGC (Kan et al. 1979). The adenine methylated in the EcoB sequence would be on the single A in the top strand and (most likely) the first A on the bottom strand (van Ormondt et al. 1973); in the case of EcoK the single A in the bottom strand was the obvious target for methylation. In the top strand, the second adenine was the site of methylation, as shown by Ineichen and Bickle: They cleverly used one of the EcoK sites in lambda DNA, where the trimeric part of the recognition sequence (AAC) overlaps with a HindII site (GTAAC) (Fig. 2 in Bickle 1982).

Reaction Mechanisms of Type I Systems

Most of the data on the reaction scheme were derived from work on EcoK, but it likely applied to the other Type I enzymes (Bickle 1982). EcoK had at least three binding sites for its cofactor SAM and no detectable affinity for DNA in the absence of SAM. Addition of SAM led to a slow allosteric change, and this form could bind DNA (Hadi et al. 1975). The activated enzyme did not require free SAM in order to restrict and could bind nonspecifically to DNA (Yuan et al. 1975). Subsequently, the enzyme could bind tightly to specific sites and, in the absence of ATP, to both modified and unmodified sites. Such complexes were very stable (like EcoRI; see above): The $t_{1/2}$ of a complex with a modified site was ~6 min, and with an unmodified even as long as 22 min (Yuan et al. 1975).

Once firmly bound to the site, one of three things would happen: if the DNA was modified, the enzyme would be released from the DNA, as shown by both gel filtration and EM studies (Bickle et al. 1978). On heteroduplex DNA (one strand modified), methylation was efficient, stimulated by ATP (Vovis et al. 1974; Burckhardt et al. 1981; Bickle 1982). Such a scenario made sense (Bickle 1982): Newly replicated DNA must be fully methylated before the next round of replication to avoid double-strand breaks. It would be shown later that EcoK does not kill *E. coli* so easily (Chapter 7)! In the third event, that of unmodified DNA, surprisingly, the enzyme underwent a large conformational change: It remained bound to the recognition site, but generated large loops visible in the EM (Bickle et al. 1978; Rosamond et al. 1979). The enzyme would pump the DNA past itself and DNA cleavage would occur somewhere between 400 and 7000 bp away (Horiuchi and Zinder 1972; Adler and Nathans 1973; Bickle et al. 1978; Rosamond et al. 1979; Yuan et al. 1980a). Here, a difference was noted between EcoK and EcoB: Loops by EcoK were bidirectional, and those of EcoB were one-directional (and away from the trimeric part of the recognition sequence). No formal explanation has presented itself until the day of writing, but it is tacitly assumed that EcoK has two HsdR subunits, and EcoB only one during the restriction process.

Whatever the exact mechanism, the cleavage reaction itself was a two-step process: The DNA was nicked in one strand and then, seconds later, in the second strand. It was already noted that in the presence of excess DNA, only single-strand breaks were found, suggesting that two enzyme molecules were required for double-strand cleavage (Meselson and Yuan 1968; Roulland-Dussoix and Boyer 1969; Lautenberger and Linn 1972; Adler and Nathans 1973).

Several other curious and unexplained observations were made, the most puzzling one that of massive continued ATP hydrolysis after restriction, with hydrolysis of an estimated 10.000 ATP/min/enzyme molecule (see Bickle 1982 for further discussion).

TYPE III ENZYMES

By 1982, only three members had been identified. In addition to the system of Joe Bertani's phage P1, EcoP1, a second system was found on a plasmid in *E. coli* 15T$^-$ (EcoP15) and the third in *Haemophilus influenzae* Rf (HinfIII, probably chromosomal) (Arber and Dussoix 1962; Arber and Wauters-Willems 1970; Piekarowicz et al. 1974). EcoP1 and EcoP15 were allelic and highly homologous. Based on their behavior after methionine starvation in the presence of ethionine, they were initially grouped with Type II enzymes, but as their substrate requirements were more complex, they were designated Type III in 1978 (Kauc and Piekarowicz 1978). The EcoP1 enzyme was the first Type III enzyme to be purified (Haberman 1974).

Genetics of Type III Systems

The genetics of the P1 system initially suggested that EcoP1 required three genes for restriction (Scott 1970; Rosner 1973; Reiser 1975), but transposon mutagenesis indicated only two genes, *mod* ahead of *res*, which were transcribed in the same direction from their own promoter (Iida et al. 1983). Both EcoP1 and EcoP15 enzymes contained two subunits of ~75 kDa (Mod) and 100 kDa (Res) (Heilmann et al. 1980; Hadi et al. 1983), which showed immunological cross-reactivity with each other (Fig. 5 in Bickle 1982).

DNA Recognition and Cleavage Sequences of Type III Systems

The recognition sequences were established for EcoP1 (5'AGACC), EcoP15 (5'CAGCAG), and HinfIII (5'CGAAT). Like EcoK and EcoB, methylation occurred on adenine (Brockes et al. 1972; Reiser 1975; Kauc and Piekarowicz 1978). It would remain a puzzle for quite some time and an interesting biological problem—that is, how the cell avoided cell death in the case of EcoP1 and EcoP15, as only the strand shown here can be modified (the opposite strand lacks adenine residues). All three enzymes cleaved the DNA ~25–27 bp away to the right of the sequence shown, with two to three base single-strand 5' extensions. It was estimated that this distance of ~9 nm could easily be spanned by the enzyme without the enzyme having to move along the DNA. A likely model for the action of the enzyme was proposed: The modification subunit would recognize the specificity site and direct the restriction subunit to the site to be cleaved (Bickle 1982).

Reaction Mechanisms of Type III Systems

The reaction mechanisms of all three enzymes were rather similar: All three required ATP for cleavage, but ATP was not hydrolyzed in the process. SAM was not essential, although it stimulated the restriction process. In the presence of both cofactors, restriction and modification competed with each other. Working with these enzymes was complicated for various reasons, making the interpretation of results often difficult (see Bickle 1982 for discussion). SAM was the methyl donor for modification (Yuan and Reiser 1978). The enzyme was able to bind DNA in complex with SAM but also in the absence of cofactors, as seen in the EM (Yuan et al. 1980b). The Type III enzymes did not cleave or methylate each of their recognition sites with equal efficiency: A good methylation site was not necessarily a poor restriction site and vice versa (Yuan et al. 1980b).

In conclusion, a lot of mysterious aspects of the biology and reaction mechanisms of the Type III enzymes remained to be explained. Why were the enzymes inefficient nucleases in vitro but not in vivo? Why was only one strand of the recognition sequence modified (similar to the Type II REase MboII) (Bachi et al. 1979). How did the cell avoid restricting these sites? It left Tom Bickle with an "uncomfortable feeling that something basic" was missing from their in vitro systems (Bickle 1982). It would take decades to find answers to the many puzzling questions with respect to the Type III enzymes. To mention just two landmark papers in the history of the Type III enzymes: Tom Bickle and his coworkers resolved one of the mysterious aspects in 1992, when they showed that the Type III enzymes needed two inversely orientated sites for restriction (Meisel et al. 1992; see Chapter 6), whereas another major breakthrough would be published by Aneel Aggarwal and colleagues in 2015, when they published the structure of EcoP15I (Gupta et al. 2015; see Chapter 8).

REFERENCES

Aaij C, Borst P. 1972. The gel electrophoresis of DNA. *Biochim Biophys Acta* **269:** 192–200.

Adler SP, Nathans D. 1973. Studies of SV 40 DNA. V. Conversion of circular to linear SV 40 DNA by restriction endonuclease from *Escherichia coli* B. *Biochim Biophys Acta* **299:** 177–188.

Arber W, Dussoix D. 1962. Host specificity of DNA produced by *Escherichia coli*. I. Host controlled modification of bacteriophage λ. *J Mol Biol* **5:** 18–36.

Arber W, Wauters-Willems D. 1970. Host specificity of DNA produced by *Escherichia coli*. XII. The two restriction and modification systems of strain 15T. *Mol Gen Genet* **108:** 203–217.

Bachi B, Reiser J, Pirrotta V. 1979. Methylation and cleavage sequences of the *Eco*P1 restriction-modification enzyme. *J Mol Biol* **128:** 143–163.

Bachmann BJ, Low KB. 1980. Linkage map of *Escherichia coli* K-12, edition 6. *Microbiol Rev* **44:** 1–56.

Baumstark BR, Roberts RJ, RajBhandary UL. 1979. Use of short synthetic DNA duplexes as substrates for the restriction endonucleases *Hpa* II and *Mno* I. *J Biol Chem* **254:** 8943–8950.

Berg OG, Winter RB, von Hippel PH. 1981. Diffusion-driven mechanisms of protein translocation on nucleic acids. 1. Models and theory. *Biochemistry* **20:** 6929–6948.

Berkner KL, Folk WR. 1977. EcoRI cleavage and methylation of DNAs containing modified pyrimidines in the recognition sequence. *J Biol Chem* **252:** 3185–3193.

Berkner KL, Folk WR. 1979. The effects of substituted pyrimidines in DNAs on cleavage by sequence-specific endonucleases. *J Biol Chem* **254:** 2551–2560.

Bickle TA. 1982. The ATP-dependent restriction endonucleases. In *Nucleases* (ed. Linn SM, Roberts RJ), pp. 85–108. Cold Spring Harbor Laboratory, Cold Spring Harbor, NY.

Bickle TA, Brack C, Yuan R. 1978. ATP-induced conformational changes in the restriction endonuclease from *Escherichia coli* K-12. *Proc Natl Acad Sci* **75:** 3099–3103.

Bigger CH, Murray K, Murray NE. 1973. Recognition sequence of a restriction enzyme. *Nature (London), New Biol* **244:** 7–10.

Bird AP, Taggart MH, Smith BA. 1979. Methylated and unmethylated DNA compartments in the sea urchin genome. *Cell* **17:** 889–901.

Bishop JO. 1979. A DNA sequence cleaved by restriction endonuclease R.EcoRI in only one strand. *J Mol Biol* **128:** 545–549.

Bishop JO, Davies JA. 1980. Plasmid cloning vectors that can be nicked at a unique site. *Mol Gen Genet MGG* **179:** 573–580.

Blakesley RW, Wells RD. 1975. 'Single-stranded' DNA from phiX174 and M13 is cleaved by certain restriction endonucleases. *Nature* **257:** 421–422.

Blakesley RW, Dodgson JB, Nes IF, Wells RD. 1977. Duplex regions in "single-stranded" φX174 DNA are cleaved by a restriction endonuclease from *Haemophilus aegyptius*. *J Biol Chem* **252:** 7300–7306.

Borck K, Beggs JD, Brammar WJ, Hopkins AS, Murray NE. 1976. The construction in vitro of transducing derivatives of phage λ. *Mol Gen Genet* **146:** 199–207.

Botterman J, Zabeau M. 1985. High-level production of the *Eco*RI endonuclease under the control of the p_L promoter of bacteriophage λ. *Gene* **37:** 229–239.

Bougueleret L. 1985. "Contribution a l'etude des systemes de restriction modification EcoRI and EcoRV." PhD thesis, Université Paris VII Fevrier 8.

Bougueleret L, Schwarzstein M, Tsugita A, Zabeau M. 1984. Characterization of the genes coding for the *Eco* RV restriction and modification system of *Escherichia coli*. *Nucleic Acids Res* **12:** 3659–3676.

Bougueleret L, Tenchini ML, Botterman J, Zabeau M. 1985. Overproduction of the *Eco*R V endonuclease and methylase. *Nucleic Acids Res* **13:** 3823–3839.

Boyer H. 1964. Genetic control of restriction and modification in *Escherichia coli*. *J Bacteriol* **88:** 1652–1660.

Boyer HW, Chow LT, Dugaiczyk A, Hedgpeth J, Goodman HM. 1973. DNA substrate site for the EcoRII restriction endonuclease and modification methylase. *Nature (London), New Biol* **244:** 40–43.

Brockes JP, Brown PR, Murray K. 1972. The deoxyribonucleic acid modification enzyme of bacteriophage P1. *Biochem J* **127:** 1–10.

Brown NL, Smith M. 1977. Cleavage specificity of the restriction endonuclease isolated from *Haemophilus gallinarum* (Hga I). *Proc Natl Acad Sci* **74:** 3213–3216.

Brown NL, Hutchison CA III, Smith M. 1980. The specific non-symmetrical sequence recognized by restriction endonuclease *Mbo*II. *J Mol Biol* **140:** 143–148.

Bullas LR, Colson C. 1975. DNA restriction and modification systems in *Salmonella*. III. SP, a *Salmonella* potsdam system allelic to the SB system in *Salmonella typhimurium*. *Mol Gen Genet* **139:** 177–188.

Bullas LR, Colson C, Neufeld B. 1980. Deoxyribonucleic acid restriction and modification systems in *Salmonella*: chromosomally located systems of different serotypes. *J Bacteriol* **141:** 275–292.

Burckhardt J, Weisemann J, Yuan R. 1981. Characterization of the DNA methylase activity of the restriction enzyme from *Escherichia coli* K. *J Biol Chem* **256:** 4024–4032.

Cheng SC, Kim R, King K, Kim SH, Modrich P. 1984. Isolation of gram quantities of *Eco*RI restriction and modification enzymes from an overproducing strain. *J Biol Chem* **259:** 11571–11575.

Chou PY, Fasman GD. 1978a. Empirical predictions of protein conformation. *Ann Rev Biochem* **47:** 251–276.

Chou PY, Fasman GD. 1978b. Prediction of the secondary structure of proteins from their amino acid sequence. *Adv Enzymol Relat Areas Mol Biol* **47:** 45–148.

Drew HR, Wing RM, Takano T, Broka C, Tanaka S, Itakura K, Dickerson RE. 1981. Structure of a B-DNA dodecamer: conformation and dynamics. *Proc Natl Acad Sci* **78:** 2179–2183.

Endow SA, Roberts RJ. 1977. Two restriction-like enzymes from *Xanthomonas malvacearum*. *J Mol Biol* **112:** 521–529.

Fuchs C, Rosenvold EC, Honigman A, Szybalski W. 1978. A simple method for identifying the palindromic sequences recognized by restriction endonucleases: the nucleotide sequence of the AvaII site. *Gene* **4:** 1–23.

Gann A, Beggs J. 2014. Noreen Elizabeth Murray CBE. 26 February 1935–12 May 2011. http:// rsbmroyalsocietypublishingorg/content/roybiogmem/60/349.

Garfin DE, Goodman HM. 1974. Nucleotide sequences at the cleavage sites of two restriction endonucleases from *Hemophilus parainfluenzae*. *Biochem Biophys Res Commun* **59:** 108–116.

Gelinas RE, Myers PA, Roberts RJ. 1977. Two sequence-specific endonucleases from *Moraxella bovis*. *J Mol Biol* **114:** 169–179.

Gingeras TR, Milazzo JP, Roberts RJ. 1978. A computer assisted method for the determination of restriction enzyme recognition sites. *Nucleic Acids Res* **5:** 4105–4127.

Glover SW. 1970. Functional analysis of host-specificity mutants in *Escherichia coli*. *Genet Res* **15:** 237–250.

Godson GN, Roberts RJ. 1976. A catalogue of cleavages of φX174, S13, G4, and ST-1 DNA by 26 different restriction endonucleases. *Virology* **73:** 561.

Greene PH, Poonian MS, Nussbaum AL, Tobias L, Garfin DE, Boyer HW, Goodman HM. 1975. Restriction and modification of a self-complementary octanucleotide containing the *Eco*RI substrate. *J Mol Biol* **99:** 237–261.

Greene PJ, Gupta M, Boyer HW, Brown WE, Rosenberg JM. 1981. Sequence analysis of the DNA encoding the *Eco* RI endonuclease and methylase. *J Biol Chem* **256:** 2143–2153.

Gromkova R, Goodgal SH. 1972. Action of haemophilus endodeoxyribonuclease on biologically active deoxyribonucleic acid. *J Bacteriol* **109:** 987–992.

Gromkova R, Goodgal SH. 1976. Biological properties of a *Haemophilus influenzae* restriction enzyme, Hind I. *J Bacteriol* **127:** 848–854.

Gunthert U, Freund M, Trautner TA. 1981. Restriction and modification in *Bacillus subtilis*: two DNA methyltransferases with BsuRI specificity. I. Purification and physical properties. *J Biol Chem* **256:** 9340–9345.

Gupta YK, Chan SH, Xu SY, Aggarwal AK. 2015. Structural basis of asymmetric DNA methylation and ATP-triggered long-range diffusion by EcoP15I. *Nat Commun* **6:** p7363. doi: 7310.1038/ncomms8363.

Haberman A. 1974. The bacteriophage P1 restriction endonuclease. *J Mol Biol* **89:** 545–563.

Hadi SM, Bachi B, Iida S, Bickle TA. 1983. DNA restriction–modification enzymes of phage P1 and plasmid p15B. Subunit functions and structural homologies. *J Mol Biol* **165:** 19–34.

Hadi SM, Bickle TA, Yuan R. 1975. The role of *S*-adenosylmethionine in the cleavage of deoxyribonucleic acid by the restriction endonuclease from *Escherichia coli* K. *J Biol Chem* **250:** 4159–4164.

Halford SE. 1980. The specificity of the *Eco*RI restriction endonuclease. *Biochem Soc Trans* **8:** 399–400.

Halford SE. 2013. http://library.cshl.edu/Meetings/restriction-enzymes/Halford.php.

Halford SE, Johnson NP. 1980. The *Eco*RI restriction endonuclease with bacteriophage λ DNA. Equilibrium binding studies. *Biochem J* **191:** 593–604.

Halford SE, Johnson NP. 1983. Single turnovers of the *Eco*RI restriction endonuclease. *Biochem J* **211:** 405–415.

Halford SE, Johnson NP, Grinsted J. 1979. The reactions of the *Eco*Ri and other restriction endonucleases. *Biochem J* **179:** 353–365.

Halford SE, Johnson NP, Grinsted J. 1980. The *Eco*RI restriction endonuclease with bacteriophage λ DNA. Kinetic studies. *Biochem J* **191:** 581–592.

Hamablet L, Chen GC, Brown A, Roberts RJ. 1989. LpnI, from *Legionella pneumophila*, is a neoschizomer of HaeII. *Nucleic Acids Res* **17:** 6417.

Heilmann H, Burkardt HJ, Puhler A, Reeve JN. 1980. Transposon mutagenesis of the gene encoding the bacteriophage P1 restriction endonuclease. Co-linearity of the gene and gene product. *J Mol Biol* **144:** 387–396.

Hendrix RW, Roberts JW, Stahl FW, Weisberg RA. ed. 1983. *Lambda II*. Cold Sping Harbor Laboratory, Cold Spring Harbor, NY.

Hershey AD. ed. 1971. *The bacteriophage lambda*. Cold Spring Harbor Laboratory, Cold Spring Harbor, NY.

Horiuchi K, Zinder ND. 1972. Cleavage of bacteriophage fl DNA by the restriction enzyme of *Escherichia coli* B. *Proc Natl Acad Sci* **69:** 3220–3224.

Horiuchi K, Zinder ND. 1975. Site-specific cleavage of single-stranded DNA by a *Hemophilus* restriction endonuclease. *Proc Natl Acad Sci* **72:** 2555–2558.

Iida S, Meyer J, Bachi B, Stalhammar-Carlemalm M, Schrickel S, Bickle TA, Arber W. 1983. DNA restriction–modification genes of phage P1 and plasmid p15B. Structure and in vitro transcription. *J Mol Biol* **165:** 1–18.

Jack WE, Rubin A, Modrich P. 1980. Structures and mechanisms of EcoRI DNA restriction and modification enzymes. In *Gene amplification and analysis volume I: restriction endonucleases* (ed. Chirikjian JG), pp. 165. Elsevier/North-Holland, NY.

Jack WE, Rubin RA, Newman A, Modrich P. 1981. Structures and mechanisms of Eco RI DNA restriction and modification enzymes. *Gene Amplif Anal* **1:** 165–179.

Jack WE, Terry BJ, Modrich P. 1982. Involvement of outside DNA sequences in the major kinetic path by which *Eco*RI endonuclease locates and leaves its recognition sequence. *Proc Natl Acad Sci* **79:** 4010–4014.

Jovin TM. 1976. Recognition mechanisms of DNA-specific enzymes. *Ann Rev Biochemy* **45:** 889–920.

Kan NC, Lautenberger JA, Edgell MH, Hutchison CA III. 1979. The nucleotide sequence recognized by the *Escherichia coli* K12 restriction and modification enzymes. *J Mol Biol* **130:** 191–209.

Kaplan DA, Nierlich DP. 1975. Cleavage of nonglucosylated bacteriophage T4 deoxyribonucleic acid by restriction endonuclease *Eco* RI. *J Biol Chem* **250:** 2395–2397.

Kauc L, Piekarowicz A. 1978. Purification and properties of a new restriction endonuclease from *Haemophilus influenzae* Rf. *Eur J Biochem/FEBS* **92:** 417–426.

Kelly TJ Jr, Smith HO. 1970. A restriction enzyme from *Hemophilus influenzae*. II. *J Mol Biol* **51:** 393–409.

Kim R, Modrich P, Kim SH. 1984. 'Interactive' recognition in *Eco*RI restriction enzyme–DNA complex. *Nucleic Acids Res* **12:** 7285–7292.

Kim YC, Grable JC, Love R, Greene PJ, Rosenberg JM. 1990. Refinement of *Eco* RI endonuclease crystal structure: a revised protein chain tracing. *Science* **249:** 1307–1309.

Langowski J, Pingoud A, Goppelt M, Maass G. 1980. Inhibition of *Eco* RI action by polynucleotides. A characterization of the non-specific binding of the enzyme to DNA. *Nucleic Acids Res* **8:** 4727–4736.

Lautenberger JA, Linn S. 1972. The deoxyribonucleic acid modification and restriction enzymes of *Escherichia coli* B. I. Purification, subunit structure, and catalytic properties of the modification methylase. *J Biol Chem* **247**: 6176–6182.

Lautenberger JA, Kan NC, Lackey D, Linn S, Edgell MH, Hutchison CA III. 1978. Recognition site of *Escherichia coli* B restriction enzyme on φ XsB1 and simian virus 40 DNAs: an interrupted sequence. *Proc Natl Acad Sci* **75**: 2271–2275.

Lautenberger JA, Edgell MH, Hutchison CA III, Godson GN. 1979. The DNA sequence on bacteriophage G4 recognized by the *Escherichia coli* B restriction enzyme. *J Mol Biol* **131**: 871–875.

Lee YH, Chirikjian JG. 1979. Sequence-specific endonuclease Bgl I. Modification of lysine and arginine residues of the homogeneous enzyme. *J Biol Chem* **254**: 6838–6841.

Loenen WA, Brammar WJ. 1980. A bacteriophage λ vector for cloning large DNA fragments made with several restriction enzymes. *Gene* **10**: 249–259.

Lu A-L, Jack WE, Modrich P. 1981. DNA determinants important in sequence recognition by *Eco* RI endonuclease. *J Biol Chem* **256**: 13200–13206.

Mann MB, Smith HO. 1979. Specificity of DNA methylases from *Haemophilus sp*. In *Transmethylation* (ed. Usdin E, Borchardt R, Kreveling C), pp. 483. Elsevier/North-Holland, New York.

Mann MB, Rao RN, Smith HO. 1978. Cloning of restriction and modification genes in *E. coli*: the HhaII system from *Haemophilus haemolyticus*. *Gene* **3**: 97–112.

Marchionni MA, Roufa RJ. 1978. Digestion of 5-bromodeoxyuridine-substituted lambda-DNA by restriction endonucleases. *J Biol Chem* **253**: 9075–9081.

Marinus MG, Morris NR. 1973. Isolation of deoxyribonucleic acid methylase mutants of *Escherichia coli* K-12. *J Bacteriol* **114**: 1143–1150.

Maxam AM, Gilbert W. 1980. Sequencing end-labeled DNA with base-specific chemical cleavages. *Methods Enzymol* **65**: 499–560.

McClarin JA, Frederick CA, Wang BC, Greene P, Boyer HW, Grable J, Rosenberg JM. 1986. Structure of the DNA–Eco RI endonuclease recognition complex at 3 Å resolution. *Science* **234**: 1526–1541.

Meisel A, Bickle TA, Kruger DH, Schroeder C. 1992. Type III restriction enzymes need two inversely oriented recognition sites for DNA cleavage. *Nature* **355**: 467–469.

Meselson M, Yuan R. 1968. DNA restriction enzyme from *E. coli*. *Nature* **217**: 1110–1114.

Middleton JH, Edgell MH, Hutchison CA III. 1972. Specific fragments of φX174 deoxyribonucleic acid produced by a restriction enzyme from *Haemophilus aegyptius*, endonuclease Z. *J Virol* **10**: 42–50.

Modrich P. 1979. Structures and mechanisms of DNA restriction and modification enzymes. *Q Rev Biophys* **12**: 315–369.

Modrich P. 1982. Studies on sequence recognition by type II restriction and modification enzymes. *CRC Crit Rev Biochem* **13**: 287–323.

Modrich P, Roberts RJ. 1982. Type II restriction and modification enzymes. In *Nucleases* (ed. Linn SM, Roberts RJ), 109–154. Cold Spring Harbor Laboratory, Cold Spring Harbor, NY.

Modrich P, Rubin RA. 1977. Role of the 2-amino group of deoxyguanosine in sequence recognition by EcoRI restriction and modification enzymes. *J Biol Chem* **252**: 7273–7278.

Modrich P, Zabel D. 1976. EcoRI endonuclease. Physical and catalytic properties of the homogenous enzyme. *J Biol Chem* **251**: 5866–5874.

Molloy PL, Symons RH. 1980. Cleavage of DNA.RNA hybrids by type II restriction enzymes. *Nucleic Acids Res* **8**: 2939–2946.

Murray K. 1973. Nucleotide sequence analysis with polynucleotide kinase and nucleotide "mapping" methods. 5'-Terminal sequences of deoxyribonucleic acid from bacteriophages λ and 424. *Biochem J* **131:** 569–582.

Murray NE. 2000. Type I restriction systems: sophisticated molecular machines (a legacy of Bertani and Weigle). *Microbiol Mol Biol Rev* **64:** 412–434.

Murray NE. 2002. 2001 Fred Griffith review lecture. Immigration control of DNA in bacteria: self versus non-self. *Microbiology* **148:** 3–20.

Murray NE, Murray K. 1974. Manipulation of restriction targets in phage λ to form receptor chromosomes for DNA fragments. *Nature* **251:** 476–481.

Murray NE, Batten PL, Murray K. 1973. Restriction of bacteriophage λ by *Escherichia coli* K. *J Mol Biol* **81:** 395–407.

Murray K, Murray NE, Bertani G. 1975. Base changes in the recognition site for *ter* functions in lambdoid phage DNA. *Nature* **254:** 262–265.

Murray NE, Gough JA, Suri B, Bickle TA. 1982. Structural homologies among type I restriction-modification systems. *EMBO J* **1:** 535–539.

Murray IA, Stickel SK, Roberts RJ. 2010. Sequence-specific cleavage of RNA by Type II restriction enzymes. *Nucleic Acids Res* **38:** 8257–8268.

Newman AK, Rubin RA, Kim SH, Modrich P. 1981. DNA sequences of structural genes for *Eco* RI DNA restriction and modification enzymes. *Journal Biol Chem* **256:** 2131–2139.

Old R, Murray K, Boizes G. 1975. Recognition sequence of restriction endonuclease III from *Hemophilus influenzae*. *J Mol Biol* **92:** 331–339.

Piekarowicz A, Kauc L, Glover SW. 1974. Host specificity of DNA in *Haemophilus influenzae*: the restriction and modification systems in strains Rb and Rf. *J Gen Microbiol* **81:** 391–403.

Polisky B, Greene P, Garfin DE, McCarthy BJ, Goodman HM, Boyer HW. 1975. Specificity of substrate recognition by the *Eco*RI restriction endonuclease. *Proc Natl Acad Sci* **72:** 3310–3314.

Ravetch JV, Horiuchi K, Zinder ND. 1978. Nucleotide sequence of the recognition site for the restriction-modification enzyme of *Escherichia coli* B. *Proc Natl Acad Sci* **75:** 2266–2270.

Reiser J. 1975. "The P1 and P15-specific restriction endonucleases: a comparative study." PhD thesis. Basel University.

Richter PH, Eigen M. 1974. Diffusion controlled reaction rates in spheroidal geometry. Application to repressor–operator association and membrane bound enzymes. *Biophys Chem* **2:** 255–263.

Roberts RJ. 1976. Restriction endonucleases. *CRC Crit Rev Biochem* **4:** 123–164.

Roberts RJ, Halford SE. 1993. Type II restriction enzymes. In *Nucleases*, 2nd ed. (ed. Linn SM, Lloyd RS, Roberts RJ), 35–88. Cold Spring Harbor Laboratory Press, Cold Spring Harbor, NY.

Roberts RJ, Macelis D. 1993. REBASE—restriction enzymes and methylases. *Nucleic Acids Res* **21:** 3125–3137.

Rosamond J, Endlich B, Linn S. 1979. Electron microscopic studies of the mechanism of action of the restriction endonuclease of *Escherichia coli* B. *J Mol Biol* **129:** 619–635.

Rosenberg JM. 2013. http://library.cshl.edu/Meetings/restriction-enzymes/v-Rosenberg.php.

Rosenberg JM, Greene P. 1982. *Eco* RI* specificity and hydrogen bonding. *DNA* **1:** 117–124.

Rosenberg JM, Dickerson RE, Greene PJ, Boyer HW. 1978. Preliminary X-ray diffraction analysis of crystalline *Eco*RI endonuclease. *J Mol Biol* **122:** 241–245.

Rosenberg JM, Boyer HW, Greene P. 1980. The structure and function of EcoRI endonuclease. In *Gene amplification and analysis, volume I: restriction endonucleases* (ed. Chirikjian JG), p. 131. Elsevier/North-Holland, New York.

Rosenberg JM, Boyer HW, Greene P. 1981. The structure and function of the Eco RI restriction endonuclease. *Gene Amplif Anal* **1:** 131–164.

Rosner JL. 1973. Modification-deficient mutants of bacteriophage P1. I. Restriction by P1 cryptic lysogens. *Virology* **52:** 213–222.

Roulland-Dussoix D, Boyer HW. 1969. The *Escherichia coli* B restriction endonuclease. *Biochim Biophys Acta* **195:** 219–229.

Roychoudhury R, Kossel H. 1971. Synthetic polynucleotides. Enzymic synthesis of ribonucleotide terminated oligodeoxynucleotides and their use as primers for the enzymic synthesis of polydeoxynucleotides. *Eur J Biochem/FEBS* **22:** 310–320.

Roychoudhury R, Jay E, Wu R. 1976. Terminal labeling and addition of homopolymer tracts to duplex DNA fragments by terminal deoxynucleotidyl transferase. *Nucleic Acids Res* **3:** 863–877.

Rubin RA, Modrich P. 1977. EcoRI methylase. Physical and catalytic properties of the homogeneous enzyme. *J Biol Chem* **252:** 7265–7272.

Rubin RA, Modrich P. 1978. Substrate dependence of the mechanism of *Eco*RI endonuclease. *Nucleic Acids Res* **5:** 2991–2997.

Rubin RA, Modrich P, Vanaman TC. 1981. Partial NH_2- and COOH-terminal sequence analyses of Eco RI DNA restriction and modification enzymes. *J Biol Chem* **256:** 2140–2142.

Sain B, Murray NE. 1980. The hsd (host specificity) genes of *E. coli* K12. *Mol Gen Genet* **180:** 35–46.

Sanger F, Coulson AR, Friedmann T, Air GM, Barrell BG, Brown NL, Fiddes JC, Hutchison CA III, Slocombe PM, Smith M. 1978. The nucleotide sequence of bacteriophage φX174. *J Mol Biol* **125:** 225–246.

Schranner R, Richter PH. 1978. Rate enhancement by guided diffusion. Chain length dependence of repressor-operator association rates. *Biophys Chem* **8:** 135–150.

Scott JR. 1970. Clear plaque mutants of phage P1. *Virology* **41:** 66–71.

Sharp PA, Sugden B, Sambrook J. 1973. Detection of two restriction endonuclease activities in *Haemophilus parainfluenzae* using analytical agarose–ethidium bromide electrophoresis. *Biochemistry* **12:** 3055–3063.

Smith HO, Wilcox KW. 1970. A restriction enzyme from *Hemophilus influenzae*. I. Purification and general properties. *J Mol Biol* **51:** 379–391.

Sommer R, Schaller H. 1979. Nucleotide sequence of the recognition site of the B-specific restriction modification system in *E. coli*. *Mol Gen Genet MGG* **168:** 331–335.

Southern EM. 1975. Detection of specific sequences among DNA fragments separated by gel electrophoresis. *J Mol Biol* **98:** 503–517.

Sussenbach JS, Monfoort CH, Schiphof R, Stobberingh EE. 1976. A restriction endonuclease from *Staphylococcus aureus*. *Nucleic Acids Res* **3:** 3193–3202.

Takanami M. 1973. Specific cleavage of coliphage fd DNA by five different restriction endonucleases from *Haemophilus* genus. *FEBS Lett* **34:** 318–322.

Thomas M, Davis RW. 1975. Studies on the cleavage of bacteriophage λ DNA with *Eco*RI restriction endonuclease. *J Mol Biol* **91:** 315–328.

Tikchonenko TI, Karamov EV, Zavizion BA, Naroditsky BS. 1978. EcoRI activity: enzyme modification or activation of accompanying endonuclease? *Gene* **4:** 195–212.

van Ormondt H, Lautenberger JA, Linn S, de Waard A. 1973. Methylated oligonucleotides derived from bacteriophage fd RF-DNA modified in vitro by *E. coli* B modification methylase. *FEBS Lett* **33:** 177–180.

von Hippel PH. 1979. On the molecular bases of the specificity of interaction of transcriptional proteins with genome DNA. *Biol Regul Dev* **1:** 279–347.

von Hippel PH, McGhee JD. 1972. DNA-protein interactions. *Ann Rev Biochem* **41:** 231–300.

Vovis GF, Horiuchi K, Zinder ND. 1974. Kinetics of methylation of DNA by a restriction endonuclease from *Escherichia coli* B. *Proc Natl Acad Sci* **71:** 3810–3813.

Waalwijk C, Flavell RA. 1978. DNA methylation at a CCGG sequence in the large intron of the rabbit β-globin gene: tissue-specific variations. *Nucleic Acids Res* **5:** 4631–4634.

Winter RB, Berg OG, von Hippel PH. 1981. Diffusion-driven mechanisms of protein translocation on nucleic acids. 3. The *Escherichia coli* lac repressor–operator interaction: kinetic measurements and conclusions. *Biochemistry* **20:** 6961–6977.

Woodbury CP Jr, Hagenbuchle O, von Hippel PH. 1980. DNA site recognition and reduced specificity of the Eco RI endonuclease. *J Biol Chem* **255:** 11534–11548.

Woodhead JL, Malcolm AD. 1980. Non-specific binding of restriction endonuclease *Eco*R1 to DNA. *Nucleic Acids Res* **8:** 389–402.

Yoo OJ, Agarwal KL. 1980a. Cleavage of single strand oligonucleotides and bacteriophage φX174 DNA by Msp I endonuclease. *J Biol Chem* **255:** 10559–10562.

Yoo OJ, Agarwal KL. 1980b. Isolation and characterization of two proteins possessing *Hpa* II methylase activity. *J Biol Chem* **255:** 6445–6449.

Yoshimori R. 1971. "A genetic and biochemical analysis of the restriction and modification of DNA by resistance transfer factors." PhD thesis. University of California, San Francisco.

Yoshimori R, Roulland-Dussoix D, Boyer HW. 1972. R factor–controlled restriction and modification of deoxyribonucleic acid: restriction mutants. *J Bacteriol* **112:** 1275–1279.

Young T, Kim S, Modrich P, Beth A, Jay E. 1981. Preliminary X-ray diffraction studies of *Eco*RI restriction endonuclease–DNA complex. *J Mol Biol* **145:** 607–610.

Yuan R, Reiser J. 1978. Steps in the reaction mechanism of the *Escherichia coli* plasmid P15-specific restriction endonuclease. *J Mol Biol* **122:** 433–445.

Yuan R, Bickle TA, Ebbers W, Brack C. 1975. Multiple steps in DNA recognition by restriction endonuclease from *E. coli* K. *Nature* **256:** 556–560.

Yuan R, Hamilton DL, Burckhardt J. 1980a. DNA translocation by the restriction enzyme from *E. coli* K. *Cell* **20:** 237–244.

Yuan R, Hamilton DL, Hadi SM, Bickle TA. 1980b. Role of ATP in the cleavage mechanism of the EcoP15 restriction endonuclease. *J Mol Biol* **144:** 501–519.

WWW RESOURCES

http://library.cshl.edu/Meetings/restriction-enzymes/Halford.php Halford SE. 2013. Type II restriction enzymes: searching for one site and then two.

http://library.cshl.edu/Meetings/restriction-enzymes/v-Rosenberg.php Rosenberg JM. 2013. EcoRI structure.

http://rebase.neb.com/rebase/rebase.html The Restriction Enzyme Database.

http://rsbmroyalsocietypublishingorg/content/roybiogmem/60/349 Gann A, Beggs J. 2014. Noreen Elizabeth Murray CBE.

CHAPTER 6

Variety in Mechanisms and Structures of Restriction Enzymes: ~1982–1993

INTRODUCTION

During the 1970s and 1980s Type II REases became the workhorses of genetic engineers, making the identification of novel enzymes important. By 1993, 188 Type II specificities had been identified, among a total of nearly 2400 enzymes, but still only a small number of Type I and III REases (Roberts and Macelis 1993a,b) rising to approximately 3700 REases by 2004 (Pingoud et al. 2005). It was the largest of any group of nucleases known at the time. It was estimated that ~25% of bacteria from all genera would carry one or more Type II systems (Roberts and Halford 1993; Roberts and Macelis 1993a,b). The presence of Type I or Type III systems in bacterial strains remained difficult to establish. One of the most important new Type II enzymes identified was EcoRV (Bougueleret et al. 1984, 1985), which would be exhaustively studied over the next three decades (reviewed in Pingoud et al. 2014). Were REase–DNA interactions truly different from those mediated by known DNA recognition modules such as zinc fingers or helix-turn-helix found in transcription factors, repressors, etc.? Was the view, held at the time, that DNA recognition involved only a few such mechanisms mistaken and naive? Also, how did REases find their specific sites among so many kilobases of DNA: Did they use one or more of the various mechanisms proposed by von Hippel and Berg (Berg and von Hippel 1985; von Hippel and Berg 1989)? And what did the long-awaited crystal structures look like?

Despite difficulties encountered with the discovery and analysis of REases, all three types of enzymes had enthusiastic followers who were fascinated with the genetics, molecular biology, and biochemistry of these highly specific enzymes (Bennett and Halford 1989; Wilson 1991; Wilson and Murray 1991; Anderson 1993; Heitman 1993). During the 1980s and early 1990s, the groups in Bristol, Pittsburgh, Edinburgh, and Basel (Chapter 5) and Alfred

Chapter doi:10.1101/restrictionenzymes_6

Pingoud and colleagues in Hannover made important progress toward the understanding of the action of EcoRI and EcoRV and the ATP-dependent Type I and III enzymes. At the same time, other REases became subject to intense study (e.g., BamHI and EcoR124). FokI was different and appeared to contain separate domains for DNA sequence recognition and cleavage (Sugisaki and Kanazawa 1981; Sugisaki et al. 1989), whereas NaeI required interaction with two recognition sequences, such as EcoRII (Van Cott and Wilson 1988; Topal et al. 1991). Like some other "oddball systems, such as BcgI and Eco57I" (Roberts and Halford 1993) these enzymes would become the prototypes of new subclasses in 2003 (Roberts et al. 2003).

Another interesting enzyme was MmeI. Isolated by Imperial Chemistry Industries as AS1 (Aggravated Sludge 1), this bacterial strain was highly resistant to changes in temperature and pressure during growth in bulk. During the oil crisis in the early 1970s, the idea was developed to grow AS1 (which proved to be *Methylophilus methylotrophus*) in batches of 500,000 liters (!) and use them dried as protein-rich feed for cows. Chris Boyd in Bill Brammar's laboratory identified the R-M system in this strain, MmeI, which he lovingly called "My Mimi," and worked out the recognition sequence using his own computer program (Boyd et al. 1986). This was still not very common at the time and sometimes Bill would ask in (mock) despair: "Yes, Chris, could you perhaps also do an experiment at the bench to back up your computer data?" The oil crisis ended, and MmeI was shelved until two decades later Richard (Rick) Morgan discovered that it was a rather interesting enzyme (Morgan et al. 2008, 2009; Callahan et al. 2011; Morgan 2013; Chapter 8).

One goal of many workers at the time would prove elusive—that of the generation of new specificities of Type II REases, particularly longer recognition sequences, by genetic manipulation of existing systems. Perhaps enzymes such as EcoRI were simply not "malleable" to reengineering (Roberts and Halford 1993) in this way.

With respect to the Type I systems, the role of these enzymes was under strong debate: Why spend so much metabolic energy on these enzymes, as the vast majority of individual bacterial cells were unlikely to experience phage infection or conjugation outside the laboratory (Bickle 1993)? Did these enzymes play a role in the cellular economy (e.g., in genetic recombination [Price and Bickle 1986]), as well as offer protection against potential phage infections in nature?

TYPE II ENZYMES

Characterization of New Specificities of Type II Enzymes

Initially, the Type II systems identified were R-M systems like EcoRI, a homodimeric REase, which required Mg^{2+} as cofactor and cleaved at a specific recognition sequence 4–8 bp in length, and a monomeric MTase, which recognized and methylated the same ("cognate") 4–8-bp site and by doing so protected the site against the REase. This view changed over the years, as the superfamily grew—for example, in some cases two MTases were associated with the REase (e.g., DpnI and DpnII [de la Campa et al. 1987], Esp3I [Bitinaite et al. 1992], and MboI [Ueno et al. 1993]).

By 1993 the rate of discovery of new specificities had dropped to about eight per year (http://library.cshl.edu/Meetings/restriction-enzymes/v-Roberts .php). More iso- and neoschizomer pairs like MboI/Sau3AI and HpaII/MspI were identified that were differentially sensitive to methylation, either within or outside their recognition site (Nelson et al. 1993; Roberts and Halford 1993). This led to progress in cloning in *E. coli* (Chapter 5) and investigations in the methylation state of the CG dinucleotide in and around genes in higher organisms, a tool toward understanding the organization and transcription of eukaryotic genes (Doerfler 1983).

The Type IIS Enzymes

In addition to the Type II REases that cut at a recognition site with dyad symmetry, a substantial number of the new enzymes cleaved at a short distance away from an asymmetric sequence. These were termed Type IIS (for shifted) (Szybalski et al. 1991) to distinguish them from the EcoRI-like enzymes (Type IIP for palindrome [Roberts et al. 2003]). By 1991, only a small percentage of the more than 1000 Type II REases belonged to the Type IIS class (35 specificities and 80 isoschizomers) (Szybalski et al. 1991).

The Type IIS REases had separate recognition and cleavage domains, somewhat like the Type III enzymes. However, the Type IIS REases appeared to be monomeric proteins, whereas Type III REases were heterodimeric with Mod and Res subunits. In support of this separation of recognition and cleavage modules, deletion of the carboxyl terminus of FokI prevented cleavage but not DNA binding (Li et al. 1993). The distance between the recognition sequence and the cleavage site varied from enzyme to enzyme: FokI cut GGATG (9/13)—that is, it cleaved 9 nt beyond 5′-GGATG on the same strand and 13 nt beyond this sequence on the complementary strand (Sugisaki and Kanazawa 1981). Isoschizomer StsI cleaved 1 nt further along: GGATG (10/14). Also, some enzymes showed variable cleavage distance, either as wild-type or mutant enzyme (Li and Chandrasegaran 1993; Roberts and Halford 1993; see below).

In the Type IIS systems, cognate methylation was due to two activities— one for methylation of each strand of the recognition sequence. In the case of M·FokI the two MTases were fused into a single protein (Looney et al. 1989; Sugisaki et al. 1989), whereas the first Type IIS REase to be discovered, HgaI (found in 1974, but not recognized as such), had two separate cognate MTases (Barsomian et al. 1990). Would the REases have two cutting domains, one for each strand?

The separation between recognition and cleavage domains made Type IIS useful for various manipulations (Szybalski et al. 1991) and would lead to engineering of FokI to obtain hybrid REases (Kim et al. 1998; Sanders et al. 2009; Guo et al. 2010; Halford et al. 2011; Li et al. 2011; Chapter 8). By 1991, it was possible to generate mutants by cutting selected DNAs with FokI, filling in or removing the single-stranded (ss) ends, then cutting again, and again, as one wished, since the recognition site remained intact. One could mix and match sequences—for example, reassemble HgaI-derived fragments of phage f1 replicative form (RF) DNA (Moses and Horiuchi 1979)—and various other applications such as precise excision and amplification (see Szybalski et al. 1991 for details).

Other Type II Enzymes

Among the Type II REases, some enzymes had unusual features. Examples were BcgI, which required SAM as cofactor and cleaved symmetrically on both sides of its asymmetric recognition sequence (GCA[N6]TCG [Kong et al. 1993]), and Sgr20I (Orekhov et al. 1982). The latter recognized the same sequence as EcoRII (CC(A/T)GG, written as CCWGG) but appeared to cleave on both sides of this sequence, as it produced bands on gels that were slightly smaller than those generated with EcoRII (Roberts and Halford 1993). These would be the first of the Type IIC and Type IIH (BcgI), Type IIF (Sgr20I), and Type IIE (EcoRII) REase subgroups that would lead to the new classification in 2003 in 11 subclasses (Roberts et al. 2003).

Determination of Cleavage Sites of Type II Enzymes

New computer programs combined with the availability of sequenced DNA from plasmid, phage, and other sources came to the aid of scientists to characterize the recognition and cleavage sites of new REases. Analysis of the size of bands produced by the REases on sequenced DNA helped to identify potential restriction sites (Gingeras et al. 1978; Tolstoshev and Blakesley 1982; Boyd et al. 1986). Experimental proof was usually obtained using primed synthesis reactions on a couple of sites (Brown and Smith 1980): A primer close to the site was extended with DNA polymerase on a given template. The product was cleaved with the REase, and polymerase and dNTPs were added to one-half of the sample. If the REase produced blunt ends, no dNTPs would be incorporated; if the enzyme produced sticky ends, then a 5′ extension would be repaired, and a 3′ extension trimmed back by the polymerase. In the second and third case the fragment would become longer or shorter than in the first case, which would not change in length. The precise sites of cleavage on both strands could then be determined by running the treated and untreated samples in parallel on a sequencing gel.

Genes and Organization of Type II Enzymes

A comprehensive survey of the cloned R-M systems was compiled by Geoff Wilson at New England Biolabs (NEB) in 1991 (Wilson 1991), showing different organizations, but always tight linkage between the REase and MTase genes. For example, the EcoRII genes were convergently transcribed from separate promoters on opposite DNA strands, in contrast to those encoding EcoRI (Kosykh et al. 1980, 1989; Som et al. 1987; Bhagwat et al. 1990; Reuter et al. 1999), which were in-line. By 1993, more than 60 REase and 100 MTase genes had been sequenced. The m5C-MTases clearly shared a common architecture (Lauster et al. 1989; Pósfai et al. 1989). The m4C- and m6A-MTases also showed similarities, although they were less pronounced (Klimasauskas et al. 1989; Lauster et al. 1989). The molecular weight (MW) of the REases, already known to vary considerably, proved to vary even more substantially: PvuII was only 18 kDa (Athanasiadis et al. 1990; Tao and Blumenthal 1992) and BsuRI was 66 kDa (Kiss et al. 1985). All this reinforced the idea that most enzymes evolved independently and used a variety of mechanisms to recognize DNA. Only in a few pairs was significant homology found (e.g., RsrI and EcoRI, which recognize the same sequence) (Aiken 1986; Stephenson et al. 1989).

The evidence that, like EcoRI, many Type II REases acted as dimers and the MTases as monomers was consolidated and supported by the first structures of Type II enzymes, those of EcoRI and EcoRV.

DNA Binding by Type II Enzymes

Virtually all Type II REases required Mg^{2+} for cleavage. However, many REases could form stable DNA–protein complexes in the absence of Mg^{2+}—for example, BamHI, EcoRI, EcoRII, EcoRV, RsrI, and TaqI (Terry et al. 1987; Aiken et al. 1991a; Taylor et al. 1991; Xu and Schildkraut 1991; Gabbara and Bhagwat 1992; Zebala et al. 1992a). Fortuitously, both EcoRI and EcoRV first bound DNA before binding Mg^{2+} (Halford 1983; Taylor and Halford 1989). Hence, it was relevant to the reaction pathway to study the complexes formed in the absence of Mg^{2+}. Gel-shift assays superseded the filter-binding method used for EcoRI (Halford and Johnson 1980; Jack et al. 1982; Chapter 5). Such assays revealed that binding of EcoRI to DNA with two recognition sites resulted in two complexes, with either one or two enzyme molecules bound to the DNA (Terry et al. 1985). Footprinting and a preferential cleavage assay were also used to monitor binding of REases to their substrates (Jack et al. 1982; Lu et al. 1983; Terry et al. 1983; Becker et al. 1988; Lesser et al. 1990; Roberts and Halford 1993).

EcoRI could bind very tightly to its recognition site, even on DNA molecules as long as 40 kb (Halford and Johnson 1980; Terry et al. 1983). Sets of synthetic oligonucleotide duplexes were used to study the effect of single-base changes in the recognition site (Lesser et al. 1990; Thielking et al. 1990). The equilibrium binding constants varied with each substitution, but these were much lower than at the recognition site—typically ~5000-fold! BamHI and RsrI gave similar results (Aiken et al. 1991a; Xu and Schildkraut 1991), but a surprise came in the shape of EcoRV. In contrast to EcoRI, in the absence of Mg^{2+}, EcoRV bound all DNA sequences with equal affinity (Fig. 1) irrespective of whether an EcoRV site was present or not. Gel shifts revealed a series of complexes due to the binding of 1, 2, 3, …, n molecules of protein per molecule of DNA (Taylor et al. 1991; Roberts and Halford 1993).

This suggested that, in the absence of Mg^{2+}, EcoRV bound the DNA indiscriminately. Also, the equilibrium constants all had the same value, even if the DNA contained an EcoRV site. (Note that although surprising, this is

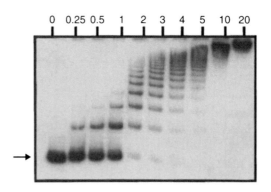

FIGURE 1. EcoRV binds all DNA sequences with equal affinity in the absence of Mg^{2+}. Gel shifts with increasing concentrations of EcoRV added to 0.1 nM 32P-labeled DNA in EDTA buffer (no Mg^{2+}). DNA (381 bp) with one EcoRV site. The enzyme concentrations (nM) are shown above each lane. The arrow marks the mobility of the free DNA. The same DNA without an EcoRV site gave the same results (not shown). The band above the arrow is the only band seen with specific DNA when Ca^{2+} was added (Vipond and Halford 1995; Halford 2013). With 50-, 100-, and 200-bp DNA, an increasing number of retarded bands appeared: 3, 6, and 12, respectively (Taylor et al. 1991; Halford 2013). (Reprinted, with permission, from Taylor et al. 1991, © American Chemical Society.)

FIGURE 2. Scheme for target site location (Fig. 1 in Halford and Marko 2004, as proposed by von Hippel and Berg in 1989 [von Hippel and Berg 1989; Halford and Marko 2004], legend adapted). Three routes for transfer of a protein from one site to another along a long DNA molecule: "sliding," "hopping," and "intersegmental transfer" (also called "jumping"). (*Top*) A protein might "slide" along the double helix from base pair to base pair without dissociation from the DNA. Repeated sliding results in 1D diffusion of the protein along the DNA. (*Middle*) If dissociation occurs, the protein might "hop" onto the same DNA a little further along. (*Bottom*) Beyond 150 bp (50 nm), the DNA can bend back on itself, and the enzyme could bind transiently to both DNAs, and then move to the other DNA, a process called "intersegmental transfer" or "jumping." (Reprinted from Halford and Marko 2004.)

not relevant in vivo, as cells have ample Mg^{2+}.) Like EcoRV, TaqI and Cfr9I (and also other DNA-binding proteins) could bind DNA without preference for their recognition sequences in the absence of Mg^{2+} (Zebala et al. 1992a; Roberts and Halford 1993).

EcoRI appeared to bind nonspecifically to the DNA and then transfer to its specificity site. In general, such transfers could occur by sliding, hopping, or jumping (Fig. 2): Sliding is linear diffusion from nonspecific to specific sites; hopping refers to tiny dissociations/reassociations within the same DNA molecule; and jumping indicates total release of the DNA (von Hippel and Berg 1989). EcoRI probably used sliding, as this is fast: The association rate was very fast, and "too fast" for 3D "intersegmental transfer" (or "hopping") (Jack et al. 1982; Terry et al. 1985; Halford 2013). Was hopping unlikely for REases with only one DNA-binding site?

DNA Cleavage by Type II Enzymes

During this decade, increasing support was obtained for variation in restriction efficiencies at different sites and the role of flanking sequences (Ehbrecht et al. 1985; Terry et al. 1985; Nardone et al. 1986). In a few cases the effects of these flanking sequences were examined systematically (Taylor and Halford 1992; Yang and Topal 1992). For kinetic studies, it was important to use DNA with single sites, and the length of DNA in contact with the protein had to be longer than the recognition sequence, so as to include at least some flanking

DNA (Lu et al. 1983; Becker et al. 1988; Rosenberg 1991). EcoRII and NaeI became the prototypes of REases that had to interact simultaneously with two or more copies of the recognition sequence before cleaving DNA (Krüger et al. 1988; Conrad and Topal 1989; Oller et al. 1991; Roberts et al. 2003). These enzymes contained two distinct binding sites for their recognition site: One proved to be an allosteric effector that activated the other for DNA cleavage. This explained why some EcoRII sites were refractory to restriction (Krüger et al. 1988). This activation was crucial for cleavage, as without this the EcoRII–DNA complex was stable even in the presence of Mg^{2+} (Gabbara and Bhagwat 1992). The two sequences could be supplied in *cis* or in *trans* (Conrad and Topal 1989; Pein et al. 1991; Topal et al. 1991; Yang and Topal 1992).

Plasmid DNA was conclusive to the determination of whether the enzyme cleaved one strand first or both strands at the same time. If the enzyme nicked a supercoiled DNA molecule, an open circle appeared on an agarose gel with a different mobility to that of a linear molecule (produced by a double-strand cut). In this way the reaction could be followed in time, providing an answer to the question of whether the enzyme first cut one strand and then the next or both strands at the same time. In the latter case no open circles would be produced. Gel-shift assays could also be used to study single turnover reactions (Halford 1983; Halford and Johnson 1983; Terry et al. 1987; Bennett and Halford 1989; Zebala et al. 1992a).

Studies on the kinetics of DNA cleavage by EcoRI yielded conflicting results (Roberts and Halford 1993), but results with EcoRV were clear: no open circles, hence double-strand cuts (Halford and Goodall 1988). However, changing normal assay conditions (low pH or low $MgCl_2$) did produce open circles. The explanation was that in these cases Mg^{2+} bound only one of the two EcoRV subunits of the homodimer. As SalI produced similar results (Maxwell and Halford 1982), these results were consistent with a general requirement for coupled reactions to have Mg^{2+} bound to both subunits and that failure to do so would alter the mode of DNA cleavage (Bennett and Halford 1989; Hensley et al. 1990; Zebala et al. 1992a). Under optimal conditions the steady state reactions produced k_{cat} values and K_m values that approached the theoretical limit for k_{cat}/K_m in enzyme-catalyzed reactions (Roberts and Halford 1993).

In addition to plasmid DNA, short synthetic DNA duplexes (8–20 nt long) were used for kinetic studies. Time courses and analysis of reaction products (by electrophoresis, chromatography, or UV spectrophotometry [Aiken and Gumport 1991; Waters and Connolly 1992]) indicated that such short substrates posed fundamentally different problems for the REases compared to those composed of longer DNA molecules (Roberts and Halford 1993).

Specificity of Type II Enzymes

During the 1980s, research continued into the remarkable ability of the REases to discriminate the recognition sequence from all DNA sites. In vivo, a double-strand break may kill the cell and must be avoided, as only cognate sites are protected by methylation by the cognate MTase (Heitman and Model 1990b; Taylor et al. 1990; Smith et al. 1992). However, in vitro, REases cleaved DNA both at their recognition sites and some other sites (usually with one different base), albeit at a low level at noncognate sites under standard reaction conditions (Taylor and Halford 1989; Lesser et al. 1990; Thielking et al. 1990). As mentioned in Chapter 5, different conditions led to "star" (*) activity of, for example, EcoRI (Polisky et al. 1975) and other enzymes (Bennett and Halford 1989). To analyze this further, the cognate site for EcoRV (GATATC) in pAT153 (a derivative of pBR322) was compared with the preferred noncognate site (GTTATG). The latter site was flanked by alternating purines and pyrimidines, which conferred flexibility to the DNA structure (Taylor and Halford 1992). Under standard reaction conditions, the activity of EcoRV at this GTTATG was a formidable 10^6 times lower than at the cognate site, but this ratio changed to 10^3 in the presence of 10% DMSO and to just 6 in the presence of Mn^{2+} (Taylor and Halford 1989; Vermote and Halford 1992). In the case of EcoRI, systematic analysis of all nine possible single base pair substitutions in the recognition sequence caused a 10^5- to 10^9-fold reduction in k_{cat}/K_m for DNA cleavage (Lesser et al. 1990; Thielking et al. 1990). The enzymes usually cleaved noncognate sites via two successive nicks, even when the enzyme produced double-strand breaks at the cognate site (Barany 1988; Taylor and Halford 1989; Thielking et al. 1990).

E. coli DNA ligase was known to rapidly seal nicks in duplex DNA but joined double-stranded (ds) breaks much more slowly (Lehman 1974). This suggested that in vivo this enzyme might repair damage by REase action at the noncognate site but not at the cognate site. Would addition of ligase to the reaction mixture prevent product formation at the noncognate site? The answer to that was yes (Taylor and Halford 1989; Roberts and Halford 1993)!

Crystallography of Type II Enzymes[1]

The structure of EcoRI was the first to be published in 1986 (McClarin et al. 1986) but carried a mistake in the chain tracing and was revised 4 years later (Kim et al. 1990). One structure of EcoRI was solved in complex

---

[1]Adapted from Roberts (1993a).

with a 12-bp duplex containing the recognition site (in the absence of Mg^{2+}, which prevents cleavage [Kim et al. 1990]). John Rosenberg's group was lucky. On soaking the crystals with either Mg^{2+} or Mn^{2+}, the enzyme became active and cleaved the duplex; in this way the postreactive enzyme-product complex could be analyzed, because the crystals remained intact (Rosenberg 2013)!

By 1993, crystal structures for EcoRV and BamHI had also been solved at high resolution (Fig. 3; Kim et al. 1990; Strzelecka et al. 1990, 1994; Winkler et al. 1993), with five others in progress (Roberts and Halford 1993). The structure of PvuII was published by the group of John Anderson at Cold Spring Harbor Laboratory in 1994 (Cheng et al. 1994), shortly after the publication of the review in the *Nucleases* book (Roberts and Halford 1993). This led to the idea of

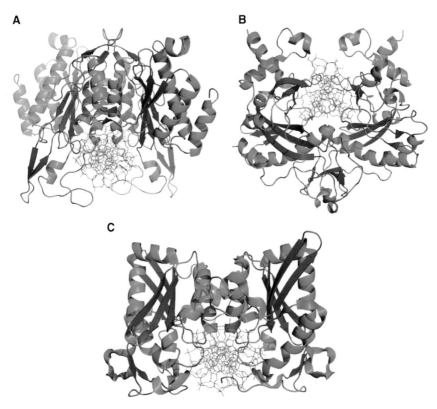

FIGURE 3. Crystal structures of (*A*) EcoRI, (*B*) EcoRV, and (*C*) BamHI (Kim et al. 1990; Strzelecka et al. 1990; Winkler et al. 1993; Strzelecka et al. 1994). (Courtesy of Aneel Aggarwal.)

two types of structures, one with EcoRI and BamHI as prototypes producing sticky ends and the other pair with EcoRV and PvuII as prototypes that generate blunt end fragments (Jack et al. 1991; Roberts and Halford 1993; Winkler et al. 1993; Newman et al. 1994; Strzelecka et al. 1994). In the case of EcoRV, three structures were solved; free protein, the protein bound to a 10-bp duplex with the EcoRV site, and the enzyme bound to noncognate DNA (Winkler et al. 1993). This provided highly valuable information, and revealed the secret to the specificity of the enzyme for cognate sites.

Protein Structures of the Type II REases EcoRI and EcoRV with DNA

The crystal structures of EcoRI and EcoRV have been extensively reviewed over the years. In 1993, the first results were summarized by Rich Roberts and Steve Halford and a shorter version of this text follows below (Roberts and Halford 1993).

a. *EcoRI.* The EcoRI–DNA enzyme complex has a single dyad symmetry relating the two subunits and the two halves of the palindromic DNA (Kim et al. 1990), as expected (Kelly and Smith 1970). At the dyad axis, the major groove of the DNA faces the protein. In each subunit, two arms extend from the main body of the protein to wrap around the DNA, but these remain within the major groove. Critical contacts to the DNA are made with a bundle of four α-helices, two from each subunit, aligned almost perpendicular to the DNA with the amino terminus of each helix poking into the major groove (Kim et al. 1990).

b. *EcoRV.* In its complexes with either specific or nonspecific DNA, EcoRV consists of two L-shaped subunits that interact with each other over a small surface area, to create a U-shaped dimer with a deep cleft between the subunits (Winkler et al. 1993). The fold of the polypeptide appeared to be completely different from that of EcoRI. In both complexes the DNA is located in the cleft, with its minor groove facing the base of the cleft (i.e., the opposite way around from EcoRI [but, as Alfred Pingoud was to point out later, the enzymes share a common β-sheet edge on to the active site]). The principal contacts to the DNA are made by two peptide loops per subunit. One loop, the R (recognition)-loop, is located toward the top of the cleft above the DNA. In the complex with the cognate site, the R-loop is positioned deep within the major groove, but in the nonspecific complex the R-loop is more distant from the DNA (Fig. 4). The second loop at the base of the cleft contains several glutamines (hence called the Q-loop) (Winkler et al. 1993). This loop approaches the minor groove of the DNA and contacts primarily phosphates rather than bases. The available structure suggests that EcoRV must undergo a series of conformational changes (Winkler et al. 1993), a prediction that would prove to be correct.

The First DNA Structures in Protein–DNA Complexes of Type II REases

In the case of the EcoRI–DNA complex (with a 12-bp duplex 5′CGCGAA TTCGCG), the DNA is distorted from the regular B-DNA structure (Fig. 4; Kim et al. 1990). The distortion is primarily an untwisting at the center of the sequence, with concomitant unstacking of the central 2 bp. This widens the major groove, thus improving access to the bases.

In the case of EcoRV, a 10-bp duplex (5′GGGATATCCC) is radically distorted from B-DNA (Winkler et al. 1993) The most marked feature of the distortion is a sharp bend, directed toward the protein, in the axis of the DNA helix at the center of the recognition site. Like EcoRI, the middle 2 bp in the recognition site are unstacked, but, in this case, the roll is in the opposite direction. The bound substrate for EcoRV has a deep and narrow major groove and a correspondingly shallow minor groove. The Winkler group would further refine the structure of EcoRV (see Winkler and Prota 2004 for review and Chapter 7).

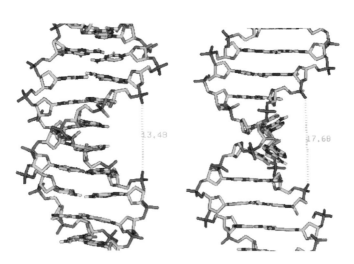

FIGURE 4. Distortion of the 12-bp duplex DNA containing the EcoRI recognition site by EcoRI (Kim et al. 1990). (*Left*) The 12-bp duplex: B-DNA on its own. (*Right*) The 12-bp duplex + EcoRI: The enzyme kinks and distorts the DNA (Rosenberg 2013). (Courtesy of John Rosenberg.)

DNA–Protein Interfaces of Type II REases
EcoRI and EcoRV

Each base pair in the duplex DNA possesses, on its edge facing the major groove, a unique array of three hydrogen-bonding functions (Seeman et al. 1976). The 5-methyl group of thymidine can also be used to distinguish DNA sequences. EcoRI uses 16 (out of 18 possible) hydrogen-bonding functions in its 6-bp recognition sequence and makes van der Waals' contacts with all of the thymidine methyl groups (it was at that time unique to have so many contacts). In contrast, EcoRV only uses four amino acids for sequence-specific binding (see Roberts and Halford 1993 for further details).

DNA Cutting by Type II REases EcoRI and EcoRV

The overall structure of EcoRI bound to its recognition sequence was radically different from that of EcoRV. But one striking similarity was already noted, raising the possibility that these enzymes did use the same mechanism to hydrolyze the phosphodiester bond. In both enzymes the bond cleaved was surrounded by a proline, two acidic residues, and a lysine in the same relative positions (Selent et al. 1992; Winkler 1992). The sequence motif PD$\cdots$(D/E)XK was noted in several other REases, but its significance still had to be established (Anderson 1993); by 2001, seven out of 12 REases had been shown to have the PD motif and nine out of 12 REases the (D/E)XK motif in their catalytic centers (Pingoud and Jeltsch 2001). The precise mechanism of the chemical reaction catalyzed by EcoRI, EcoRV, and almost all other REases remained to be determined (namely, deprotonation of a H_2O molecule, in-line attack by the resulting OH^- leading to formation of an unstable pentacovalent phosphate intermediate, and dissolution of the $3'$ bond by protonation is well established, but the details, and perhaps the order of these events, likely vary from enzyme to enzyme).

DNA Recognition Functions

To dissect recognition and catalysis by REases, one could mutate either the protein or the DNA. To alter the DNA, base analogs (Brennan et al. 1986), or replacement of phosphates with phosphorothioates (Connolly et al. 1984), were employed. In this way it was shown that EcoRI was less active toward GAAUTC than GAATTC, suggesting a role for the methyl group on the inner thymidine (Brennan et al. 1986). Replacing Gln-115 in the protein gave a mutant with the same (lower) activity toward both sequences (Jeltsch et al. 1993). This tied in with the crystal structure: The methylene side chain of Gln-115 interacted hydrophobically with the methyl group on the inner thymidine (Rosenberg 1991).

Altered Enzymes

Mutations in the R-loop of EcoRV all concurred with the crystal structure (Thielking et al. 1991; Vermote and Halford 1992; Halford et al. 1993). Initially, this was not the case for EcoRI, and this necessitated a revision of the original structure, as mentioned above (McClarin et al. 1986; Kim et al. 1990). However, these "wrong" mutants that retained specificity for the EcoRI site were useful, as they did affect nuclease activity ("secondary" functions; Rosenberg 1991; Heitman 1992).

In addition to specific mutagenesis, random mutagenesis followed by genetic selection yielded useful information about several REases (Yanofsky et al. 1987; King et al. 1989; Xu and Schildkraut 1991). Although many mutations affected the protein's affinity for DNA, for some the specificity remained unchanged. One interesting mutant of EcoRI had a Glu-111 → Gly-111 substitution, identifying this residue as a key amino acid in catalysis (King et al. 1989). A clever strategy to isolate mutants was devised by Joe Heitman and Peter Model (Heitman and Model 1990a). A strain carrying M·EcoRI and the *lacZ* gene under the control of an SOS-inducible promoter was transformed with mutagenized DNA. Induction of the SOS response indicated cutting at noncognate sites. This procedure yielded promiscuous mutants that were more active at noncognate sites than the wild-type enzyme (Heitman and Model 1990a). Moreover, their properties could be accounted for by reference to the crystal structure (Heitman 1992).

Altered Substrates

Base analogs alter the hydrogen-bonding interactions between bases and protein (Seeman et al. 1976). This allowed analysis of alterations in DNA cleavage by EcoRI, EcoRV, RsrI, and TaqI (Brennan et al. 1986; McLaughlin et al. 1987; Newman et al. 1990; Aiken et al. 1991b; Zebala et al. 1992b; Lesser et al. 1993; Waters and Connolly 1994). In the case of EcoRI and EcoRV, the loss of almost any one of the functional groups in the DNA (i.e., those that interacted in the crystal structure with the protein) reduced DNA cleavage rates relative to the cognate oligonucleotide. But other changes could also alter enzyme activity, indicating as yet unpredictable cooperativity between regions within the enzyme.

Incorporation of phosphorothioates in the DNA allowed the study of protein–DNA backbone interactions. Using three dNTPs and one dNTPαS

(Potter and Eckstein 1984) on a ss template produced duplex DNA with the phosphorothioates in one strand only (the new strand). A phosphorothioate at the scissile bond reduced or abolished DNA cleavage, and thus these substrates amplified the difference between cleaving the first and second strands of the DNA (Potter and Eckstein 1984). A drawback was that phosphorothioates elsewhere in the recognition sequence or flanking DNA also reduced enzyme activity (Olsen et al. 1990; Lesser et al. 1992). Therefore, another method used chemical synthesis of oligonucleotides with the phosphorothioate placed at a specified site in the chain (Connolly et al. 1984). This method yielded valuable information about the stereochemistry of phosphodiester hydrolysis and interaction with phosphates in the distorted backbone (Connolly et al. 1984; Grasby and Connolly 1992; Lesser et al. 1992).

Coupling Recognition to Catalysis

Taking all results together, the question arose: Could one now account for the ability of REases to distinguish their recognition sites from all other DNA sequence?

a. *EcoRI.* In the case of EcoRI, several processes appeared to be involved in the recognition of the cognate site and the rejection of noncognate sites (Lesser et al. 1990; Heitman 1992). The combined data suggested that the conformation of the DNA (and/or protein) in the specific complex differed from that in the nonspecific complex (Lesser et al. 1990, 1993; Thielking et al. 1990; Heitman 1992), and that EcoRI thus generated its specificity by a subtle combination of both direct and indirect readouts (Roberts and Halford 1993).

b. *EcoRV.* In contrast, EcoRV showed no preference for binding to its recognition site (Taylor et al. 1991), although in this case also, distortion of the enzyme-bound DNA played a key role. The finding that EcoRV had a high affinity for Mg^{2+} when bound to the cognate, but not at nonspecific sites, was probably the main factor determining the different rates of DNA cleavage (Vipond and Halford 1993). This high affinity for Mg^{2+} at only cognate sites could be explained on the structures of the enzyme DNA complexes (Winkler et al. 1993; see Vermote and Halford 1992; Halford et al. 1993; Roberts and Halford 1993 for further discussion). The crucial role for Mg^{2+} at cognate sites was supported by the lack of discrimination by EcoRV in the presence of Mn^{2+} (Vermote and Halford 1992). Both cognate and noncognate complexes with EcoRV had high affinities for Mn^{2+} (with a ratio of DNA cleavage of 6, whereas it had been ~10^6 with Mg^{2+} as cofactor).

The Structure of the PvuII REase with Cognate DNA

As mentioned above, the structure of PvuII was published shortly after the comprehensive review in the *Nucleases* book by Roberts and Halford (Roberts and Halford 1993). The enzyme binds as a dimer to the DNA (Cheng et al. 1994). The enzyme has three domains for dimerization, catalysis, and DNA recognition, respectively (Fig. 5). The catalytic domain resembles that of other REases and appears to share a conserved sequence with the active sites of EcoRI and EcoRV, whereas the direct contacts between the protein and the base pairs of the PvuII recognition site occur exclusively in the major groove via two antiparallel β-strands from the sequence recognition region of the protein.

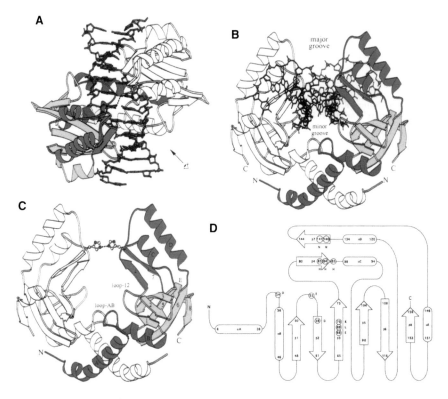

FIGURE 5. Structure of PvuII. Two subunits are shown in gray or in color, respectively, with a ball-and-stick model of the bound cognate DNA segment. Three regions are colored in red, green, and blue for dimer interaction, catalysis, and DNA recognition, respectively. (*A*) Front view of the protein–DNA complex. (*B*) Side view of the protein–DNA complex from an angle as indicated in *A*. (*C*) Same view as in *B* of dimer structure without DNA. The interaction between two H85 side chains closes off the DNA-binding cleft. (*D*) Locations of secondary structural elements in the amino acid sequence. (Reprinted from Cheng et al. 1994, with permission from EMBO.)

The catalytic regions of these REases appear to have been conserved in evolution (presumably reflecting common cleavage to yield 5'P and 3'OH groups); the subunit interface and DNA sequence recognition domains apparently are not conserved. The fact that EcoRI and BamHI produce four-base ssDNA overhangs on the major groove side of the DNA, whereas both EcoRV and PvuII generate blunt ends on the minor groove side of the DNA, may be the reason that with EcoRI and BamHI their DNA-binding cleft of the protein dimer faces the major groove, whereas in the case of EcoRV and PvuII the binding cleft faces the minor groove—thus, the difference in the directions in which EcoRI and EcoRV may be related to the positions of their scissile bonds (Anderson 1993). "The orientation of the catalytic region may need to be stabilized by the dimerization region for efficient cleavage, requiring that the DNA binding cleft face the side of the DNA the scissile bonds are on. This hypothesis is supported by the structure of PvuII-DNA, in which the enzyme binds to DNA from the minor groove side" (Cheng et al. 1994). The prediction would be that enzymes with 3' extensions would approach DNA from the minor groove side as well (Cheng et al. 1994).

With a MW of only 18 kDa, PvuII was the shortest of these four structurally characterized REases, which could explain why the B-form DNA is not distorted after protein binding, whereas the same region in EcoRV is much bulkier. In the latter case, this might require a kink in the DNA to ensure similar distances between the catalytic residue(s) and the target, while at the same time avoiding steric collision between the protein and DNA (Cheng et al. 1994).

TYPE I ENZYMES

Structural Genes and Family Relationships of Type I Enzymes

During the 1980s, various Type I enzymes were cloned and sequenced, both from *E. coli* and *Salmonella* and from other species such as *Citrobacter freundii* (Bickle 1993). As the number of REases grew, Roman numerals were added to the enzymes (e.g., EcoK became EcoKI). In all cases, the results of genetic analysis corroborated the earlier findings: three *hsd* genes, tightly clustered, and transcribed from two promoters, one in front of *hsdM* (cotranscribing *hsdS*), and the other in front of *hsdR* (Sain and Murray 1980; Gough and Murray 1983; Suri and Bickle 1985; Loenen et al. 1987; Cowan et al. 1989; Kannan et al. 1989; Price et al. 1989). Although tight linkage was conserved in all cases, the relative order of the transcription units was not. Moreover, many of the new systems showed little or no homology with EcoKI or EcoBI, or with each other, whether by complementation analysis, DNA hybridization, or immunological cross-reactivity (Murray et al. 1982; Price et al.

1987a). This led to their classification into three families with 19 members in total by 1993 (Bickle 1993): Type IA (prototypes EcoKI and EcoBI), Type IB (prototype EcoAI), and Type IC (prototype R124I, variously called StyR124I and EcoR124I). At the time, all Type IA and Type IB genes were located on the bacterial chromosome near *serB*; the Type IC genes were plasmid-encoded. But all proteins were similar in structure and required ATP and SAM for activity. In later years, two more families were added (Type ID [prototype StyBLI and KpnAI (Titheradge et al. 2001; Murray 2002; Kasarjian et al. 2004)]) and Type IE (prototype KpnBI [Chin et al. 2004]), whereas the location on the chromosome or episome would prove not as strict as originally thought.

Sequence Homologies within and between Families of Type I Enzymes

Within a family, DNA sequences of the *hsdM* and *hsdR* genes are quite highly conserved (Murray et al. 1982; Daniel et al. 1988; Gubler et al. 1992; Sharp et al. 1992) with the strongest sequence identity ~95% for *hsdM* (Sharp et al. 1992). In contrast, the *hsdS* genes contained two extensive regions of nonhomology, one at the 5' end of the genes and the other toward the 3' end, flanked by homologous regions (Gough and Murray 1983; Cowan et al. 1989; Gubler et al. 1992). The two nonhomologous "hypervariable" regions encoded protein domains that each recognized one-half of the recognition sequence. Those enzymes with the same 5' moiety of their recognition sequence showed ~50% identity in the amino-terminal hypervariable region (e.g., StySBI, EcoAI, and EcoEI all recognized 5'GAG) (Cowan et al. 1989). This was the first formal evidence for the current model of how Type I enzymes specifically recognize DNA. The conserved regions outside these nonhomologous regions were thought to provide protein–protein interactions with the HsdM and HsdR proteins.

Between families, the *hsdM* genes encoded ~26%–33% identical amino acids, a degree of homology high enough to exclude an independent origin of the different Type I families. Some of these residues were found to be conserved in all m6A MTases (Lauster et al. 1987; Guschlbauer 1988; Narva et al. 1988; Smith et al. 1990).

Evolution of DNA Specificity of Type I Enzymes

Evolution of DNA Specificity by Homologous Recombination within the hsdS Gene of Type I Enzymes

In 1976, Len Bullas in the laboratory of Stuart Glover identified a new specificity, StySQ, after transduction of the *hsd* genes of *S. potsdam* (StySP) into

S. typhimurium (StySB; a Roman I was later added to the names of all three enzymes) (Fig. 6A; Bullas et al. 1976). DNA heteroduplex analysis and DNA sequencing showed that indeed the StySQ system arose by recombination within the central conserved region of the parental *hsdS* genes (Fuller-Pace et al. 1984; Fuller-Pace and Murray 1986). The recognition sequences of the three proteins GAG(N6)RTAYG (StySBI), AAC(N6)GTRC (StySPI), and AAC(N6)RTAYG (StySQI) confirmed the hybrid nature of StySQI (Nagaraja et al. 1985a,b). These findings immediately led to the idea that *hsdS* genes encoded two DNA-binding domains: an amino-terminal domain that recognized the 5′ part of the recognition sequences and a carboxy-terminal domain that recognized the 3′ part. Thus, recombination in the central conserved fragment would allow domain swapping generating new sequence specificity. In line with this result, the reciprocal recombinant (StySJIb) between the StySBI and StySPI *hsdS* genes recognized GAG(N6)GTRC (Gann et al. 1987).

Evolution of DNA Specificity by Unequal Crossing-Over within the hsdS Gene of Type I Enzymes

In addition to domain swaps via homologous recombination, another spontaneous change in specificity was found with plasmid EcoR124. Cells carrying this plasmid could express either EcoR124 (renamed EcoR124I) or EcoR124/3 (renamed EcoR124II), with recognition sequences GAA(N6) RTCG and GAA(N7)RTCG, respectively (Price et al. 1987b). This proved to be due to unequal crossing-over in the nonspecific spacer separating the two specific parts (Fig. 6B; Price et al. 1989). Surprisingly, this crossing-over occurred at a specific site in the conserved central region, where a 12-bp sequence (encoding four amino acids, TAEL) was repeated twice in EcoR124I and three times in EcoR124II. This increased the spacer from 6 to 7 bp, rotating the two domains by 36°, a far from trivial matter with respect to enzyme recognition. This effect of the increase in length of the conserved region on that of the spacer separating the two DNA recognizing domains led to the model in which the length, but not the exact amino acid sequence, was important for function. Mutations in the DNA encoding the TAEL repeats indeed did not affect activity or specificity, whereas altering the length of the repeated region did have drastic effects (Gubler and Bickle 1991).

Variants with either one or four copies of the repeat, for example, were virtually inactive in restriction (10^5–10^6 times less active than the wild type). However, they were still efficient MTases. The mutant with a single copy of

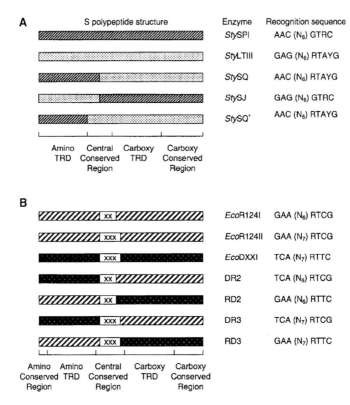

FIGURE 6. Evolution of Type I enzymes with new specificities. (*A*) Recombination between *hsdS* genes produces hybrid genes and chimeric S polypeptides. StySP1 and StyLTIII are naturally occurring Type I R-M systems. StySQ and StySJ have hybrid *hsdS* genes (Fuller-Pace et al. 1984; Gann et al. 1987). The regions originating from StySP1 are hatched and those originating from StyLTIII are stippled. Reassortment of the target recognition domains (TRDs) accordingly gave rise to recombinant recognition sequences (Nagaraja et al. 1985a; Gann et al. 1987). Site-directed mutagenesis of the central conserved region of the StySQ *hsdS* gene produced StySQ*, comprising only the amino-terminal variable region from StySP1 and the remainder from StyLTIII. The StySQ* target sequence confirms that the amino-terminal variable region is in fact a TRD responsible for recognition of the trinucleotide component of the sequence (Cowan et al. 1989). (*B*) Sequence specificity may also be altered by changing the length of the nonspecific spacer of the target sequence. The S polypeptides of EcoRI241 and EcoRI2411 differ only in the number of times a short amino acid motif (X = TAEL) is repeated within their central conserved regions (Price et al. 1989), resulting in extension of the spacer in the target sequence from 6 nt (N_6) for EcoRI241 to N_7 for EcoRI2411. The recognition sequence of EcoDXXI also contains a nonspecific spacer of 7 nt, corresponding to three TAEL repeats in its HsdS polypeptide (Gubler et al. 1992). Chimeric S polypeptides recognize the predicted target sequences (Gubler et al. 1992). (Reprinted from Murray 2002, with permission from Microbiology Society.)

the repeat methylated the EcoR124I recognition sequence, whereas a mutant with four repeats methylated both the EcoR124I and the EcoR124II sequences but would not methylate a putative recognition sequence with 8 bp in the nonspecific spacer (Gubler and Bickle 1991). It was speculated that the severe effects on restriction but not on modification might relate to the structure of the central conserved regions of the *hsdS* gene products. This region of the protein most likely had a dual function: a spacer between the DNA-binding domains of the protein, but it was also needed for interaction with the HsdR restriction subunit.

The idea that *hsdS* genes contained two DNA-binding domains separated by a spacer region whose length determined the number of base pairs separating the two components of the recognition sequences was tested using the Type IC enzyme, EcoDXXI (Piekarowicz et al. 1985; Skrzypek and Piekarowicz 1989). This enzyme recognized TCA(N7)RTTC and contained three copies of the 12-bp repeated sequence in the conserved region (Meister et al. 1993). Hybrids between the two halves of the *hsdS* genes of *ecoDXXI* and *ecoR124* with either two or three copies of the 12-bp repeated sequence all were active in restriction and recognized DNA sequences consistent with this model.

Evolution of DNA Specificity by Transposition within the hsdS Gene of Type I Enzymes

A third mode of changing specificity of a Type I enzyme was also discovered in EcoDXXI. A Tn5 derivative in the *ecoDXXI hsd* region appeared to have an altered DNA sequence specificity. It turned out that Tn5 had inserted into the *hsdS* gene, just 3′ to the central conserved region. The *hsdS* gene product produced by the mutant was much shorter but retained the amino-terminal part of the protein. The sequence recognized was TCA(N8)TGA (Fig. 6B; Meister et al. 1993). The rotational symmetry of the site and the length of the truncated HsdS protein led to the inescapable conclusion that the enzyme apparently assembled two copies of the truncated HsdS protein (called EcoDXXsI) (Loenen 2003).

Enzyme Structures and Mechanisms of Type I Enzymes

As mentioned earlier, purification of the EcoKI and EcoBI enzymes resulted in various oligomeric protein complexes. New Type I enzymes of any of the three families behaved similarly (Suri et al. 1984a,b; Price et al. 1987a; Gubler and

Bickle 1991; Taylor et al. 1992). Enzymes could be MTases with or without the REase, with stoichiometries reported for the HsdM and HsdS subunits as 2:1 or 1:1 for the methylase on its own, and most likely 2:2:1 for the pentameric REase, although some enzymes were unstable and aggregated or fell apart upon storage (Bickle 1993). The HsdS subunit of EcoKI could not be purified on its own, but that of EcoR124I was shown to be a sequence-specific DNA-binding protein without enzymatic activity (Kusiak et al. 1992).

EcoKI

The genes encoding the modification and restriction subunits of EcoKI were the first Type I HsdR and HsdM proteins to be sequenced and had sequence motifs typical of SAM- and ATP-binding proteins (Loenen et al. 1987). EcoKI had no affinity for DNA in the absence of cofactor SAM. After allosteric activation, the enzyme bound with high affinity to both modified and unmodified recognition sites (Bickle et al. 1978). In the absence of ATP, these complexes with DNA were relatively stable on both modified ($t_{1/2}$ = 6 min), and unmodified sites ($t_{1/2}$ = 22 min) (Yuan et al. 1975). The enzyme would modify the second strand of hemimethylated DNA with overall first-order reaction kinetics with rate constants of 3×10^{-3} sec^{-1} in the presence of ATP and 4×10^{-4} sec^{-1} in its absence (the enzyme could also modify unmethylated DNA, with a rate constant of 6×10^{-5} sec^{-1}; Suri et al. 1984a; Bickle 1993). The addition of ATP to complexes with unmethylated DNA set the cleavage mode in action. After a massive conformational change, the enzyme remained bound to the recognition site, but cleaved randomly far from the recognition sites (400–7000 bp; Chapter 5). This was followed by massive ATP hydrolysis. This last aspect of the reaction mechanism was an utter mystery and generated considerable controversy (Bickle 1993).

Cleavage Models for Type I Enzymes

Different models were proposed for the mechanism whereby the enzyme cleaves DNA at loci distant from the recognition sequence. Based mainly on EM data on the EcoBI and EcoKI enzymes, the enzymes tracked along the DNA, forming ever-larger loops until the cleavage site was reached. Although EcoBI cleaved only to one side of the asymmetric recognition site (Rosamond et al. 1979), EcoKI translocated and cleaved the DNA in both directions (Yuan et al. 1980a). The "collision" model, which stated that restriction required at least two recognition sites in the DNA, was proposed in 1988 by William (Bill) Studier. An enzyme molecule would bind to each site, and the molecules would move along the DNA until they bumped into each other, at which point the DNA would be cut (Fig. 7; Studier and Bandyopadhyay 1988; Studier 2013).

How did Bill Studier arrive at this model? He studied phage T7 and tried to find out why this phage was resistant to cleavage by EcoKI (details about this and other restriction evasion strategies by plasmids and phages [extensively reviewed in Krüger and Bickle 1983; Bickle and Krüger 1993] will follow later in this book). Bill identified an early function (the product of gene 0.3, called Ocr [overcoming restriction]) that blocked EcoKI. Mutations in this gene led to restriction of T7 DNA by EcoKI. However, he found distinct restriction fragments on the gel, rather than a smear caused by random fragmentation. Did this mean that two enzymes would bind two recognition sites and translocate the DNA until they met in the middle? That cleavage occurred when they met, stalled, and cleaved (Studier 2013)? The model was very appealing, but if this collision model were true, how did one explain cleavage of DNA molecules with single recognition sites?

Whatever the exact mechanism, DNA cleavage was a two-step process: a nick in one strand, followed by a second cut, probably by another EcoKI molecule. In addition to the curious ATP hydrolysis, the DNA ends produced by Type I enzymes were a mystery too. They could not be labeled with polynucleotide kinase and had long 3′ protrusions (Eskin and Linn 1972; Murray et al. 1973; Endlich and Linn 1985).

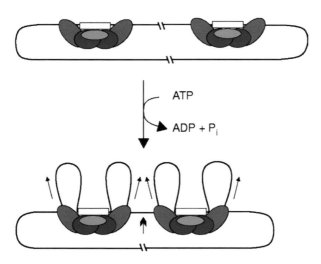

FIGURE 7. The collision model for DNA breakage (Studier and Bandyopadhyay 1988). EcoKI bound to target sequences translocates DNA toward itself. Collision blocks translocation and stimulates the nicking of both DNA strands. REase activity may be stimulated when translocation is impeded by some other protein or structure (Janscak et al. 1999). (Reprinted from Murray 2002, with permission from Microbiology Society.)

Maintenance versus De Novo Methylation by Type I Enzymes

There were additional unexplained observations and differences between different Type I enzymes. Whereas EcoKI and EcoBI and other Type IA enzymes preferentially methylated hemimethylated substrates, the reaction with nonmethylated DNA was slow (see above), and the reaction with both substrates was inhibited by ATP (Suri and Bickle 1985). Type 1C enzymes showed the same substrate preferences as Type IA enzymes; however, for these enzymes, the reaction was stimulated by ATP (Price et al. 1987a). The Type IB enzyme EcoAI showed a completely different pattern. Hemimethylated and nonmethylated substrates were modified equally well, but the reaction was completely dependent on ATP (Suri and Bickle 1985).

An interesting finding was the small Ral (restriction alleviation) protein of lambda, which rescued superinfecting phages from restriction by EcoKI (Zabeau et al. 1980; Loenen and Murray 1986). (An analogous protein Lar is present on the cryptic lambdoid prophage, Rac [King and Murray 1995].) Ral appeared to switch EcoKI from a maintenance to a de novo MTase by enhancing methylation of unmodified sites (Loenen and Murray 1986). Perhaps even more interesting were the Ral-independent EcoKI m* mutants isolated in Noreen Murray's laboratory (Kelleher et al. 1991). They mapped to a few specific places in the HsdM subunit, in line with later evidence that assigned the discriminatory capacity with respect to the methylation state of the DNA to the M_2S complex (see, e.g., Loenen 2003 for details).

Biochemical experiments were carried out with m* mutants (LL113Q, L134V [Winter 1997])—for example, the kinetic constants for L113Q versus wild type were determined on unmethylated DNA and on hemimethylated DNA with the methyl group on either the top or bottom strand (Tables 4.3 and 4.4 in Winter 1997).

TYPE III ENZYMES

Occurrence and Genetics of Type III Enzymes

By 1993, only four members of this family had been identified. In addition to EcoP1I, EcoP15I, and HinfIII (Chapter 5), the chromosomal StyLTI system was present in many *Salmonella* strains (Arber and Dussoix 1962; Arber and Wauters-Willems 1970; Colson et al. 1970; Piekarowicz and Kalinowska 1974; Bullas et al. 1980). The enzymes were encoded by the *mod* and *res* genes (Iida et al. 1983). *mod* encoded the MTase and recognized the DNA specificity site. The *res* gene product was essential for restriction in a complex with the MTase but lacked enzymatic activity on its own. The data on transcription were controversial. Were both genes transcribed from a single promoter located in front of the *mod* gene, as judged from transposon insertion experiments (Iida et al. 1983)? Or did other in vitro and in vivo studies suggest a more complex situation (Iida et al. 1983; Sharrocks and Hornby 1991)?

The sequences of the *mod* genes of EcoP1I, EcoP15I, and StyLTI, as well as the *res* genes of EcoP1I and StyLTI, were known (Humbelin et al. 1988; Dartois et al. 1993). DNA heteroduplex analysis indicated strong homology between the *res* genes of EcoP15I and EcoP1I (Iida et al. 1983). The *mod* genes were mosaics of conserved and nonconserved regions, a structure reminiscent of that of the Type I *hsdS* genes: a totally dissimilar central region and conserved 5′ and 3′ regions. It was thought that the conserved regions encoded protein domains that interacted with the *res* gene product and that the variable regions encoded sequence-specific DNA-binding domains. In line with this, mutations that led to loss of modification without affecting restriction (Rosner 1973) mapped in this region (Humbelin et al. 1988). These mutants were shown to have lost the ability to bind cofactor SAM (Rao et al. 1989a). The EcoP1I and StyLTI *res* sequences showed surprisingly little homology apart from a stretch of 50 amino acids toward the center, where the two proteins were virtually identical—a good candidate region for the interaction between the restriction and modification subunits (Dartois et al. 1993).

Enzyme Mechanisms of Type III Enzymes

All four REases recognized asymmetric DNA sequences and cut 25–26 bp downstream from the sequence: EcoP1I (AGACC), EcoP15I (CAGCAG), HinfIII (CGAAT), and StyLTI (CAGAG) (Bachi et al. 1979; Hadi et al. 1979; Piekarowicz et al. 1981; De Backer and Colson 1991). Surprisingly, these sequences had only adenine in the strand shown, and thus half of the sites following DNA replication completely lack modification. Yet, this was apparently not lethal to the cell. Experiments with phage T7 would solve that mystery in 1992, as detailed later.

The available data on the reaction mechanism of the Type III enzymes came mainly from studies on EcoP1I and EcoP15I. They were quite different from the Type I enzymes: They required ATP for restriction but did not hydrolyze as much ATP as the Type I enzymes. SAM was not essential, but stimulated cleavage, leading to competition between restriction and modification (Haberman 1974; Risser et al. 1974; Reiser and Yuan 1977; Kauc and Piekarowicz 1978). Both ATP and SAM were allosteric effectors for DNA cleavage, and nonhydrolyzable ATP analogs only weakly supported cleavage (Yuan and Reiser 1978; Yuan et al. 1980b).

Also in contrast to the Type I enzymes, the Type III MTase could bind either DNA or SAM first. An unusual feature of the kinetics of methylation was that the enzyme was inhibited by SAM concentrations of only slightly more than the K_m values for SAM (Rao et al. 1989b). This suggested nonproductive binding of SAM to the methylated DNA–enzyme complex. A mutant, S240A, supported this idea, being more active than wild-type enzyme because it could no longer be inhibited by substrate SAM. This serine was important for activity, as a S240P mutation led to loss of SAM binding (Rao et al. 1989a).

DNA Cleavage by Type III Enzymes

As mentioned above, EcoP1I, EcoP15I, and StyLTI have a methylatable adenine in the top strand only (Hadi et al. 1979; Meisel et al. 1991). How did cells survive after DNA replication? The answer to this mystery came from the sequence of phage T7. This phage was not restricted by EcoP15I, although it contained 36 EcoP15I sites (Dunn and Studier 1983). Interestingly, these sites were all in the same orientation: the CAGCAG sequence in one DNA strand and its CTGCTG complement in the other strand (Schroeder et al. 1986). Did this make T7 DNA refractory to EcoP15I cleavage? If so, this would mean that EcoP15I restriction should require two recognition sites, and also that these two recognition sites should be in inverse orientation! M13 constructs with different numbers and orientations of EcoP15I sites were made and proved this idea to be correct. Single or multiple sites with any orientation could be methylated, but only unmodified sites in inverse orientation could be restricted (Meisel et al. 1992). Hence, the true recognition site for the EcoP15I REase complex consisted of a twofold rotationally symmetrical sequence interrupted by a nonspecific spacer of variable length. As all newly replicated sites would be in the same orientation, unmodified sites were not cleaved, but modified.

Although little evidence was available at the time, there was good reason to believe that this phenomenon would be a common characteristic of the Type III enzymes. This would be in line with some earlier observations with EcoP1I on phage lambda (Arber et al. 1963; Hattman et al. 1978; Bickle 1993).

Foresight

Thomas (Tom) Bickle ends his 1993 review (Bickle 1993) with: "I believe that the prevalence of DNA restriction systems is a sign of genetic selection operating at the population level: A population whose individual members can prevent phage propagation (even if the infected individual is killed in the process) is fitter than one that cannot. Paradoxically, selection for function takes place in those cells in which the function is not used."

REFERENCES

Aiken CR, Gumport RI. 1991. Base analogs in study of restriction enzyme–DNA interactions. *Methods Enzymol* **208:** 433–457.

Aiken C, Milarski-Brown K, Gumport RI. 1986. RsrI and EcoRI are two different restriction endonucleases that recognise the same DNA sequence. *Fed Proc* **45:** 1914.

Aiken CR, Fisher EW, Gumport RI. 1991a. The specific binding, bending, and unwinding of DNA by RsrI endonuclease, an isoschizomer of EcoRI endonuclease. *J Biol Chem* **266:** 19063–19069.

Aiken CR, McLaughlin LW, Gumport RI. 1991b. The highly homologous isoschizomers RsrI endonuclease and EcoRI endonuclease do not recognize their target sequence identically. *J Biol Chem* **266:** 19070–19078.

Anderson JE. 1993. Restriction endonucleases and modification methylases. *Curr Opin Struct Biol* **3:** 24–30.

Arber W, Dussoix D. 1962. Host specificity of DNA produced by *Escherichia coli*. I. Host controlled modification of bacteriophage λ. *J Mol Biol* **5:** 18–36.

Arber W, Wauters-Willems D. 1970. Host specificity of DNA produced by *Escherichia coli*. XII. The two restriction and modification systems of strain 15T. *Mol Gen Genet* **108:** 203–217.

Arber W, Hattman S, Dussoix D. 1963. On the host-controlled modification of bacteriophage λ. *Virology* **21:** 30–35.

Athanasiadis A, Gregoriu M, Thanos D, Kokkinidis M, Papamatheakis J. 1990. Complete nucleotide sequence of the *Pvu*II restriction enzyme gene from *Proteus vulgaris*. *Nucleic Acids Res* **18:** 6434.

Athanasiadis A, Vlassi M, Kotsifaki D, Tucker PA, Wilson KS, Kokkinidis M. 1994. Crystal structure of *Pvu*II endonuclease reveals extensive structural homologies to *Eco*RV. *Nat Struct Biol* **1:** 469–475.

Bachi B, Reiser J, Pirrotta V. 1979. Methylation and cleavage sequences of the *Eco*P1 restriction-modification enzyme. *J Mol Biol* **128:** 143–163.

Barany F. 1988. The *Taq*I 'star' reaction: strand preferences reveal hydrogen-bond donor and acceptor sites in canonical sequence recognition. *Gene* **65:** 149–165.

Barsomian J, Landry D, Moran L, Slatko B, Feehery G, Jack W, Nwanko D, Wilson G. 1990. Organisation and evolution of Type IIS MTases. In *Second NEB Workshop on Biological DNA Modification, Berlin*, Vol. 11.

Becker MM, Lesser D, Kurpiewski M, Baranger A, Jen-Jacobson L. 1988. "Ultraviolet footprinting" accurately maps sequence-specific contacts and DNA kinking in the *Eco*RI endonuclease–DNA complex. *Proc Natl Acad Sci* **85:** 6247–6251.

Bennett SP, Halford SE. 1989. Recognition of DNA by type II restriction enzymes. *Curr Top Cell Regul* **30:** 57–104.

Berg OG, von Hippel PH. 1985. Diffusion-controlled macromolecular interactions. *Annu Rev Biophys Biophys Chem* **14:** 131–160.

Bhagwat AS, Johnson B, Weule K, Roberts RJ. 1990. Primary sequence of the *Eco*RII endonuclease and properties of its fusions with β-galactosidase. *J Biol Chem* **265:** 767–773.

Bickle TA. 1993. The ATP-dependent restriction enzymes. In *Nucleases*, 2nd ed. (ed. Linn SM, LLoyd SR, Roberts RJ), pp. 88–109. Cold Spring Harbor Laboratory Press, Cold Spring Harbor, NY.

Bickle TA, Krüger DH. 1993. Biology of DNA restriction. *Microbiol Rev* **57:** 434–450.

Bickle TA, Brack C, Yuan R. 1978. ATP-induced conformational changes in the restriction endonuclease from *Escherichia coli* K-12. *Proc Natl Acad Sci* **75:** 3099–3103.

Bitinaite J, Maneliene Z, Menkevicius S, Klimasauskas S, Butkus V, Janulaitis A. 1992. *Alw*26I, *Eco*31I and *Esp*3I–type IIs methyltransferases modifying cytosine and adenine in complementary strands of the target DNA. *Nucleic Acids Res* **20**: 4981–4985.

Bougueleret L, Schwarzstein M, Tsugita A, Zabeau M. 1984. Characterization of the genes coding for the *Eco* RV restriction and modification system of *Escherichia coli*. *Nucleic Acids Res* **12**: 3659–3676.

Bougueleret L, Tenchini ML, Botterman J, Zabeau M. 1985. Overproduction of the *Eco*R V endonuclease and methylase. *Nucleic Acids Res* **13**: 3823–3839.

Boyd AC, Charles IG, Keyte JW, Brammar WJ. 1986. Isolation and computer-aided characterization of *Mme*I, a type II restriction endonuclease from *Methylophilus methylotrophus*. *Nucleic Acids Res* **14**: 5255–5274.

Brennan CA, Van Cleve MD, Gumport RI. 1986. The effects of base analogue substitutions on the methylation by the *Eco*RI modification methylase of octadeoxyribonucleotides containing modified *Eco*RI recognition sequences. *J Biol Chem* **261**: 7279–7286.

Brown NL, Smith M. 1980. A general method for defining restriction enzyme cleavage and recognition sites. *Methods Enzymol* **65**: 391–404.

Bullas LR, Colson C, Van Pel A. 1976. DNA restriction and modification systems in *Salmonella*. SQ, a new system derived by recombination between the SB system of *Salmonella typhimurium* and the SP system of *Salmonella potsdam*. *J Gen Microbiol* **95**: 166–172.

Bullas LR, Colson C, Neufeld B. 1980. Deoxyribonucleic acid restriction and modification systems in *Salmonella*: chromosomally located systems of different serotypes. *J Bacteriol* **141**: 275–292.

Callahan SJ, Morgan RD, Jain R, Townson SA, Wilson GG, Roberts RJ, Aggarwal AK. 2011. Crystallization and preliminary crystallographic analysis of the type IIL restriction enzyme MmeI in complex with DNA. *Acta Crystallogr, Sect F: Struct Biol Cryst Commun* **67**: 1262–1265.

Cheng X, Balendiran K, Schildkraut I, Anderson JE. 1994. Structure of *Pvu*II endonuclease with cognate DNA. *EMBO J* **13**: 3927–3935.

Chin V, Valinluck V, Magaki S, Ryu J. 2004. KpnBI is the prototype of a new family (IE) of bacterial type I restriction-modification system. *Nucleic Acids Res* **32**: e138.

Colson C, Colson AM, Van Pel A. 1970. Chromosomal location of host specificity in *Salmonella typhimurium*. *J Gen Microbiol* **60**: 265–271.

Connolly BA, Eckstein F, Pingoud A. 1984. The stereochemical course of the restriction endonuclease *Eco*RI-catalyzed reaction. *J Biol Chem* **259**: 10760–10763.

Conrad M, Topal MD. 1989. DNA and spermidine provide a switch mechanism to regulate the activity of restriction enzyme *Nae* I. *Proc Natl Acad Sci* **86**: 9707–9711.

Cowan GM, Gann AA, Murray NE. 1989. Conservation of complex DNA recognition domains between families of restriction enzymes. *Cell* **56**: 103–109.

Daniel AS, Fuller-Pace FV, Legge DM, Murray NE. 1988. Distribution and diversity of *hsd* genes in *Escherichia coli* and other enteric bacteria. *J Bacteriol* **170**: 1775–1782.

Dartois V, De Backer O, Colson C. 1993. Sequence of the *Salmonella typhimurium* StyLT1 restriction-modification genes: homologies with *Eco*P1 and *Eco*P15 type-III R-M systems and presence of helicase domains. *Gene* **127**: 105–110.

De Backer O, Colson C. 1991. Identification of the recognition sequence for the M.StyLTI methyltransferase of *Salmonella typhimurium* LT7: an asymmetric site typical of type-III enzymes. *Gene* **97**: 103–107.

de la Campa AG, Kale P, Springhorn SS, Lacks SA. 1987. Proteins encoded by the *Dpn*II restriction gene cassette. Two methylases and an endonuclease. *J Mol Biol* **196**: 457–469.

Doerfler W. 1983. DNA methylation and gene activity. *Ann Rev Biochem* **52:** 93–124.

Dunn JJ, Studier FW. 1983. Complete nucleotide sequence of bacteriophage T7 DNA and the locations of T7 genetic elements. *J Mol Biol* **166:** 477–535.

Ehbrecht HJ, Pingoud A, Urbanke C, Maass G, Gualerzi C. 1985. Linear diffusion of restriction endonucleases on DNA. *J Biol Chem* **260:** 6160–6166.

Endlich B, Linn S. 1985. The DNA restriction endonuclease of *Escherichia coli* B. II. Further studies of the structure of DNA intermediates and products. *J Biol Chem* **260:** 5729–5738.

Eskin B, Linn S. 1972. The deoxyribonucleic acid modification and restriction enzymes of *Escherichia coli* B. II. Purification, subunit structure, and catalytic properties of the restriction endonuclease. *J Biol Chem* **247:** 6183–6191.

Fuller-Pace FV, Murray NE. 1986. Two DNA recognition domains of the specificity polypeptides of a family of type I restriction enzymes. *Proc Natl Acad Sci* **83:** 9368–9372.

Fuller-Pace FV, Bullas LR, Delius H, Murray NE. 1984. Genetic recombination can generate altered restriction specificity. *Proc Natl Acad Sci* **81:** 6095–6099.

Gabbara S, Bhagwat AS. 1992. Interaction of *Eco*RII endonuclease with DNA substrates containing single recognition sites. *J Biol Chem* **267:** 18623–18630.

Gann AA, Campbell AJ, Collins JF, Coulson AF, Murray NE. 1987. Reassortment of DNA recognition domains and the evolution of new specificities. *Mol Microbiol* **1:** 13–22.

Gingeras TR, Milazzo JP, Roberts RJ. 1978. A computer assisted method for the determination of restriction enzyme recognition sites. *Nucleic Acids Res* **5:** 4105–4127.

Gough JA, Murray NE. 1983. Sequence diversity among related genes for recognition of specific targets in DNA molecules. *J Mol Biol* **166:** 1–19.

Grasby JA, Connolly BA. 1992. Stereochemical outcome of the hydrolysis reaction catalyzed by the *Eco*RV restriction endonuclease. *Biochemistry* **31:** 7855–7861.

Gubler M, Bickle TA. 1991. Increased protein flexibility leads to promiscuous protein–DNA interactions in type IC restriction-modification systems. *EMBO J* **10:** 951–957.

Gubler M, Braguglia D, Meyer J, Piekarowicz A, Bickle TA. 1992. Recombination of constant and variable modules alters DNA sequence recognition by type IC restriction-modification enzymes. *EMBO J* **11:** 233–240.

Guo J, Gaj T, Barbas CF III. 2010. Directed evolution of an enhanced and highly efficient *Fok*I cleavage domain for zinc finger nucleases. *J Mol Biol* **400:** 96–107.

Guschlbauer W. 1988. The DNA and *S*-adenosylmethionine-binding regions of *Eco*Dam and related methyltransferases. *Gene* **74:** 211–214.

Haberman A. 1974. The bacteriophage P1 restriction endonuclease. *J Mol Biol* **89:** 545–563.

Hadi SM, Bachi B, Shepherd JC, Yuan R, Ineichen K, Bickle TA. 1979. DNA recognition and cleavage by the EcoP15 restriction endonuclease. *J Mol Biol* **134:** 655–666.

Halford SE. 1983. How does *Eco*RI cleave its recognition site on DNA? *Trends Biochem Sci* **8:** 455–460.

Halford SE, Goodall AJ. 1988. Modes of DNA cleavage by the *Eco*RV restriction endonuclease. *Biochemistry* **27:** 1771–1777.

Halford SE, Johnson NP. 1980. The *Eco*RI restriction endonuclease with bacteriophage λ DNA. Equilibrium binding studies. *Biochem J* **191:** 593–604.

Halford SE, Johnson NP. 1983. Single turnovers of the EcoRI restriction endonuclease. *Biochem J* **211:** 405–415.

Halford SE, Marko JF. 2004. How do site-specific DNA-binding proteins find their targets? *Nucleic Acids Res* **32:** 3040–3052.

Halford SE, Taylor JD, Vermote CLM, Vipond IB. 1993. Mechanism of action of restriction endonuclease *Eco*RV. *Nucleic Acids Mol Biol* **7:** 47–69.

Halford SE, Catto LE, Pernstich C, Rusling DA, Sanders KL. 2011. The reaction mechanism of FokI excludes the possibility of targeting zinc finger nucleases to unique DNA sites. *Biochem Soc Trans* **39:** 584–588.

Hattman S, Brooks JE, Masurekar M. 1978. Sequence specificity of the P1 modification methylase (M·*Eco* P1) and the DNA methylase (M·*Eco* dam) controlled by the *Escherichia coli* dam gene. *J Mol Biol* **126:** 367–380.

Heitman J. 1992. How the EcoRI endonuclease recognizes and cleaves DNA. *Bioessays* **14:** 445–454.

Heitman J. 1993. On the origins, structures and functions of restriction-modification enzymes. *Genet Eng* **15:** 57–108.

Heitman J, Model P. 1990a. Mutants of the *Eco*RI endonuclease with promiscuous substrate specificity implicate residues involved in substrate recognition. *EMBO J* **9:** 3369–3378.

Heitman J, Model P. 1990b. Substrate recognition by the *Eco*RI endonuclease. *Proteins* **7:** 185–197.

Hensley P, Nardone G, Chirikjian JG, Wastney ME. 1990. The time-resolved kinetics of superhelical DNA cleavage by BamHI restriction endonuclease. *J Biol Chem* **265:** 15300–15307.

Humbelin M, Suri B, Rao DN, Hornby DP, Eberle H, Pripfl T, Kenel S, Bickle TA. 1988. Type III DNA restriction and modification systems *Eco*P1 and *Eco*P15. Nucleotide sequence of the *Eco*P1 operon, the *Eco*P15 mod gene and some *Eco*P1 mod mutants. *J Mol Biol* **200:** 23–29.

Iida S, Meyer J, Bachi B, Stalhammar-Carlemalm M, Schrickel S, Bickle TA, Arber W. 1983. DNA restriction–modification genes of phage P1 and plasmid p15B. Structure and in vitro transcription. *J Mol Biol* **165:** 1–18.

Jack WE, Terry BJ, Modrich P. 1982. Involvement of outside DNA sequences in the major kinetic path by which *Eco*RI endonuclease locates and leaves its recognition sequence. *Proc Natl Acad Sci* **79:** 4010–4014.

Jack WE, Greenough L, Dorner LF, Xu SY, Strzelecka T, Aggarwal AK, Schildkraut I. 1991. Overexpression, purification and crystallization of BamHI endonuclease. *Nucleic Acids Res* **19:** 1825–1829.

Janscak P, MacWilliams MP, Sandmeier U, Nagaraja V, Bickle TA. 1999. DNA translocation blockage, a general mechanism of cleavage site selection by type I restriction enzymes. *EMBO J* **18:** 2638–2647.

Jeltsch A, Alves J, Oelgeschläger T, Wolfes H, Maass G, Pingoud A. 1993. Mutational analysis of the function of Gln115 in the *Eco*RI restriction endonuclease, a critical amino acid for recognition of the inner thymidine residue in the sequence -GAATTC- and for coupling specific DNA binding to catalysis. *J Mol Biol* **229:** 221–234.

Kannan P, Cowan GM, Daniel AS, Gann AA, Murray NE. 1989. Conservation of organization in the specificity polypeptides of two families of type I restriction enzymes. *J Mol Biol* **209:** 335–344.

Kasarjian JK, Hidaka M, Horiuchi T, Iida M, Ryu J. 2004. The recognition and modification sites for the bacterial type I restriction systems KpnAI, StySEAI, StySENI and StySGI. *Nucleic Acids Res* **32:** e82.

Kauc L, Piekarowicz A. 1978. Purification and properties of a new restriction endonuclease from *Haemophilus influenzae* Rf. *Eur J Biochem/FEBS* **92:** 417–426.

Kelleher JE, Daniel AS, Murray NE. 1991. Mutations that confer de novo activity upon a maintenance methyltransferase. *J Mol Biol* **221:** 431–440.

Kelly TJ Jr, Smith HO. 1970. A restriction enzyme from *Hemophilus influenzae* II. *J Mol Biol* **51:** 393–409.

Kim YC, Grable JC, Love R, Greene PJ, Rosenberg JM. 1990. Refinement of Eco RI endonuclease crystal structure: a revised protein chain tracing. *Science* **249:** 1307–1309.

Kim YG, Smith J, Durgesha M, Chandrasegaran S. 1998. Chimeric restriction enzyme: Gal4 fusion to FokI cleavage domain. *Biol Chem* **379:** 489–495.

King G, Murray NE. 1995. Restriction alleviation and modification enhancement by the Rac prophage of *Escherichia coli* K-12. *Mol Microbiol* **16:** 769–777.

King K, Benkovic SJ, Modrich P. 1989. Glu-111 is required for activation of the DNA cleavage center of *Eco*RI endonuclease. *J Biol Chem* **264:** 11807–11815.

Kiss A, Posfai G, Keller CC, Venetianer P, Roberts RJ. 1985. Nucleotide sequence of the BsuRI restriction-modification system. *Nucleic Acids Res* **13:** 6403–6421.

Klimasauskas S, Timinskas A, Menkevicius S, Butkiene D, Butkus V, Janulaitis A. 1989. Sequence motifs characteristic of DNA[cytosine-N4]methyltransferases: similarity to adenine and cytosine-C5 DNA-methylases. *Nucleic Acids Res* **17:** 9823–9832.

Kong H, Morgan RD, Maunus RE, Schildkraut I. 1993. A unique restriction endonuclease, BcgI, from *Bacillus coagulans*. *Nucleic Acids Res* **21:** 987–991.

Kosykh VG, Buryanov YI, Bayev AA. 1980. Molecular cloning of EcoRII endonuclease and methylase genes. *Mol Gen Genet MGG* **178:** 717–718.

Kosykh VG, Repik AV, Kaliman AV, Bur'ianov Ia I, Baev AA. 1989. [Primary structure of the gene of restriction endonuclease EcoRII]. *Dokl Akad Nauk SSSR* **308:** 1497–1499.

Krüger DH, Bickle TA. 1983. Bacteriophage survival: multiple mechanisms for avoiding the deoxyribonucleic acid restriction systems of their hosts. *Microbiol Rev* **47:** 345–360.

Krüger DH, Barcak GJ, Reuter M, Smith HO. 1988. *Eco*RII can be activated to cleave refractory DNA recognition sites. *Nucleic Acids Res* **16:** 3997–4008.

Kusiak M, Price C, Rice D, Hornby DP. 1992. The HsdS polypeptide of the type IC restriction enzyme EcoR124 is a sequence-specific DNA-binding protein. *Mol Microbiol* **6:** 3251–3256.

Lauster R, Kriebardis A, Guschlbauer W. 1987. The GATATC-modification enzyme *Eco*RV is closely related to the GATC-recognizing methyltransferases *Dpn*II and *dam* from *E. coli* and phage T4. *FEBS Lett* **220:** 167–176.

Lauster R, Trautner TA, Noyer-Weidner M. 1989. Cytosine-specific type II DNA methyltransferases. A conserved enzyme core with variable target-recognizing domains. *J Mol Biol* **206:** 305–312.

Lehman IR. 1974. DNA ligase: structure, mechanism, and function. *Science* **186:** 790–797.

Lesser DR, Kurpiewski MR, Jen-Jacobson L. 1990. The energetic basis of specificity in the Eco RI endonuclease–DNA interaction. *Science* **250:** 776–786.

Lesser DR, Grajkowski A, Kurpiewski MR, Koziolkiewicz M, Stec WJ, Jen-Jacobson L. 1992. Stereoselective interaction with chiral phosphorothioates at the central DNA kink of the EcoRI endonuclease-GAATTC complex. *J Biol Chem* **267:** 24810–24818.

Lesser DR, Kurpiewski MR, Waters T, Connolly BA, Jen-Jacobson L. 1993. Facilitated distortion of the DNA site enhances *Eco*RI endonuclease-DNA recognition. *Proc Natl Acad Sci* **90:** 7548–7552.

Li L, Chandrasegaran S. 1993. Alteration of the cleavage distance of *Fok* I restriction endonuclease by insertion mutagenesis. *Proc Natl Acad Sci* **90:** 2764–2768.

Li L, Wu LP, Clarke R, Chandrasegaran S. 1993. C-terminal deletion mutants of the FokI restriction endonuclease. *Gene* **133:** 79–84.

Li T, Huang S, Jiang WZ, Wright D, Spalding MH, Weeks DP, Yang B. 2011. TAL nucleases (TALNs): hybrid proteins composed of TAL effectors and FokI DNA-cleavage domain. *Nucleic Acids Res* **39:** 359–372.

Loenen WAM. 2003. Tracking EcoKI and DNA fifty years on: a golden story full of surprises. *Nucleic Acids Res* **31**: 7059–7069.

Loenen WA, Murray NE. 1986. Modification enhancement by the restriction alleviation protein (Ral) of bacteriophage λ. *J Mol Biol* **190**: 11–22.

Loenen WA, Daniel AS, Braymer HD, Murray NE. 1987. Organization and sequence of the hsd genes of *Escherichia coli* K-12. *J Mol Biol* **198**: 159–170.

Looney MC, Moran LS, Jack WE, Feehery GR, Benner JS, Slatko BE, Wilson GG. 1989. Nucleotide sequence of the *Fok*I restriction-modification system: separate strand-specificity domains in the methyltransferase. *Gene* **80**: 193–208.

Lu A-L, Jack WE, Modrich P. 1983. DNA determinants important in sequence recognition by EcoRI endonuclease. *J Biol Chem* **256**: 13200–13206.

Maxwell A, Halford SE. 1982. The SalGI restriction endonuclease. Mechanism of DNA cleavage. *Biochem J* **203**: 85–92.

McClarin JA, Frederick CA, Wang BC, Greene P, Boyer HW, Grable J, Rosenberg JM. 1986. Structure of the DNA-Eco RI endonuclease recognition complex at 3 Å resolution. *Science* **234**: 1526–1541.

McLaughlin LW, Bensler F, Graeser E, Pile N, Scholtissek S. 1987. Effects of functional group changes in the EcoRI recognition site on the cleavage reaction catalysed by the endonuclease. *Biochemistry* **26**: 7238–7245.

Meisel A, Krüger DH, Bickle TA. 1991. M.*Eco*P15 methylates the second adenine in its recognition sequence. *Nucleic Acids Res* **19**: 3997.

Meisel A, Bickle TA, Krüger DH, Schroeder C. 1992. Type III restriction enzymes need two inversely oriented recognition sites for DNA cleavage. *Nature* **355**: 467–469.

Meister J, MacWilliams M, Hubner P, Jutte H, Skrzypek E, Piekarowicz A, Bickle TA. 1993. Macroevolution by transposition: drastic modification of DNA recognition by a type I restriction enzyme following Tn5 transposition. *EMBO J* **12**: 4585–4591.

Morgan RD, Bhatia TK, Lovasco L, Davis TB. 2008. MmeI: a minimal Type II restriction-modification system that only modifies one DNA strand for host protection. *Nucleic Acids Res* **36**: 6558–6570.

Morgan RD, Dwinell EA, Bhatia TK, Lang EM, Luyten YA. 2009. The MmeI family: type II restriction-modification enzymes that employ single-strand modification for host protection. *Nucleic Acids Res* **37**: 5208–5221.

Moses PB, Horiuchi K. 1979. Specific recombination in vitro promoted by the restriction endonuclease HgaI. *J Mol Biol* **135**: 517–524.

Murray NE. 2002. 2001 Fred Griffith review lecture. Immigration control of DNA in bacteria: self versus non-self. *Microbiology* **148**: 3–20.

Murray NE, Batten PL, Murray K. 1973. Restriction of bacteriophage λ by *Escherichia coli* K. *J Mol Biol* **81**: 395–407.

Murray NE, Gough JA, Suri B, Bickle TA. 1982. Structural homologies among type I restriction-modification systems. *EMBO J* **1**: 535–539.

Nagaraja V, Shepherd JC, Bickle TA. 1985a. A hybrid recognition sequence in a recombinant restriction enzyme and the evolution of DNA sequence specificity. *Nature* **316**: 371–372.

Nagaraja V, Shepherd JC, Pripfl T, Bickle TA. 1985b. Two type I restriction enzymes from *Salmonella* species. Purification and DNA recognition sequences. *J Mol Biol* **182**: 579–587.

Nardone G, George J, Chirikjian JG. 1986. Differences in the kinetic properties of BamHI endonuclease and methylase with linear DNA substrates. *J Biol Chem* **261**: 12128–12133.

Narva KE, Van Etten JL, Slatko BE, Benner JS. 1988. The amino acid sequence of the eukaryotic DNA [N^6-adenine]methyltransferase, M·CviBIII, has regions of similarity with the prokaryotic isoschizomer M·*Taq*I and other DNA [N^6-adenine] methyltransferases. *Gene* 74: 253–259.

Nelson M, Raschke E, McClelland M. 1993. Effect of site-specific methylation on restriction endonucleases and DNA modification methyltransferases. *Nucleic Acids Res* 21: 3139–3154.

Newman PC, Williams DM, Cosstick R, Seela F, Connolly BA. 1990. Interaction of the EcoRV restriction endonuclease with the deoxyadenosine and thymidine bases in its recognition hexamer d(GATATC). *Biochemistry* 29: 9902–9910.

Newman M, Strzelecka T, Dorner LF, Schildkraut I, Aggarwal AK. 1994. Structure of restriction endonuclease bamhi phased at 1.95 A resolution by MAD analysis. *Structure* 2: 439–452.

Oller AR, Vanden Broek W, Conrad M, Topal MD. 1991. Ability of DNA and spermidine to affect the activity of restriction endonucleases from several bacterial species. *Biochemistry* 30: 2543–2549.

Olsen DB, Kotzorek G, Eckstein F. 1990. Investigation of the inhibitory role of phosphorothioate internucleotidic linkages on the catalytic activity of the restriction endonuclease EcoRV. *Biochemistry* 29: 9546–9551.

Orekhov AV, Rebentish BA, Debabov VG. 1982. [New site-specific endonuclease from *Streptomyces*: SGRII]. *Dokl Akad Nauk SSSR* 263: 217–220.

Pein CD, Reuter M, Meisel A, Cech D, Krüger DH. 1991. Activation of restriction endonuclease *Eco*RII does not depend on the cleavage of stimulator DNA. *Nucleic Acids Res* 19: 5139–5142.

Piekarowicz A, Kalinowska J. 1974. Host specificity of DNA in *Haemophilus influenzae*: similarity between host-specificity types of *Haemophilus influenzae* Re and Rf. *J Gen Microbiol* 81: 405–411.

Piekarowicz A, Bickle TA, Shepherd JC, Ineichen K. 1981. The DNA sequence recognised by the *Hinf*III restriction endonuclease. *J Mol Biol* 146: 167–172.

Piekarowicz A, Goguen JD, Skrzypek E. 1985. The *Eco*DXX1 restriction and modification system of *Escherichia coli* ET7. Purification, subunit structure and properties of the restriction endonuclease. *Eur J Biochem/FEBS* 152: 387–393.

Pingoud A, Jeltsch A. 2001. Structure and function of type II restriction endonucleases. *Nucleic Acids Res* 29: 3705–3727.

Pingoud A, Fuxreiter M, Pingoud V, Wende W. 2005. Type II restriction endonucleases: structure and mechanism. *Cell Mol Life Sci* 62: 685–707.

Pingoud A, Wilson GG, Wende W. 2014. Type II restriction endonucleases—a historical perspective and more. *Nucleic Acids Res* 42: 7489–7527.

Polisky B, Greene P, Garfin DE, McCarthy BJ, Goodman HM, Boyer HW. 1975. Specificity of substrate recognition by the *Eco*RI restriction endonuclease. *Proc Natl Acad Sci* 72: 3310–3314.

Pósfai J, Bhagwat AS, Posfai G, Roberts RJ. 1989. Predictive motifs derived from cytosine methyltransferases. *Nucleic Acids Res* 17: 2421–2435.

Potter BV, Eckstein F. 1984. Cleavage of phosphorothioate-substituted DNA by restriction endonucleases. *J Biol Chem* 259: 14243–14248.

Price C, Bickle TA. 1986. A possible role for DNA restriction in bacterial evolution. *Microbiol Sci* 3: 296–299.

Price C, Pripfl T, Bickle TA. 1987a. *Eco*R124 and *Eco*R124/3: the first members of a new family of type I restriction and modification systems. *Eur J Biochem/FEBS* 167: 111–115.

Price C, Shepherd JC, Bickle TA. 1987b. DNA recognition by a new family of type I restriction enzymes: a unique relationship between two different DNA specificities. *EMBO J* 6: 1493–1497.

Price C, Lingner J, Bickle TA, Firman K, Glover SW. 1989. Basis for changes in DNA recognition by the EcoR124 and EcoR124/3 type I DNA restriction and modification enzymes. *J Mol Biol* **205:** 115–125.

Rao DN, Eberle H, Bickle TA. 1989a. Characterization of mutations of the bacteriophage P1 *mod* gene encoding the recognition subunit of the *Eco*P1 restriction and modification system. *J Bacteriol* **171:** 2347–2352.

Rao DN, Page MG, Bickle TA. 1989b. Cloning, over-expression and the catalytic properties of the EcoP15 modification methylase from *Escherichia coli*. *J Mol Biol* **209:** 599–606.

Reiser J, Yuan R. 1977. Purification and properties of the P15 specific restriction endonuclease from *Escherichia coli*. *J Biol Chem* **252:** 451–456.

Reuter M, Schneider-Mergener J, Kupper D, Meisel A, Mackeldanz P, Krüger DH, Schroeder C. 1999. Regions of endonuclease *Eco*RII involved in DNA target recognition identified by membrane-bound peptide repertoires. *J Biol Chem* **274:** 5213–5221.

Risser R, Hopkins N, Davis RW, Delius H, Mulder C. 1974. Action of *Escherichia coli* P1 restriction endonuclease on simian virus 40 DNA. *J Mol Biol* **89:** 517–544.

Roberts RJ, Halford S.E. 1993. Type II restriction enzymes. In *Nucleases* (2nd ed.) (ed. Linn SM, Lloyd RS, Roberts RJ), pp. 35–88. Cold Spring Harbor Laboratory Press, Cold Spring Harbor, NY.

Roberts RJ, Macelis D. 1993a. REBASE—restriction enzymes and methylases. *Nucleic Acids Res* **21:** 3125–3137.

Roberts RJ, Macelis D. 1993b. Appendix A. The restriction enzymes. In *Nucleases* 2nd ed. (ed. Linn S, Lloyd RS, Roberts RJ), pp. 439–444.

Roberts RJ, Belfort M, Bestor T, Bhagwat AS, Bickle TA, Bitinaite J, Blumenthal RM, Degtyarev S, Dryden DT, Dybvig K, et al. 2003. A nomenclature for restriction enzymes, DNA methyltransferases, homing endonucleases and their genes. *Nucleic Acids Res* **31:** 1805–1812.

Rosamond J, Endlich B, Linn S. 1979. Electron microscopic studies of the mechanism of action of the restriction endonuclease of *Escherichia coli* B. *J Mol Biol* **129:** 619–635.

Rosenberg JM. 1991. Structure and function of restriction endonucleases. *Curr Opin Struct Biol* **1:** 104–113.

Rosner JL. 1973. Modification-deficient mutants of bacteriophage P1. I. Restriction by P1 cryptic lysogens. *Virology* **52:** 213–222.

Sain B, Murray NE. 1980. The hsd (host specificity) genes of *E. coli* K 12. *Mol Gen Genet* **180:** 35–46.

Sanders KL, Catto LE, Bellamy SR, Halford SE. 2009. Targeting individual subunits of the FokI restriction endonuclease to specific DNA strands. *Nucleic Acids Res* **37:** 2105–2115.

Schroeder C, Jurkschat H, Meisel A, Reich JG, Krüger D. 1986. Unusual occurrence of EcoP1 and *Eco*P15 recognition sites and counterselection of type II methylation and restriction sequences in bacteriophage T7 DNA. *Gene* **45:** 77–86.

Seeman NC, Rosenberg JM, Rich A. 1976. Sequence-specific recognition of double helical nucleic acids by proteins. *Proc Natl Acad Sci* **73:** 804–808.

Selent U, Ruter T, Kohler E, Liedtke M, Thielking V, Alves J, Oelgeschläger T, Wolfes H, Peters F, Pingoud A. 1992. A site-directed mutagenesis study to identify amino acid residues involved in the catalytic function of the restriction endonuclease EcoRV. *Biochemistry* **31:** 4808–4815.

Sharp PM, Kelleher JE, Daniel AS, Cowan GM, Murray NE. 1992. Roles of selection and recombination in the evolution of type I restriction-modification systems in enterobacteria. *Proc Natl Acad Sci* **89:** 9836–9840.

Sharrocks AD, Hornby DP. 1991. Transcriptional analysis of the restriction and modification genes of bacteriophage P1. *Mol Microbiol* **5:** 685–694.

Skrzypek E, Piekarowicz A. 1989. The EcoDXX1 restriction and modification system: cloning the genes and homology to type I restriction and modification systems. *Plasmid* **21:** 195–204.

Smith HO, Annau TM, Chandrasegaran S. 1990. Finding sequence motifs in groups of functionally related proteins. *Proc Natl Acad Sci* **87:** 826–830.

Smith DW, Crowder SW, Reich NO. 1992. *In vivo* specificity of *Eco*RI DNA methyltransferase. *Nucleic Acids Res* **20:** 6091–6096.

Som S, Bhagwat AS, Friedman S. 1987. Nucleotide sequence and expression of the gene encoding the *Eco*RII modification enzyme. *Nucleic Acids Res* **15:** 313–332.

Stephenson FH, Ballard BT, Boyer HW, Rosenberg JM, Greene PJ. 1989. Comparison of the nucleotide and amino acid sequences of the *Rsr*I and *Eco*RI restriction endonucleases. *Gene* **85:** 1–13.

Strzelecka T, Dorner L, Schildkraut I, Agarwal A. 1990. Structural studies of the BamHI restriction enzyme. *Biophys J* **57:** 68.

Strzelecka T, Newman M, Dorner LF, Knott R, Schildkraut I, Aggarwal AK. 1994. Crystallization and preliminary X-ray analysis of restriction endonuclease BamHI-DNA complex. *J Mol Biol* **239:** 430–432.

Studier FW, Bandyopadhyay PK. 1988. Model for how type I restriction enzymes select cleavage sites in DNA. *Proc Natl Acad Sci* **85:** 4677–4681.

Sugisaki H, Kanazawa S. 1981. New restriction endonucleases from *Flavobacterium okeanokoites* (FokI) and *Micrococcus luteus* (MluI). *Gene* **16:** 73–78.

Sugisaki H, Kita K, Takanami M. 1989. The *Fok*I restriction-modification system. II. Presence of two domains in *Fok*I methylase responsible for modification of different DNA strands. *J Biol Chem* **264:** 5757–5761.

Suri B, Bickle TA. 1985. EcoA: the first member of a new family of type I restriction modification systems. Gene organization and enzymatic activities. *J Mol Biol* **186:** 77–85.

Suri B, Nagaraja V, Bickle TA. 1984a. Bacterial DNA modification. *Curr Top Microbiol Immunol* **108:** 1–9.

Suri B, Shepherd JC, Bickle TA. 1984b. The *Eco*A restriction and modification system of *Escherichia coli* 15T-: enzyme structure and DNA recognition sequence. *EMBO J* **3:** 575–579.

Szybalski W, Kim SC, Hasan N, Podhajska AJ. 1991. Class-IIS restriction enzymes—a review. *Gene* **100:** 13–26.

Tao T, Blumenthal RM. 1992. Sequence and characterization of *pvuIIR*, the *Pvu*II endonuclease gene, and of *pvuIIC*, its regulatory gene. *J Bacteriol* **174:** 3395–3398.

Taylor JD, Halford SE. 1989. Discrimination between DNA sequences by the EcoRV restriction endonuclease. *Biochemistry* **28:** 6198–6207.

Taylor JD, Halford SE. 1992. The activity of the EcoRV restriction endonuclease is influenced by flanking DNA sequences both inside and outside the DNA–protein complex. *Biochemistry* **31:** 90–97.

Taylor JD, Goodall AJ, Vermote CL, Halford SE. 1990. Fidelity of DNA recognition by the EcoRV restriction/modification system in vivo. *Biochemistry* **29:** 10727–10733.

Taylor JD, Badcoe IG, Clarke AR, Halford SE. 1991. EcoRV restriction endonuclease binds all DNA sequences with equal affinity. *Biochemistry* **30:** 8743–8753.

Taylor I, Patel J, Firman K, Kneale G. 1992. Purification and biochemical characterisation of the *Eco*R124 type I modification methylase. *Nucleic Acids Res* **20:** 179–186.

Terry BJ, Jack WE, Rubin RA, Modrich P. 1983. Thermodynamic parameters governing interaction of EcoRI endonuclease with specific and nonspecific DNA sequences. *J Biol Chem* **258:** 9820–9825.

Terry BJ, Jack WE, Modrich P. 1985. Facilitated diffusion during catalysis by *Eco*RI endonuclease. Nonspecific interactions in *Eco*RI catalysis. *J Biol Chem* **260:** 13130–13137.

Terry BJ, Jack WE, Modrich P. 1987. Mechanism of specific site location and DNA cleavage by EcoR I endonuclease. *Gene Amplif Anal* **5:** 103–118.

Thielking V, Alves J, Fliess A, Maass G, Pingoud A. 1990. Accuracy of the EcoRI restriction endonuclease: binding and cleavage studies with oligodeoxynucleotide substrates containing degenerate recognition sequences. *Biochemistry* **29:** 4682–4691.

Thielking V, Selent U, Kohler E, Wolfes H, Pieper U, Geiger R, Urbanke C, Winkler FK, Pingoud A. 1991. Site-directed mutagenesis studies with EcoRV restriction endonuclease to identify regions involved in recognition and catalysis. *Biochemistry* **30:** 6416–6422.

Titheradge AJ, King J, Ryu J, Murray NE. 2001. Families of restriction enzymes: an analysis prompted by molecular and genetic data for type ID restriction and modification systems. *Nucleic Acids Res* **29:** 4195–4205.

Tolstoshev CM, Blakesley RW. 1982. RSITE: a computer program to predict the recognition sequence of a restriction enzyme. *Nucleic Acids Res* **10:** 1–17.

Topal MD, Thresher RJ, Conrad M, Griffith J. 1991. NaeI endonuclease binding to pBR322 DNA induces looping. *Biochemistry* **30:** 2006–2010.

Ueno T, Ito H, Kimizuka F, Kotani H, Nakajima K. 1993. Gene structure and expression of the MboI restriction–modification system. *Nucleic Acids Res* **21:** 2309–2313.

Van Cott EM, Wilson GG. 1988. Cloning the *Fnu*DI, *Nae*I, *Nco*I and *Xba*I restriction-modification systems. *Gene* **74:** 55–59.

Vermote CL, Halford SE. 1992. EcoRV restriction endonuclease: communication between catalytic metal ions and DNA recognition. *Biochemistry* **31:** 6082–6089.

Vipond IB, Halford SE. 1993. Structure-function correlation for the *Eco*RV restriction enzyme: from non-specific binding to specific DNA cleavage. *Mol Microbiol* **9:** 225–231.

Vipond IB, Halford SE. 1995. Specific DNA recognition by EcoRV restriction endonuclease induced by calcium ions. *Biochemistry* **34:** 1113–1119.

von Hippel PH, Berg OG. 1989. Facilitated target location in biological systems. *J Biol Chem* **264:** 675–678.

Waters TR, Connolly BA. 1992. Continuous spectrophotometric assay for restriction endonucleases using synthetic oligodeoxynucleotides and based on the hyperchromic effect. *Anal Biochem* **204:** 204–209.

Waters TR, Connolly BA. 1994. Interaction of the restriction endonuclease *Eco*RV with the deoxyguanosine and deoxycytidine bases in its recognition sequence. *Biochemistry* **33:** 1812–1819.

Wilson GG. 1991. Organization of restriction-modification systems. *Nucleic Acids Res* **19:** 2539–2566.

Wilson GG, Murray NE. 1991. Restriction and modification systems. *Ann Rev Genet* **25:** 585–627.

Winkler FK. 1992. Structure and function of restriction endonucleases. *Curr Opin Struct Biol* **2:** 93–99.

Winkler FK, Banner DW, Oefner C, Tsernoglou D, Brown RS, Heathman SP, Bryan RK, Martin PD, Petratos K, Wilson KS. 1993. The crystal structure of *Eco*RV endonuclease and of its complexes with cognate and non-cognate DNA fragments. *EMBO J* **12:** 1781–1795.

Winkler FK, Prota AE. 2004. Structure and function of EcoRV endonuclease. In *Restriction endonucleases* (ed. Pingoud A), pp. 179–214. Springer, Berlin.

Winter M. 1997. "Investigation of de novo methylation activity in mutants of the EcoKI methyltransferase." PhD thesis, University of Edinburgh, UK.

Xu SY, Schildkraut I. 1991. Isolation of *Bam*HI variants with reduced cleavage activities. *J Biol Chem* **266:** 4425–4429.

Yang CC, Topal MD. 1992. Nonidentical DNA-binding sites of endonuclease NaeI recognize different families of sequences flanking the recognition site. *Biochemistry* **31:** 9657–9664.

Yanofsky SD, Love R, McClarin JA, Rosenberg JM, Boyer HW, Greene PJ. 1987. Clustering of null mutations in the EcoRI endonuclease. *Proteins* **2:** 273–282.

Yuan R, Reiser J. 1978. Steps in the reaction mechanism of the *Escherichia coli* plasmid P15-specific restriction endonuclease. *J Mol Biol* **122:** 433–445.

Yuan R, Bickle TA, Ebbers W, Brack C. 1975. Multiple steps in DNA recognition by restriction endonuclease from *E. coli* K. *Nature* **256:** 556–560.

Yuan R, Hamilton DL, Burckhardt J. 1980a. DNA translocation by the restriction enzyme from *E. coli* K. *Cell* **20:** 237–244.

Yuan R, Hamilton DL, Hadi SM, Bickle TA. 1980b. Role of ATP in the cleavage mechanism of the *Eco*P15 restriction endonuclease. *J Mol Biol* **144:** 501–519.

Zabeau M, Friedman S, Van Montagu M, Schell J. 1980. The *ral* gene of phage λ. I. Identification of a non-essential gene that modulates restriction and modification in *E. coli. Mol Gen Genet MGG* **179:** 63–73.

Zebala JA, Choi J, Barany F. 1992a. Characterization of steady state, single-turnover, and binding kinetics of the *Taq*I restriction endonuclease. *J Biol Chem* **267:** 8097–8105.

Zebala JA, Choi J, Trainor GL, Barany F. 1992b. DNA recognition of base analogue and chemically modified substrates by the *Taq*I restriction endonuclease. *J Biol Chem* **267:** 8106–8116.

WWW RESOURCES

http://library.cshl.edu/Meetings/restriction-enzymes/v-Aggarwal.php Aggarwal A. 2013.

http://library.cshl.edu/Meetings/restriction-enzymes/Halford.php Halford SE. 2013.

http://library.cshl.edu/Meetings/restriction-enzymes/v-Morgan.php Morgan RD. 2013.

http://library.cshl.edu/Meetings/restriction-enzymes/v-Roberts.php Roberts RJ. 2013.

http://library.cshl.edu/Meetings/restriction-enzymes/v-Rosenberg.php Rosenberg JM. 2013.

http://library.cshl.edu/Meetings/restriction-enzymes/v-Studier.php Studier FW. 2013.

CHAPTER 7

Crystal Structures of Type II Restriction Enzymes and Discovery of the Common Core of the Catalytic Domain: ~1993–2004

INTRODUCTION

Chapter 6 used as starting material two reviews on the Type II and ATP-dependent (Type I and III) REases, published in the 1993 *Nucleases* book (Bickle 1993; Roberts and Halford 1993). This chapter, covering the next 10 years, is based on more than 25 reviews, reflecting the tremendous progress made during this period. Forty years after the first paper on EcoKI and EcoP1I (Bertani and Weigle 1953), reviews mentioned only about 20 Type I and Type III REases (Bickle and Krüger 1993; Roberts and Halford 1993), and because of lack of commercial interest, most research continued on EcoKI, EcoR124I, EcoP1I, and EcoP15I. Together, these two types comprised <2% of the enzymes identified in REBASE (Table 1); the newly named Type IV REases featured even less at 0.1% (and will be discussed in Chapter 8). This changed dramatically as, by the end of the century, improved sequencing and computer prediction programs showed R-M systems to be ubiquitous among Eubacteria and Archaea, with almost one-half of these genomes containing candidate Type II genes (Table 1; Roberts and Macelis 1991, 1993; Titheradge et al. 2001; Roberts et al. 2003b). To summarize these findings briefly, for which evidence had already become apparent (Chapter 6), there appeared to be extensive horizontal transfer of all R-M systems. Closely related systems were often present in unrelated organisms, and codon usage of R-M genes was often different from other host genes. In addition, the genes were found in different places on the genome in different strains of the same organism (Jeltsch and Pingoud 1996; Chinen et al. 2000; Nobusato et al. 2000a,b).

Chapter doi:10.1101/restrictionenzymes_7

TABLE 1. *Identified and candidate REases in REBASE in 2004*

Enzymes[a]	% Total	Genomes[b]	% Total
Type I	1.6	Type I	39.5
Type II	98.0	Type II	43.0
Type III	0.3	Type III	8.3
Type IV	0.1	Type IV	9.2

Courtesy of Rich Roberts; REBASE stats presented by Noreen Murray at the 5th NEB Meeting (Bristol 2004).

Total ~3700 REases.

[a] Identified REase genes and characterized by biochemical assays.

[b] Candidate genes for REases in sequenced genomes.

By 2004, the number of REases had risen to ~3700, making the family of REases a very large one indeed (Pingoud and Jeltsch 2001; Roberts et al. 2004). The dictum that Type II REases were simple dimeric enzymes, requiring Mg^{2+} (and no ATP or SAM) came under further scrutiny (Stasiak 1980a,b; Bennett and Halford 1989; Bujnicki 2000b; Murray 2000; Sapranauskas et al. 2000; Pingoud and Jeltsch 2001). Many new Type II enzymes recognized nonpalindromic DNA sites as monomers, tetramers, or higher-order complexes. Recognition sequences were often not unique, but could be discontinuous, degenerate, or asymmetric, whereas cleavage did not necessarily occur at the recognition site. The distinction into Type I, II, and III was still useful, but many REases clearly had intermediate properties, with no simple way to classify or predict DNA–protein interactions (Luscombe et al. 2001). What should one make of the many functional similarities, but also surprising diversity in DNA recognition and cleavage, and the positioning of metal cofactors (Aggarwal 1995; Wah et al. 1997; Viadiu and Aggarwal 1998; Pingoud and Jeltsch 2001)? Did this indicate highly subtle protein–DNA–cofactor interactions? The discovery of REases that had to interact with two copies of their recognition sequence before they could cleave DNA (Halford 2001) was exciting news. This process involved DNA looping, as reported for enzymes involved in replication, recombination, and transcription (Schleif 1992; Rippe et al. 1995). How could other REases that needed two sites in order to cleave be identified, and how did this cleavage occur? Would these REases perhaps be good tools to analyze interactions between distant DNA sites?

The rapid expansion of the REase family led to a book solely dedicated to "restriction enzymes" in 2004 (Pingoud 2004). The first chapter is a reprint of the Survey and Summary on the novel nomenclature of REases, MTases, Homing Endonucleases and their genes, published a year earlier (Roberts et al. 2003a). Edited by Alfred Pingoud, this book contains 16 additional chapters, including progress on EcoRI and EcoRV, as well as novel and often unexpected data on other Type II enzymes. Only one chapter is dedicated to the ATP-dependent Type I and III "molecular motors" (McClelland and Szczelkun 2004). A matter of strong debate was the issue of "selfishness," especially that of the Type I and II systems (e.g., O'Neill et al. 1997; Kobayashi 1998). The nature of the catalytic core, the role of water and metal ions in mediating both the interaction of REases with their DNA recognition sites, and hydrolysis of the phosphodiester backbone (Cowan 2004; Sidorova and Rau 2004) were subject to intense study and fierce debate: How did the kind of metal cofactor (Mg^{2+}, Mn^{2+}, or Ca^{2+}) influence the reactivity of the enzymes, and how many metal ions were needed? Another big question was whether it was possible to alter the specificity of EcoRI, EcoRV, BamHI, and other enzymes (Alves and Vennekohl 2004)? The elucidation of the structures of BamHI and BglII, which recognized sequences that differed only in the outer 2 base pairs, provided one explanation (Scheuring Vanamee et al. 2004). This disappointing immutability of REases contrasted with that of the mutability of transcription factors (Scheuring Vanamee et al. 2004). Expectations, however, were high with respect to the novel applications with chimeric REases, such as fusions with zinc fingers (Kandavelou et al. 2004).

Most data in the previous chapter concerned the mechanism of DNA specificity and cleavage of EcoRI and EcoRV, and reported the first crystal structures of EcoRI (Kim et al. 1990), EcoRV (Winkler et al. 1993), BamHI (Newman et al. 1994a,b), and PvuII with cognate DNA (Athanasiadis et al. 1994; Cheng et al. 1994). Twenty years after its initial characterization (Kuz'-min et al. 1984; Schildkraut et al. 1984), by 2004, EcoRV was the most thoroughly studied REase (with the exception of EcoRI) through an elegant "pas de deux" of structural and mechanistic studies (Jen-Jacobson 1997; Winkler and Prota 2004). Toward the end of the century, there were reports on ~1000 new Type II REases, and many biochemical and novel crystal studies (http://rebase. neb.com/rebase/rebase.html; Pingoud and Jeltsch 1997, 2001). By 2004, 16 Type II REases structures had been solved, plus those of four other nucleases and two resolvases (summarized in Table 1 of Horton et al. 2004a, p. 362). This included cocrystals of FokI, BglI, MunI, BglII, NgoMIV, BsoBI, and HincII (Wah et al. 1997; Newman et al. 1998; Deibert et al. 1999, 2000; Lukacs et al. 2000; van der Woerd et al. 2001; Horton et al. 2002). These new structures questioned the view held until the mid-1990s that the baffling lack of common features between most REases (in contrast to the MTases) suggested independent convergence, and not divergence from a common ancestor (Wilson 1991; Heitman 1993; Bujnicki 2004).

This chapter gives an overview of the period roughly from 1993 to 2004, during which the groups in Bristol, Pittsburgh, Edinburgh, and Basel continued their research into the biochemistry, structure, and relationships of the

Type I, II, and III REases. NEB and Fermentas International, Inc. continued their search for novel enzymes, and investigations into the mechanisms of, and relationships between, these enzymes. In Germany, Alfred Pingoud moved from Hannover to Giessen, where he and Albert Jeltsch studied the structure, mechanism, and evolution of Type II REases (e.g., Pingoud and Jeltsch 1997, 2001; Jeltsch and Urbanke 2004). In Tokyo, Ichizo Kobayashi worked on his concept of R-M systems as "selfish" elements and minimal forms of life (Kobayashi 2004). Aneel Aggarwal in New York elucidated the surprisingly different structures of BamHI and BglII, mentioned previously (Scheuring Vanamee et al. 2004). In Berlin, Detlev Krüger and Monika Reuter investigated the reported, and puzzling, refractory EcoRII sites, even though these sites could be modified by M·EcoRII (Reuter et al. 2004). Virginijus (Virgis) Šikšnys in Vilnius started to unravel the structural and molecular mechanisms of sequence discrimination by REases recognizing closely related sequences (Šikšnys et al. 2004). In Warsaw, Janusz Bujnicki used the nine available crystal structures in combination with database searches to build evolutionary trees of the REase and nuclease superfamilies (Bujnicki 2004). Other aspects of the ATP-dependent Type I and III enzymes (Murray 2000, 2002; Dryden et al. 2001; Loenen 2003) and the Type II enzymes (Jeltsch and Pingoud 1996; Kovall and Matthews 1998, 1999; Bujnicki 2000a,b; Halford 2001; Mucke et al. 2003; Halford and Marko 2004; Kirsanova et al. 2004; Pingoud 2004; Pingoud et al. 2005) yielded valuable information and led to the new classification into 11 Type II subtypes in 2003, mentioned previously (Roberts et al. 2003a). Was the original idea of the function of REases too narrow, which had been based on the "arms race" of phages and conjugative plasmids to avoid restriction by counterattacks (Krüger and Bickle 1983; Bickle and Krüger 1993)? Would some of these REases perhaps have an additional role in recombination and transposition, rather than simply protect their host against foreign invaders (Arber 1979; Heitman 1993; McKane and Milkman 1995)?

For reasons of space, only a few examples can be discussed in this chapter and the reader is referred to the reviews for detailed information. Although the emphasis of this book is on REases, it should be mentioned that research into the MTases led to one of the most exciting discoveries of the 1990s—that is, base flipping. In the wake of the structure of M·HhaI (Cheng et al. 1993a, b), the M·HhaI–DNA complex revealed that the enzyme flipped the target base out of the DNA helix (Klimasauskas et al. 1994; Horton et al. 2004b). Base flipping would prove to be a more general property: Other MTases, endonucleases, and RNA enzymes "do it" (Winkler 1994; Mernagh et al. 1998; Roberts and Cheng 1998; Blumenthal and Cheng 2001; Cheng and Roberts 2001; Cheng and Blumenthal 2002; Su et al. 2005; Bochtler et al. 2006; Horton et al. 2006, 2014; Hashimoto et al. 2008). Other emerging interesting features of MTases such as the molecular evolution by circular permutations (e.g., Jeltsch 1999; Vilkaitis et al. 2002) would also be found in the HsdS subunits of Type I families (e.g., Loenen et al. 2014a).

TYPE II ENZYMES

Subtypes of Type II REases

In Chapter 6, several REases (e.g., EcoRII, NaeI, FokI, BcgI, and Sgr10I) were mentioned that were clearly not Type I or Type III, but also differed from the conventional Type II REases like EcoRI, EcoRV, and BamHI (Pingoud and Jeltsch 2001). With the discovery of many such new enzymes, Rich Roberts took the initiative to subdivide the Type II REases into 11 subtypes in 2003, being different from Type IIP ("P" for palindrome: EcoRI and EcoRV), Type IIS (FokI), and Type IIE (EcoRII, NaeI) (Table 2; Roberts et al. 2003a).

The initial definition of a Type II REase was that it cleaved at, or close to, the recognition site in an ATP-independent manner. The cleavage site could have a 5′ or 3′ sticky end (EcoRI, BglI) (Hedgpeth et al. 1972; Van Heuverswyn and Fiers 1980) or blunt/flush end (EcoRV) (Schildkraut et al. 1984). Type IIS ("S" for shifted) (e.g., FokI) was the first new subtype, named in 1991 (Szybalski et al. 1991). FokI had separate recognition and catalytic domains; the recognition domain had three smaller subdomains with helix-turn-helix (HTH) motifs, with the catalytic domain involved in potential dimerization (Wah et al. 1998). Initially thought to act as a monomer, FokI later proved to dimerize on the DNA (Bitinaite et al. 1998; Wah et al. 1998), now known to be not so unusual. EcoRII and NaeI were the first Type IIE subtypes, which interacted with two copies of their recognition sequence, one serving as allosteric effector (Krüger et al. 1988, 1995; Mucke et al. 2003). The Type IIF REases (e.g., SfiI and NgoMIV) also interacted with two copies of the recognition sequence but, in contrast to Type IIE, cleaved both sequences (Halford et al. 1999). Type IIT REases were heterodimeric proteins (e.g., Bpu10I and BslI) (Stankevicius et al. 1998; Hsieh et al. 2000). Type IIB were SAM-dependent heterodimeric REases, cleaving on both sides of an asymmetric recognition sequence (e.g., BcgI and BplI) (Kong and Smith 1997; Vitkute et al. 1997). In the case of "oddball" BcgI (Chapter 6), the catalytic centers for restriction and modification were located in the α subunit, with the DNA recognition domain in the β subunit (Kong 1998). Type IIG were single-chain SAM-dependent REases (e.g., Eco57I) (Janulaitis et al. 1992). Type IIM recognized methylated DNA (e.g., DpnI) (Lacks and Greenberg 1975). Other methylation-dependent REases were grouped as Type IV REases, the best-known enzyme being McrBC of *E. coli* (Raleigh and Wilson 1986; Stewart et al. 2000), which

TABLE 2. *Division of Type II REases in 11 subtypes in 2003*

Subtype[a]	Defining feature	Examples	Recognition sequence
A	Asymmetric recognition sequence	FokI	GGATG (9/13)
		AciI	CCGC (−3/−1)
B	Cleaves both sides of target on both strands	BcgI	(10/12) CGANNNNNNTGC (12/10)
C	Symmetric or asymmetric target. R and M functions in one polypeptide	GsuI	CTGGAG (16/14)
		HaeIV	(7/13) GAYNNNNNRTC (14/9)
		BcgI	(10/12) CGANNNNNNTGC (12/10)
E	Two targets; one cleaved, one an effector	EcoRII	↓CCWGG
		NaeI	GCC↓GGC
F	Two targets; both cleaved coordinately	SfiI	GGCCNNNN↓NGGCC
		SgrAI	CR↓CCGGYG
G	Symmetric or asymmetric target. Affected by AdoMet	BsgI	GTGCAG (16/14)
		Eco57I	CTGAAG (16/14)
H	Symmetric or asymmetric target. Similar to Type I gene structure	BcgI	(10/12) CGANNNNNNTGC (12/10)
		AhdI	GACNNN↓NNGTC
M	Subtype IIP or IIA. Require methylated target	DpnI	Gm6A↓TC
P	Symmetric target and cleavage sites	EcoRI	G↓AATTC
		PpuMI	RG↓GWCCY
		BslI	CCNNNNN↓NNGG
S	Asymmetric target and cleavage sites	FokI	GGATG (9/13)
		MmeI	TCCRAC (20/18)
T	Symmetric or asymmetric target. R genes are heterodimers	Bpu10I	CCTNAGC (−5/−2)[b]
		BslI	CCNNNNN↓NNGG

Reprinted from Roberts et al. 2003a.

[a]Note that not all subtypes are mutually exclusive. E.g. BslI is of subtype P and T.

[b]The abbreviation indicates double strand cleavage as shown below:

5′ C C↓T N A G C
3′ G G A N T↑C G

caused so much trouble in DNA cloning experiments (Raleigh et al. 1988). These enzymes were NTP-dependent (usually ATP, but GTP in the case of McrBC) for cleavage, like Type I and III REases, and cleavage occurred between sites (Stewart and Raleigh 1998; Panne et al. 1999). McrB was responsible for DNA recognition and GTP hydrolysis and McrC for catalysis (Pieper et al. 1999; Pieper and Pingoud 2002). Although a helpful subdivision, some enzymes fit into more than one category (for recent details, see http://rebase.neb.com/rebase/rebase.html; Loenen et al. 2014b) or in none of these properly (e.g., HaeIV) (Piekarowicz et al. 1999; Pingoud and Jeltsch 2001).

Two Types of Readout of Type II REases

The availability of more than 100 protein–DNA complex structures revealed two types of readout: direct readout of sequence via contacts with bases in the major (usually the most important) groove and in the minor groove, and indirect readout of sequence via interactions with the DNA backbone (Luscombe et al. 2001; Winkler and Prota 2004). In the presence of Mg^{2+}, all Type II enzymes cleaved DNA with extremely high specificity (Roberts and Halford 1993). DNA sequences differing from the recognition site by just 1 bp were usually cleaved $>10^6$ times more slowly (Taylor and Halford 1989), far more than expected from the loss of a few H-bond interactions with a single base pair in a DNA–enzyme complex. Studies on EcoRI and other REases showed that total discrimination was always large (Lesser et al. 1990; Pingoud and Jeltsch 1997).

Type IIP: EcoRI and EcoRV

The wealth of crystal structures and biochemical studies with respect to EcoRI and EcoRV were extensively reviewed and are briefly summarized here (Pingoud and Jeltsch 2001; Grigorescu et al. 2004; Pingoud 2004; Winkler and Prota 2004).

EcoRI

John Rosenberg's group refined their data on the EcoRI structure (Kim et al. 1990), based on a high-resolution (1.85 Å) initial EcoRI–DNA recognition complex (Choi 1994), and a postreactive EcoRI–DNA complex at 2.7 Å resolution (using Mn^{2+}) (reviewed in Grigorescu et al. 2004). The latter cocrystals were possible, as, fortuitously, in situ cleavage in the crystals could take place. More than a dozen different crystals with different oligonucleotides in the complex showed that EcoRI had a strong tendency to associate in sheets, and that the complex had twofold rotational symmetry axes perpendicular to the three-fold axes (i.e., in the plane of the sheets) (Grable et al. 1984; Samudzi 1990; Wilkosz et al. 1995; Grigorescu et al. 2004). The protein became much more ordered after binding DNA, like EcoRV (see below). From earlier biochemical studies it was already known that the structure at the DNA–protein interface changed in response to even minor changes that were hard to predict (Lesser et al. 1990, 1993; Jen-Jacobson et al. 1991; Jen-Jacobson et al. 2000). Although EcoRI did bend the DNA like several other enzymes (Kim et al. 1994; Deibert et al. 1999; Lukacs et al. 2000; Pingoud and Jeltsch 2001), this was apparently not a general rule, as BamHI did not bend, kink, or unwind the DNA to any extent (Newman et al. 1995). None of the many EcoRI mutants altered the specificity of EcoRI for its recognition site, suggesting a general rule that mutations never led to a change of specificity, although catalytic activity might be severely impaired (Wolfes et al. 1986; Alves et al. 1989; Geiger et al. 1989; King et al. 1989; Needels et al. 1989; Wright et al. 1989; Hager et al. 1990; Heitman and Model 1990; Oelgeschläger et al. 1990; Jeltsch et al. 1993a; Flores et al. 1995; Grabowski et al. 1995; Muir et al. 1997; Windolph and Alves 1997; Fritz et al. 1998; Ivanenko et al. 1998; Kuster 1998; Rosati 1999; Grigorescu et al. 2004). Many amino acids in the main protein domain appeared essential for maintenance of the correct 3D structure of the dimer. In contrast, the promiscuous mutants (with reduced sequence specificity) (Chapter 6) localized to regions with low structural stability in the free enzyme (Heitman and Model 1990; Muir et al. 1997). Together with the available data on other REases, John Rosenberg concluded that no generalization could be made for the kind and extent of distortion Type II REases induced in their DNA substrate. In general, however, the DNA in the specific complex differed from ideal B-DNA, and distortions appeared to be part of the recognition process, as supported by the use of modified substrates or base analogs (e.g., Blattler et al. 1998).

EcoRV

Fritz Winkler's group refined the initial structure of EcoRV (Winkler et al. 1993) using many different crystals, revealing the initial EcoRV–DNA recognition complex, and the transition from nonspecific to specific complex with Mg^{2+}, followed by DNA cleavage (Fig. 1; reviewed in Pingoud and Jeltsch 2001; Winkler and Prota 2004). The cleavage rate proved to be highly sensitive to interactions far from the active site. Major groove interactions were probed extensively using mutant enzymes, oligonucleotides, and modified bases (Fliess et al. 1988; Alves et al. 1989; Newman et al. 1990a,b; Thielking et al. 1991; Vermote and Halford 1992; Waters and Connolly 1994; Martin et al. 1999). Conflicting results were obtained with gel-shift, filter-binding,

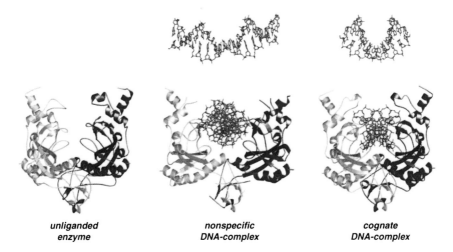

FIGURE 1. Structure of EcoRV, free and in complex with nonspecific and specific DNA. The two subunits are shown in yellow and blue, respectively, and the DNA in red. On top of the complexes the DNA is shown at a right angle from the view below to illustrate the different degree of bending. (Reprinted from Winkler and Prota 2004, with permission from Springer Nature.)

and steady-state fluorescence anisotropy techniques (Connolly et al. 2001). The gel shifts showed that EcoRV bound all sequences with equal affinity (Chapter 6), in contrast to EcoRI, suggesting a fundamentally different mechanism (Lesser et al. 1990; Thielking et al. 1990, 1992; Taylor et al. 1991; Vipond and Halford 1993, 1995; Winkler et al. 1993; Szczelkun and Connolly 1995). Conflict arose because the effect of Mg^{2+} on binding could not be analyzed directly because of rapid cleavage, a problem partly solved by using catalytically inactive mutants, Ca^{2+} as cofactor (which blocks cleavage), or using poor, or noncleavable, substrate analogs (Thielking et al. 1992; Winkler and Prota 2004). Was the interpretation of the EcoRV crystals correct? In the nonspecific complex, EcoRV bound not *one*, but *two* short DNA duplexes (of the self-complementary octamer CGAGCTCG) stacked end-to-end at the twofold axis, presumed to be representative for nonspecific binding (Winkler et al. 1993). This idea was challenged (Engler et al. 1997) and the issue reexamined (Erskine and Halford 1998; Reid et al. 2001). A decade later Winkler was still convinced that "the structure yields a very plausible explanation why no cleavage can occur in this binding mode" (Winkler and Prota 2004, p. 194).

The DNA-binding domain of EcoRV contained three segments, two of which interacted with the recognition site. This region contained a glutamine-rich "Q"-loop recognizing bases in the minor groove and a recognition "R"-loop making base-specific contacts in the major groove, presumably involved in cleaving both strands in one binding event (Thielking et al. 1991; Selent et al. 1992; Winkler et al. 1993; Kostrewa and Winkler 1995; Stahl et al. 1996, 1998b; Wenz et al. 1996; Thomas et al. 1999). The floor of the DNA-binding site appeared critical for coupling recognition and cleavage (Garcia et al. 1996). Some residues were involved in indirect readout, and others were relevant for conformational changes (Kostrewa and Winkler 1995; Thorogood et al. 1996; Wenz et al. 1996; Stahl et al. 1998a,b; Martin et al. 1999; Stanford et al. 1999). EcoRV had to open the DNA-binding site for the DNA to enter the cleft, similar to BamHI (Schulze et al. 1998; Viadiu and Aggarwal 2000).

This meant considerable conformational changes involving DNA bending and rotation of the DNA-binding domains, which wrapped around the DNA (Stover et al. 1993; Winkler et al. 1993; Kostrewa and Winkler 1995; Vipond and Halford 1995; Garcia et al. 1996; Horton and Perona 1998, 2000; Martin et al. 1999; Jones et al. 2001). EcoRV differed from EcoRI in the relative orientation of DNA and protein along the twofold symmetry axis, supporting the idea of "EcoRI-like" and "EcoRV-like" branches (Anderson 1993; Pingoud and Jeltsch 1997, 2001; Bujnicki 2000b; and below). What was the exact nature of this remarkable coupling of recognition and catalysis? The study of such changes warranted other tools and Winkler wondered whether single-molecule spectroscopy might lead to exciting new information on the relevance of the different structural states along the reaction path (Winkler and Prota 2004).

Type IIP: BamHI and BglII

The genes encoding BamHI from *Bacillus amyloliquefaciens* and BglII from *B. subtilis* subsp. globigii were finally cloned in the early 1990s (Brooks et al. 1991; Anton et al. 1997). Aneel Aggarwal's group set out to answer a key question: Did these two enzymes interact with DNA in the same way, as their recognition sequences differed by only the outer base pair (5′ G/GATCC and 5′ A/GATCT, respectively)? This analysis brought several surprises. The primary protein sequences proved to be unrelated, and, in contrast to all other known Type II REases, BamHI contained a critical glutamate as the third essential residue in the catalytic core (Selent et al. 1992; Dorner and Schildkraut 1994; Newman et al. 1994b; Grabowski et al. 1995; Lukacs et al. 2000). BglII proved unusual, too, because the catalytic residues were sequestered in a way not seen in any of the other REases.

BamHI

Like EcoRI and EcoRV, BamHI derived its specificity from both binding and catalysis, and single base pair changes in the recognition site affected binding as much as a random sequence (Lesser et al. 1990; Thielking et al. 1990; Engler 1998; Engler et al. 2001). The preliminary structure (Chapter 6; Strzelecka et al. 1994) was followed by the structure of free enzyme, and with specific and nonspecific DNA (Newman et al. 1994b, 1995; Viadiu and Aggarwal 1998, 2000). Like EcoRI, BamHI could cleave the DNA in the crystals (Viadiu and Aggarwal 1998). The BamHI–DNA complex before cleavage was obtained using Ca^{2+}, and after cleavage using Mn^{2+} (Fig. 2A; Scheuring Vanamee et al. 2004). As in most cases studied, the DNA was held in a tight-binding cleft (Aggarwal 1995; Pingoud and Jeltsch 1997, 2001). In the specific complex (Fig. 2Ac), DNA–protein interactions occurred both in the major and minor grooves (Newman et al. 1995; Scheuring Vanamee et al. 2004). Interestingly, the specific complex was asymmetrical, in contrast to the protein in the nonspecific complex. The carboxy-terminal arm of one subunit (called R) went into the DNA minor groove, whereas the arm from the other (L) subunit followed the DNA backbone. DNA cleavage occurred only in the R active site that contained two Mn^{2+} ions. In the nonspecific complex (Fig. 2Ab), the DNA protruded out of the cleft at the bottom of the BamHI dimer (Scheuring Vanamee et al. 2004). This complex would be highly competent for linear diffusion by sliding, as there were no base-specific contacts, and only a few water-mediated contacts to the phosphate backbone (Scheuring Vanamee et al. 2004). Therefore, this would prevent cleavage, like EcoRV in this situation (Winkler et al. 1993). However, in EcoRV the active site residues were displaced because of a change in DNA conformation, whereas in BamHI it was mainly because of a change in the protein conformation (Scheuring Vanamee et al. 2004).

Overall, in the specific complex, the BamHI subunits clamped onto the DNA by an ~10° rotation around the DNA axis moving in a tongs-like motion (Newman et al. 1995; Scheuring Vanamee et al. 2004). It was obvious why a DNA sequence with even a single wrong base pair would force the enzyme into a more open mode increasing the distance between the active site and scissile phosphate bonds. Thus, the enzyme could still bind to the nonspecific site (down by 10^2 to 10^3) but rarely cleave it (down by 10^7 to 10^{10}) (Scheuring Vanamee et al. 2004). Based on the complexes with Ca^{2+} and Mn^{2+} ions, a two-metal mechanism was proposed for BamHI (Scheuring Vanamee et al. 2004), as discussed for *E. coli* DNA polymerase I (Beese and Steitz 1991). This proposal fit in with the finding that the metal binding sites in BamHI were superimposable on those of NgoMIV and with other calculations (Deibert et al. 2000; Fuxreiter and Osman 2001; Mordasini et al. 2003; Scheuring Vanamee et al. 2004). The structure of the nonspecific BamHI–DNA complex was the first to provide such a detailed picture of how an enzyme selected its specific site from the multitude of nonspecific sites (Scheuring Vanamee et al. 2004).

A

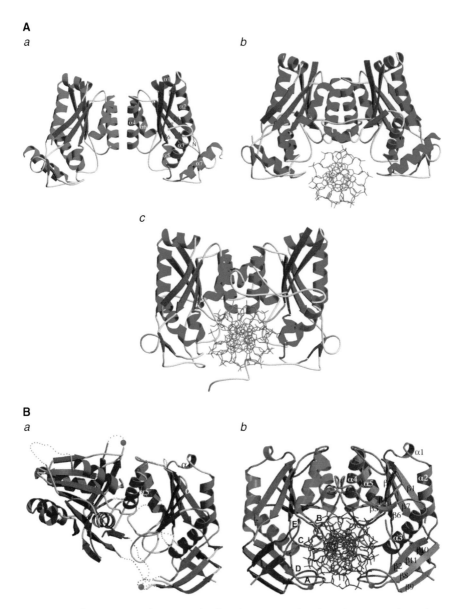

FIGURE 2. The structures of BamHI and BglII. (*A*) Structure of (*a*) free, (*b*) nonspecific, and (*c*) DNA-bound forms of BamHI, respectively. Secondary structural elements, along with the amino terminus and carboxyl terminus, are labeled on the right monomer. Overall structure looking down the DNA axis. (*B*) Structure of (*a*) free and (*b*) DNA-bound forms of BglII: The enzyme is shown with its right subunit in the same orientation as the right subunit of the complex. Loops A and D and a part of loop E are disordered in the free enzyme and are drawn with dotted lines, corresponding to the conformation seen in the enzyme–DNA complex.

(Legend continued on following page.)

An issue that remained to be resolved was how BamHI would move along the DNA: Would a "corkscrew" motion of the enzyme along the DNA major groove follow initial nonspecific binding, or would the enzyme move along one face of the DNA (Sun et al. 2003)?

BglII

The structure of BglII turned out to be a big surprise (Scheuring Vanamee et al. 2004). In the specific complex, the DNA was completely encircled by the enzyme (Fig. 2Bb). The surface area buried upon DNA binding was much larger than in the BamHI complex. Another difference was that BglII distorted the DNA by bending ~22° and by local unwinding and overwinding, similar to DNA complexes of EcoRI, EcoRV, and MunI (Kim et al. 1990; Winkler et al. 1993; Deibert et al. 1999). BglII opened up with a novel scissor-like motion to allow entry of the DNA, rather than binding the DNA in a tight cleft (Aggarwal 1995; Pingoud and Jeltsch 1997, 2001). This motion of the subunits was in a direction parallel rather than perpendicular to the DNA axis, as in the case of BamHI. To do this, the BglII monomers had to undergo a large motion to loosen their grip on the DNA. Interestingly, PvuII also completely encircled the DNA, but instead of a scissor-like motion, it opened with a tongs-like motion (Athanasiadis et al. 1994; Cheng et al. 1994; Scheuring Vanamee et al. 2004).

The large conformational change meant a change from a wedge-shaped bundle of α-helices in the free enzyme to a parallel four-helix bundle in the specific complex, which affected a so-called "lever" region. In the free enzyme, this lever was "down" hiding the catalytic site, whereas in the enzyme–DNA complex, this lever was "up," exposing the catalytic residues for cleavage (Scheuring Vanamee et al. 2004). This was in contrast with BamHI, EcoRV, and PvuII, in which most of the active residues faced the solvent, but similar to free FokI, in which enzyme the cleavage domain was hidden by the recognition domain (Winkler et al. 1993; Athanasiadis et al. 1994; Newman et al. 1994a; Wah et al. 1997; Scheuring Vanamee et al. 2004).

FIGURE 2. (*Continued.*) The complex is viewed down the DNA axis. Secondary structural elements, along with the amino terminus and carboxyl terminus, are labeled on one monomer. Blue spheres mark the respective positions of Lys188 in the free and DNA-bound dimers. Each monomer swings by as much as ~50°, like the blades of a pair of scissors, to open and close the binding cleft. The sheer magnitude of this motion is reflected by the dramatic increase in distances across the binding cleft. For example, the distance between symmetrically related Lys188 residues at the rim of the cleft increases from ~17 Å in the complex to ~61 Å in the free enzyme. (Reprinted from Scheuring Vanamee et al. 2004, with permission from Springer Nature.)

Taken together, despite these differences in the way BamHI and BglII recognized the common base pairs, importantly, in both cases the whole protein contributed to the specificity. The structures explained why attempts to change the specificity of BamHI to that of BglII did not yield viable mutants (for discussion, see Scheuring Vanamee et al. 2004). Did this immutability reflect an evolutionary pressure not to look too much alike? With the benefit of hindsight, this would make sense: A simple change of specificity of the REase through a few point mutations would mean that the cognate MTase could no longer protect the host DNA against restriction. This would put pressure on the REases to develop an intimate relationship with the recognition site that could not easily be changed. By 2004, only EcoRV and BamHI had been analyzed in a specific and nonspecific complex, showing obvious common features (Figs. 1 and 2A). Although tempted, Aneel Aggarwal cautioned that it was still too early to make general statements regarding structural changes accompanying the transition from nonspecific to specific binding, based on only these two enzymes (Scheuring Vanamee et al. 2004).

Type IIE REases: EcoRII and NaeI

EcoRII was one of the first Type II R-M systems to be discovered, and cleaved 5′ C/CWGG (Arber and Morse 1965; Bannister and Glover 1968, 1970; Takano et al. 1968; Yoshimori et al. 1972; Bigger et al. 1973; Boyer et al. 1973). But surprisingly, phage T3 DNA was resistant to EcoRII cleavage although not modified by Dcm and cut by isoschizomer BstNI (5′ CC/WGG) (Krüger et al. 1985,1988). What was going on?

The answer to this question came from an unusual experiment: The addition of pBR322 DNA (a plasmid with six EcoRII sites) to the refractory T3-EcoRII digestion mixture allowed restriction of T3 DNA (Fig. 3, lanes 4 and 6). Complete cleavage required a molar ratio of 2:1 (Pein et al. 1989). Synthetic 14-bp oligonucleotide duplexes (but not ssDNA) with a single EcoRII site could also activate cleavage of T3 DNA by EcoRII, at a molar ratio of 140:1 (Pein et al. 1989). The 14-bp duplexes were cleaved themselves as well, indicating that EcoRII could simultaneously bring together two molecules in an enzyme–DNA complex.

Similar resistance was found for phage T7 DNA, whereas that of phage f1 RF dsDNA was incomplete, although modification of ssDNA was possible (Arber 1966; Hattman 1973; Vovis et al. 1975; Krüger et al. 1988). EcoRII was the first example of a REase that could bind two copies of its DNA recognition sequence in *trans*. The next question was, if EcoRII interacted simultaneously with two DNA sites, did the activating DNA molecules necessarily have to be cleavable themselves? To address this issue, pBR322 DNA was cut

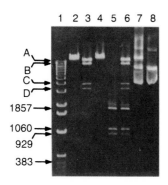

FIGURE 3. Activation in *trans* by pBR322 DNA for cleavage of refractory sites in T3 DNA by the first Type IIE REase, EcoRII, an enzyme that needs two sites for cleavage (Krüger et al. 1988). From *left to right:* 1 kb ladder (lane *1*), T3 DNA (~40 kb, lane *2*), T3 DNA treated with BstNI (lane *3*) or EcoRII (lane *4*), pBR322 DNA treated with EcoRII (lane *5*), and a mixture of T3 DNA and pBR322 DNA treated with EcoRII (lane *6*). T3 bands are labeled A, B, C, D; pBR322 band in base pairs. The fragments were separated on 0.7% agarose gel (adapted from Krüger et al. 1988; for technical reasons Dcm⁻ DNA was used, which will not be explained here). (Reprinted from Krüger et al. 1988.)

with EcoRII, and after phenol extraction and ethanol precipitation, this DNA was incubated with T3 DNA and EcoRII. The results were clear: The EcoRII-derived pBR322 cleavage products stimulated T3 cleavage, but those derived from isoschizomer BstNI (or MvaI) did not. This proved that (1) cleavage of the activating pBR322 EcoRII sites themselves was not necessary, and, interestingly, (2) the nature of the sticky ends mattered (Pein et al. 1991).

A second enzyme requiring activation in *trans* was NaeI from *Nocardia aerocolonigenes* (Conrad and Topal 1989). Both EcoRII and NaeI were dimers in solution and became the prototypes of the Type IIE REases, whose subtype also includes an enzyme of recent interest, SgrAI (Kosykh et al. 1982; Krüger et al. 1988; Conrad and Topal 1989; Vinogradova et al. 1990; Baxter and Topal 1993; Bitinaite and Schildkraut 2002; Roberts et al. 2003a; Dryden 2013). In both cases, separate catalytic and DNA recognition domains bound the two copies of the recognition site simultaneously (Krüger et al. 1988; Conrad and Topal 1989; Colandene and Topal 1998). In the case of EcoRII, the amino-terminal domain controlled the need for two sites to allow catalytic activity of the carboxy-terminal domain: The latter domain could be separately purified as a dimer, and then could cleave DNA with a single EcoRII site (Mucke et al. 2002; Zhou et al. 2004). EcoRII independently cleaved both strands in a single binding event (Yolov et al. 1985; Petrauskene et al. 1998). Activator duplexes of 14 bp were poorly cleaved at low concentrations, but were good activators and substrates at high concentrations, indicating positive cooperativity (Gabbara and Bhagwat 1992). Oligonucleotide duplexes with modified bases (phosphorothioate at the cleavage position) could still activate without being cleaved themselves (Pein et al. 1991; Conrad and Topal 1992; Senesac and Allen 1995). In the case of NaeI, this finding led to the commercial use of NaeI as "Turbo NaeI" (Senesac and Allen 1995; Reuter et al. 2004).

There appeared to be three types of sites: resistant, slow, and cleavable substrate sites; the former two were stimulated by either activator DNA or spermidine (Oller et al. 1991). Also, EcoRII cleaved duplexes of increasing length (14, 30, and 71 bp) with decreasing efficiency, in contrast to isoschizomer MvaI (Cech et al. 1988; Pein et al. 1991).

DNA loops in *cis* could be seen by EM (Topal et al. 1991; Mucke et al. 2000), but they also occurred in *trans* (Krüger et al. 1988; Conrad and Topal 1989; Pein et al. 1989, 1991; Gabbara and Bhagwat 1992; Piatrauskene et al. 1996). Interaction with two recognition sites could be achieved either by one dimer alone or by binding of one dimer per site and subsequent formation of an active tetrameric protein–DNA complex. Although able to act in *trans,* EcoRII and NaeI both preferred interactions in *cis*, forming loops with DNA molecules with two or more sites <1 kb away from each other (Krüger et al. 1988; Pein et al. 1991; Schleif 1992). This tied in with other reports that interactions between DNA sites in *cis* were usually favored over those in *trans* (Schleif 1992; Rippe et al. 1995).

NaeI had two nonequivalent DNA-binding sites: The recognition domain would bind one recognition site—the activator site; this activated the catalytic domain, enabling cleavage of the other site; the second DNA recognition site was required for efficient cleavage (Oller et al. 1991; Gabbara and Bhagwat 1992; Yang and Topal 1992; Kupper et al. 1995; Colandene and Topal 1998; Reuter et al. 1999; Huai et al. 2000, 2001; Mucke et al. 2002). This was shown using different plasmids with one or two sites (Embleton et al. 2001; Mucke et al. 2003). Catenanes (Fig. 4) could be used to test whether an enzyme used sliding along the DNA (1D tracking) to find its recognition site or used 3D looping (Adzuma and Mizuuchi 1989). Such interlinked rings with a copy of the target site in each of the two rings could be generated using resolvase on a plasmid with two *res* sites interspersed with two targets for the enzyme under study. Although a looping enzyme can move from its site to the site on the other ring, a tracking enzyme will be unable to transfer to the other ring. Type IIE enzymes cleaved only one ring of the catenanes (Embleton et al. 2001), in contrast to the Type IIF REases (Szczelkun and Halford 1996; see next subsection).

The crystal structure of NaeI revealed a structural motif for the DNA-binding site occurring in the catabolite activator protein that was not present in EcoRII (Huai et al. 2000, 2001; Zhou et al. 2004). In the absence of DNA, only the catalytic domain of NaeI contributed to dimerization, whereas, in the presence of activator DNA, the DNA-binding domain also contributed, resulting in a more compact dimer (Huai et al. 2000, 2001). Apparently, NaeI changed conformation after binding activator DNA, which promoted binding of, and cleavage by, the catalytic domain of the DNA, the active complex being

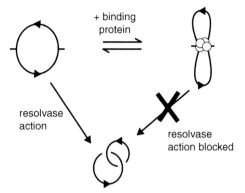

FIGURE 4. Use of catenanes to study tracking and looping by REases. A plasmid contains two target sites for resolvase in direct repeat, the res sites (black triangles), and two recognition sites (hatch marks) for a protein that can bind concurrently to two sites (e.g., a Type IIE REase). The binding of this protein (shown as four spheres) to both sites sequesters the res sites into separate loops. In the absence of this binding protein, resolvase acts on the plasmid to yield a DNA catenane. However, resolvase is blocked if the res sites are sequestered in separate loops. (Reprinted from Welsh et al. 2004, with permission from Springer Nature; originally adapted from Milsom et al. 2001, with permission from Elsevier.)

a protein dimer bound to two DNA recognition sites (Petrauskene et al. 1994; Reuter et al. 1998; Huai et al. 2000, 2001; Mucke et al. 2002).

The differences between the DNA-binding domains of EcoRII and NaeI suggested that different Type IIE enzymes had independently (i.e., "convergently") found a similar solution to the same problem of binding two identical DNA sites.

Type IIF REases: SfiI, Cfr10I, and NgoMIV

Like Type IIE REases, SfiI, Cfr10I, and NgoMIV bound simultaneously to two DNA sites, but were classed as subtype IIF as they acted at both sites at the same time, and converted catenanes with two sites directly to two linear products, unlike Type IIE (Wentzell et al. 1995; Szczelkun and Halford 1996; Šiksnys et al. 1999; Embleton et al. 2001; Roberts et al. 2003a). To align two sites for cleavage, these enzymes may require a certain length of the intervening DNA. Would it have consequences for the activity of the protein, if one altered the length of the nonspecific spacer by 5–6 bp, or 10–11 bp? In the former case, such a change would rotate one recognition surface relative to the other by about 180°, requiring (under- or over-)twisting the intervening DNA to align the sites, an energetically costly event. In contrast, an alteration of 10–11 bp would simply add an additional helical turn. If so, would variations in the length of the spacer show a cyclical response characteristic of DNA looping, with a periodicity expected for the helical repeat (Schleif 1992)? Indeed, SfiI (GGCC[N5]GGCC) did show such a cyclical response when tested against plasmids with two SfiI sites separated by various lengths of DNA of <300 bp (Wentzell and Halford 1998; Welsh et al. 2004).

NgoMIV, Bse634I, and Cfr10I

Around 1995, the group of Virgis Šikšnys started the analysis of NgoMIV, and several other REases that recognized 5′ CCGG in different contexts, but belonged to different subtypes. SsoII and StyD4I (IIP), EcoRII (IIE), and Ngo-MIV (IIF), appeared to possess a similar DNA-binding motif and catalytic center: Were these subfamilies perhaps evolutionary related, and did they share a common ancestor, probably a homodimer (Pingoud et al. 2002; Tamulaitis et al. 2002)? The first three crystal structures analyzed were NgoMIV from *Neisseria gonorrhoeae* (recognition site 5′ G/CCGGC) (Stein et al. 1992) and two isoschizomers that shared ~30% identity: Cfr10I from *C. freundii* and Bse634I from *Bacillus stearothermophiles* (recognition site 5′ Pu/CCGGPy) (Janulaitis et al. 1983; Repin et al. 1995; Grazulis et al. 2002). The structures of Cfr10I and Bse634I without DNA (Bozic et al. 1996; Grazulis et al. 2002) and that of NgoMIV with DNA (Deibert et al. 2000) proved that these enzymes acted as tetramers (although Cfr10I was initially thought to be a dimer as the dimer–dimer interface was considered to be due to crystal packing). Two monomers formed a primary dimer similar to that of Type IIP enzymes such as EcoRI (Fig. 5A; Rosenberg 1991). This similarity supported the notion of a common core and active site but also the idea that perhaps the cleavage pattern rather than the recognition sequence played a key role in the structure of the dimer (Anderson 1993; Aggarwal 1995).

The specific complex of NgoMIV with two 10-bp oligonucleotide duplexes showed two primary dimers back-to-back with the DNA on the opposite sides of a tetramer, with the major groove contacts between the dimer and the recognition site. The DNA recognition and dimerization interfaces in Ngo-MIV were intertwined, and the tetramer was fixed by contacts between both subunits and primary dimers (Šikšnys et al. 2004). The most extensive contacts were located in the "tetramerization" loop. A single mutation in this region, and parallel experiments in solution, showed that being a tetramer was important for restriction (Bilcock and Halford 1999; Šikšnys et al. 1999; Deibert et al. 2000; Milsom et al. 2001; Pingoud and Jeltsch 2001; Grazulis et al. 2002; Šikšnys et al. 2004). Using the above catenane assays, simultaneous cleavage of all four bonds by NgoMIV, Bse634I, and Cfr10I was confirmed (Bilcock et al. 1999; Bath et al. 2002; Šikšnys et al. 2004). Slow cleavage of single-site

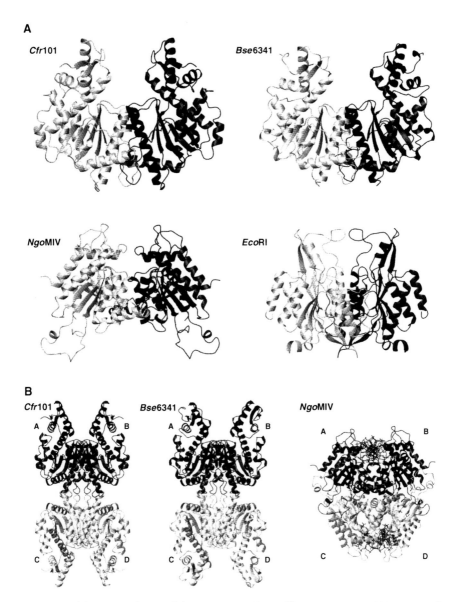

FIGURE 5. (*A*) Primary dimers of the tetrameric REases Cfr10I, Bse634I, and NgoMIV and comparison with EcoRI. Individual subunits are shown in gray and black. (*B*) Tetramers of Cfr10I, Bse634I, and NgoMIV. Two back-to-back primary dimers are shown in gray and black. The monomers are labeled A, B, C, D, respectively. DNA molecules bound to NgoMIV are shown in the stick presentation. (Reprinted from Šikšnys et al. 2004, with permission from Springer Nature.)

plasmids could be speeded up by oligonucleotide duplexes with (but not without) the recognition sequence, similar to transactivation of SfiI (Nobbs and Halford 1995).

Modeling of the structures of free Bse634I and Cfr10I on the NgoMIV–DNA complex indicated similar recognition of the central CCGG, but not the outer base pair (Grazulis et al. 2002; Šikšnys et al. 2004). This resembled the pattern shown for EcoRI (G/AATTC) and MunI (C/AATTG) (Kim et al. 1990; Šikšnys et al. 1994; Jen-Jacobson et al. 1996; Deibert et al. 1999; Lukacs and Aggarwal 2001), but not BamHI and BglII (see above, and Lukacs et al. 2000; Lukacs and Aggarwal 2001; Scheuring Vanamee et al. 2004). Interestingly, in the crystals, the amino acids contacting the two Mg^{2+} ions were in the same relative location, although they were derived from different regions of the polypeptide (Skirgaila et al. 1998; Deibert et al. 2000; Grazulis et al. 2002). This suggested plasticity of the active sites as long as the structure was conserved, rather than the primary sequence, as also reported for other enzymes (Todd et al. 2002). This was supported by a residue swapping experiment of Cfr10I, which created a reengineered metal binding site with significant catalytic activity (Skirgaila et al. 1998).

Computer and sequence analysis suggested that NgoMIV, Cfr10I, and Bse634I shared a common ancestor with SsoII and PspGI, but in these enzymes the orientation of the monomers in the dimer had changed, and their pentanucleotide 5' sticky ends were the result of an increased distance between the two catalytic sites compared to NgoMIV (Pingoud et al. 2002; Bujnicki 2004). Was this perhaps a common evolutionary mechanism for the generation of new specificities?

NgoMIV, Cfr10I, and Bse634I were the first Type IIF enzymes to be crystallized, ahead of the structure of the archetype Type IIF enzyme, SfiI (Viadiu et al. 2003; Vanamee et al. 2005). Sfi I serves as a model enzyme to study how two DNA molecules can be sequestered in a synaptic complex, an event that is used for many reactions in the cell. Analysis of how this enzyme recognizes and cleaves its target DNA would provide insight into the sequential binding events that result in such a complex. The sequence of Sfi appeared to be totally unrelated to other proteins, but its mode of DNA recognition is similar to that of the dimeric Type IIP BglI enzyme, even though SfiI is a tetramer (Vanamee et al. 2005). Bioinformatic analysis supported the notion that SfiI was more closely related to BglI than to any other REase, including other Type IIF REases with known structures, such as NgoMIV·NgoMIV and BglI were judged to belong to two different, very remotely related branches of the PD·(D/E)XK superfamily: the α class (EcoRI-like) and the β class (EcoRV-like), respectively. This analysis provided "evidence that the ability to tetramerize and cut the two DNA sequences in a concerted manner was developed independently at least two times in the evolution of the PD…(D/E)XK super-family of REases." The model of SfiI would be useful for further experimental analyses.

The Common Core of the Majority of the Type II REases

The studies in Vilnius provided the first formal evidence for the conserved sequence motif in the active site of REases, now known as the PD…(D/E) XK site (Fig. 6).

This motif appeared to be common to 11 other REases belonging to Type IIP, IIE, and IIF (Fig. 7; Kovall and Matthews 1999; Pingoud et al. 2002; Šikšnys et al. 2004). This core had already been noted, when the structures of EcoRI and EcoRV were compared (Venclovas et al. 1994). A five-stranded mixed β-sheet was flanked by α-helices, also present in four other endonu-cleases: lambda exonuclease, MutH, Vsr endonuclease, and TnsA (Ban and Yang 1998; Kovall and Matthews 1999; Tsutakawa et al. 1999a; Hickman et al. 2000). Of the four β-strands, three β-strands were absolutely conserved. Within this common core, two β-strands would be directly involved in catalysis, and the others would be important for the structure itself. Was there a common nuclease ancestor, which had been subject to divergent evolution, as long as this structure remained intact (Huai et al. 2000)?

Mutants in the PD ⋯ (D/E)XK motif of EcoRV confirmed the essential role of these residues (Thielking et al. 1991; Winkler 1992). Outside this PD…(D/E)XK fold, different REases would have acquired different additional

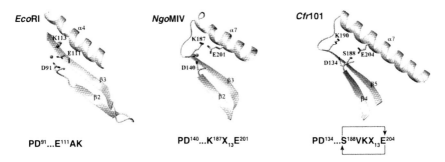

FIGURE 6. Structural localization of the active site residues of EcoRI, NgoMIV, and Cfr10I. Conserved structural elements are shown in stick representation and labeled. Mn^{2+} ion present in the active site of EcoRI (PDB entry 1 qps) is shown as a gray sphere. Two Mg^{2+} ions present in NgoMIV are shown as gray spheres. Active site motifs corresponding to the first metal ion binding site are shown below each figure. Arrows indicate Cfr10I active site residues subjected to swapping (see text; Skirgaila et al. 1998). (Reprinted from Šikšnys et al. 2004, with permission from Springer Nature.)

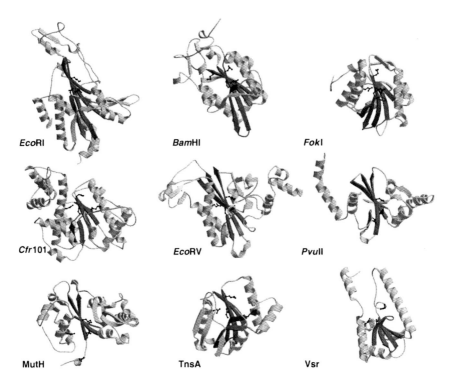

FIGURE 7. Comparison of the REase folds in some members of the REase superfamily. The conserved central β-sheet fold is highlighted in dark gray. The catalytic residues are in ball-and-stick and colored black. Single molecules of BamHI, EcoRI, EcoRV, PvuII, and Cfr10I and only the amino-terminal catalytic domains of FokI and TnsA are shown. (Reprinted from Horton et al. 2004a, with permission from Springer Nature.)

elements (e.g., Type IIF REases a tetramerization region and Type IIE REases a domain for binding two sites). This idea was supported by mutations in the respective regions (Reuter et al. 1999; Šikšnys et al. 1999, 2004; Deibert et al. 2000; Mucke et al. 2002; Bujnicki 2004; Zaremba et al. 2005, 2006, 2012). Did the idea of evolutionary relationships between different subtypes suggest an abrupt or continuous transition? Was this independent of different higher-order tertiary and quaternary structures and despite different substrate requirements for DNA cleavage (Pingoud et al. 2002; Tamulaitis et al. 2002)? Similarity at the tertiary structure was strongest between REases with a similar cleavage pattern—for example, BamHI and EcoRI (four-base 5′ overhang, DNA binding from the major groove side) or EcoRV and PvuII (blunt end, DNA binding from the minor groove side) (Anderson 1993; Aggarwal 1995). Again, this indicated that the nature of the DNA cleavage site was important, rather than the recognition sequence (Anderson 1993; Aggarwal 1995). An exception was BglI: It had a fold similar to EcoRV and PvuII, but it cleaved DNA to leave a 3′ overhang (Newman et al. 1998). This difference could be "explained away" by relatively minor modifications of the protein surface (Newman et al. 1998). It was concluded that two families of enzymes could be distinguished that were structurally very similar: EcoRI-like enzymes and EcoRV-like enzymes. The EcoRI-like REases usually recognized specific bases in the DNA mainly via residues from an α-helix, whereas the EcoRV-like REases usually recognized specific bases in the DNA mainly via residues from an additional β-sheet (Fig. 8; Aggarwal 1995; Bujnicki 2000b, 2001b; Huai et al. 2000; Pingoud and Jeltsch 2001). Despite this, the two families of enzymes were structurally very similar (see Table 1 in Horton et al. 2004a for details, pp. 362–363).

The fact that Type II subtypes shared the PD···(D/E)XK motif suggested a basically similar reaction mechanism, but there was at least one exception: BfiI was Type IIS like FokI, but used a "zero-metal" mechanism (Sapranauskas et al. 2000; Zaremba et al. 2004). Surprisingly, BfiI belonged to the phospholipase D (PLD) superfamily and resembled NucA from *S. typhimurium,* whose crystal structure was known (Stuckey and Dixon 1999). As NucA was a homodimer with one catalytic center formed by the two subunits, perhaps BfiI would also form a tetramer for double-strand cleavage (Pingoud and Jeltsch 2001). More exceptions like BfiI were likely to follow based on other sequence comparisons, which suggested that some Type II REases belonged to the HNH and GIY-YIG families of endonucleases (Aravind et al. 2000; Bujnicki et al. 2001).

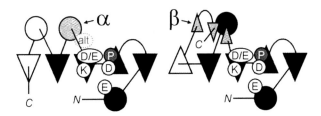

FIGURE 8. Diagrams showing the major structural differences between the α (EcoRI-like) and β (EcoRV-like) subclasses of the PD···(D/E)XK enzymes (Huai et al. 2000; Bujnicki 2001b). Common secondary structures are shown in black. Key elements involved in DNA recognition are shown in gray (in α class it is a universally conserved α-helix B; in β class it is an additional small β-sheet). Other elements specific for α and β subclasses (including the topologically fifth β-strand) are shown in white. The alternative site in α-helix B, to which the D/E carboxylate migrated in some of the enzymes from the α class, is indicated as "alt." (Modified with permission of Bentham Science Publishers, Ltd., from Bujnicki 2003.)

Evolutionary Relationships between REases

Janusz Bujnicki used the method of Johnson and coworkers (Johnson et al. 1990) and the atomic coordinates of the nine available REase structures to propose an evolutionary tree (Fig. 9; Bujnicki 2000b).

A comparison of crystal structures of REases with other proteins suggested that they were related to other DNA processing proteins, including DNA recombinases and transcription factors, which formed loops in the DNA depending on the length of the DNA between recognition sites, as first reported in 1984 (Dunn et al. 1984; Topal et al. 1991; Ban and Yang 1998; Wentzell

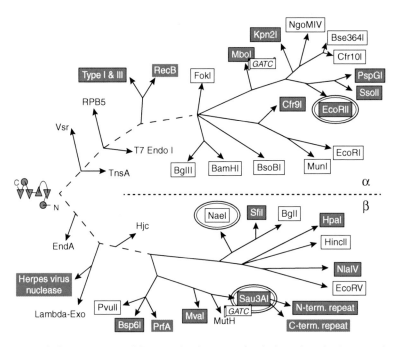

FIGURE 9. Phylogenetic tree of the PD···(D/E)XK superfamily, based on the "structural tree" (Bujnicki 2004, p. 76), and expanded to include additional members, identified by sequence analyses and protein-fold recognition (Šikšnys et al. 1995; Aravind et al. 2000; Bujnicki 2001a; Bujnicki and Rychlewski 2001a,b; Friedhoff et al. 2001; Pingoud et al. 2002; Rigden et al. 2002). REases are shown in black frames. PD···(D/E)XK domains identified by bioinformatics and not by crystallography are shown in white on gray background. Subtype IIE enzymes from three different lineages are indicated by circles. Isoschizomers MboI and Sau3A that originated from two different lineages are indicated by a label with their recognition site GATC. Parts of the tree that could not be confidently resolved based on either sequence or structural analysis are shown in broken lines. (Reprinted from Bujnicki 2004, with permission from Springer Nature.)

and Halford 1998; Kovall and Matthews 1999; Tsutakawa et al. 1999b; Hickman et al. 2000). EcoRII and NaeI shared motifs with some site-specific recombinases, and, in line with this, a single mutation in EcoRII and NaeI turned these enzymes into topoisomerases (Topal and Conrad 1993; Jo and Topal 1995; Nunes-Duby et al. 1998; Carrick and Topal 2003). Extensions or insertions in regions outside the common PD···(D/E)XK fold of SsoII, PspGI, NgoMIV, and EcoRII appeared to determine the features characteristic for the different subtypes mentioned previously (IIP, IIF, or IIE) (Reuter et al. 1999; Deibert et al. 2000; Bujnicki 2004). Sau3AI was a highly unusual enzyme with similarity to MutH: It bound two recognition sites and formed DNA loops, like Type IIE and IIF, but used two copies of a duplicated PD···(D/E)XK domain, of which only the amino-terminal copy retained the conserved catalytic residues (Bujnicki 2001a; Friedhoff et al. 2001). The Type IIS REase FokI interacted with two sites via a domain resembling that of Tn7 transposase (TnsA) (Hickman et al. 2000), whereas DNA excision by SfiI was reminiscent of recombinases that simultaneously cleaved four DNA strands (Wentzell and Halford 1998). This structural and functional similarity with recombinases and transposases led to speculations that REases might promote genetic rearrangements and enhance genome diversity (Carlson and Kosturko 1998; Hickman et al. 2000; Mucke et al. 2002). REases would thus benefit the population as a whole, rather than individual organisms or the R-M systems themselves, as Tom Bickle had already pondered (Chapter 6; Bickle 1993; Arber 2000). In the absence of in vivo or in vitro evidence (Petrauskene et al. 1998; Šikšnys et al. 1999; Deibert et al. 2000), by 2004, the tentative general conclusion seemed to be that R-M systems could integrate DNA fragments into the genome of the host but could also shuffle protein domains around as a neat way to create new functions. Interestingly, this was apparently not limited to bacteria and their plasmids, viruses, and transposons, as genes could also pass from vertebrates into bacteria (Ponting and Russell 2002).

The Role of Water in Specific and Nonspecific Recognition by Type II REases

Nina Sidorova and Donald Rau analyzed the role of H_2O in specific and nonspecific recognition by EcoRI (Sidorova and Rau 1996, 2004; Pingoud 2004), important in light of the effect of hydrostatic and osmotic pressure on the activity of REases (e.g., that on Type IV Mrr; Ghosh et al. 2014). As these studies require further analysis, the reader is referred to the original review for further information on this topic (Sidorova and Rau 2004).

Role of Mg^{2+} and Other Metal Cofactors
of Type II REases

Most REase cleavage studies were consistent with a direct attack by H_2O and absence of an intermediate species, as shown by inversion of the stereochemistry at phosphorus (Connolly et al. 1984; Grasby and Connolly 1992; Mizuuchi et al. 1999), in contrast to BfiI that goes via a covalent complex and thus retention of stereo configuration as mentioned previously. Different groups proposed models to explain the metal dependence in this process (Jeltsch et al. 1992, 1995b; Baldwin et al. 1995; Vipond et al. 1995; Vipond and Halford 1995; Horton et al. 1998a). Usually, REase activity first increased with increasing $[Mg^{2+}]$, but then dropped, possibly as a result of substrate inhibition, most likely because of general ionic strength effects or competition between the metal ions (Demple et al. 1986; Black and Cowan 1994; Friedhoff et al. 1996). Interestingly, polyamines diminished this inhibition by displacing Mg^{2+} ions (Friedhoff et al. 1996). But what made Mg^{2+} so special? According to Cowan in Columbus, Ohio, Mg^{2+} was "a good choice due to its high abundance, and a favorable combination of physical and chemical properties" (Cowan 2004). These properties included a tendency to bind H_2O molecules rather than bulkier ligands (Cowan 2004). Effectively, Mg^{2+} usually interacted with two to three oxygens on REase side chains, unlike other metals such as Mn^{2+} (Cowan 1998, 2004). In strong contrast, Ca^{2+} blocked cleavage, a useful property in the crystal studies discussed previously (Jose et al. 1999; Conlan and Dupureur 2002a; Cowan 2004).

Everybody agreed that the elucidation of the mechanism of DNA cleavage critically depended on the number of Mg^{2+} ions directly involved in catalysis. Alfred Pingoud and coworkers proposed a one-metal mechanism for EcoRV catalysis but could not exclude a two-metal mechanism (Jeltsch et al. 1992, 1993b). A heated debate continued for decades: How many metal ions were needed for catalysis: one, two, or three (Jeltsch et al. 1992; Vipond et al. 1995; Cowan 1998; Lukacs et al. 2000; Chevalier et al. 2001)? The issue remains unresolved until today (The Seventh NEB Meeting 2015). This was not helped by the "perplexing" observation that Mg^{2+}, Ca^{2+}, and Mn^{2+} did not necessarily bind to the same amino acid side chains within the catalytic core (Pingoud and Jeltsch 2001). For EcoRV, at least three different mechanisms were proposed, based on combined data from different crystal structures (Kostrewa and Winkler 1995; Pingoud and Jeltsch 1997, 2001; Horton et al. 1998b; Kovall and Matthews 1999; Cowan 2004; Horton et al. 2004a). How misleading were these structures? Was the three-metal catalytic model the result of movement of the metal(s) during the transition from nonspecific to specific binding, and positioning the catalytic site of the enzyme near the bond to cleave

(Kostrewa and Winkler 1995; Cowan 2004)? Perhaps the crystal structures of EcoRV were "snapshots" along the reaction pathway (Horton et al. 2004a)!

In the case of PvuII, Dupureur and coworkers showed a conformational change after metal ion binding and decided on a two-metal mechanism (Conlan et al. 1999; Dupureur and Hallman 1999; Dupureur and Conlan 2000; Dominguez et al. 2001; Dupureur and Dominguez 2001; Conlan and Dupureur 2002a,b). PvuII–DNA binding was only promoted by metal ions for specific recognition (Conlan and Dupureur 2002a,b), suggesting that "the placement of the metal cofactor is optimal to promote specific contacts with the cognate sequence, either through direct binding interactions or an indirect influence on enzyme structure" (Cowan 2004). Did the enzyme need two metal ions located close together ($<\sim4$ Å) on two sides of the substrate, as proposed for DNA polymerase I (Beese and Steitz 1991; Cowan 1998)? And did the overall data support the notion that one metal ion would promote cleavage, whereas the other one (or both) would serve a structural role and/or influence substrate binding (Cowan 2004)? In this process each metal ion would influence the binding of the other (Cowan 2004). Evidently more kinetic studies were needed under solution turnover conditions with Mg^{2+} as cofactor; not an easy task, but vital to solve this problem and put an end to the metal debate (Cowan 2004).

Engineering and Applications of Chimeric Type II REases

A wide variety of mutations were introduced in various REases, ranging from mutant enzymes with enhanced cleavage or relaxed specificity, recognition of altered or modified sequences, and lengthening of the recognition site to changed subunit composition and single-chain nucleases (summarized in Table 1, pp. 394–395 of Alves and Vennekohl 2004). Unfortunately, REases appeared to require many changes in order to generate REases with new specificities (Jeltsch et al. 1995a; Anton et al. 1997; Bujnicki 2001b; Bitinaite et al. 2002; Pingoud et al. 2002). This was especially disappointing with respect to attempts to increase the length of the recognition site. Why was this hunt for longer recognition sites so important? The answer was their potential applications in gene therapy! To create a single unique dsDNA break within a mammalian genome, the REase would need to recognize DNA sequences of 16 bp or more (occurrence once every 4^{16} bp = 4.3×10^9 bp). Expectations were high with the discovery of FokI. This R-M system from *Flavobacterium okeanokoites* was cloned in 1989, had separate DNA recognition and cleavage domains, required dimerization to produce a double-strand break, and cleaved 9/13 nucleotides downstream from the recognition site

(Sugisaki and Kanazawa 1981; Kita et al. 1989; Looney et al. 1989; Kandavelou et al. 2004). Could one create a chimeric nuclease by fusing the nonspecific cleavage domain of FokI to zinc finger (ZF) domains to obtain a ZF nuclease (ZFN)? At the time, the structure of ZF domains made them the most versatile recognition motifs for the development of such artificial DNA-binding proteins (Pabo et al. 2001; Beerli and Barbas 2002; Kandavelou et al. 2004). Each ZF would bind a 3-bp DNA sequence, and tandem ZF motifs could increase the length of the sequence recognized (Kandavelou et al. 2004). FokI/3ZF would recognize a 9-bp inverted site, hence an effective recognition site of 18 bp, which hopefully would cut only once in a mammalian genome and stimulate homologous recombination at that single unique site. This should be feasible, as several laboratories had already reported homologous recombination at the cleavage site by ZFN (Bibikova et al. 2001, 2002; Porteus and Baltimore 2003). Srinivasan Chandrasegaran and colleagues in Baltimore saw "a glimpse of potential future therapeutic applications of ZFN in modifying and rewiring the human genome itself" (Kandavelou et al. 2004). Could chimeric nucleases be the new molecular scissors for research in stem cells? Would this technique eventually make correction of a genetic defect feasible, especially in treating single-gene diseases? In a decade or two, would gene therapy become routine in a clinical setting? In 2004, these studies were still in their infancy as gathered from the summary of the data on homologous recombination in frog oocytes, fruit flies (very inefficient), gene targeting in murine embryonic stem cells, and studies with CCR5 (the HIV[s] chemokine receptor) and CFTR (involved in cystic fibrosis) (Kandavelou et al. 2004). There would be many more hurdles to overcome in the years ahead (see Durai et al. 2005; Kandavelou et al. 2005, 2009; Mani et al. 2005a,b; Wu et al. 2007; Kandavelou and Chandrasegaran 2009; Ramalingam et al. 2011 for further details), some of them now no longer of consequence because of the arrival of CRISPR–Cas technology (Chapter 8).

TYPE I AND III ENZYMES

Initially identified only in enterobacteriaceae because of limited detection methods (Bickle 1993; Bickle and Krüger 1993; King and Murray 1994; Barcus and Murray 1995), whole-genome sequencing had by the end of the century revealed candidate Type I R-M systems to be as abundant as Type II (Table 1). Using ATP as energy source, the intriguing Type I "molecular motor" proteins required 1D translocation along the DNA from the sequence-specific DNA site to the site of nonspecific cleavage (Murray 2000; Rao et al. 2000; Dryden et al. 2001; Bourniquel and Bickle 2002). Rather than biological tests, high sequence identity of putative HsdR and HsdM polypeptides allowed assignment of these to the Type IA, IB, IC, and ID subclasses, with 80%–99% identity even from different species (Barcus and Murray 1995). Between subclasses, identity was only ~20%–35%, irrespective of the host. Despite this apparent lack of common ancestry, early coevolution was likely because of the unusual diversity of putative genes in distantly related species with intermediate levels of identity and the unlikelihood that Type I R-M systems would have evolved more than once (Sharp et al. 1992; Barcus and Murray 1995). This was reminiscent of other genes that discriminate "self" from "non-self" (e.g., the mammalian major histocompatibility complex), which led to the concept of "primitive bacterial immune system" (Barcus and Murray 1995).

Modeling the DNA Recognition Complex of the Type I M·EcoKI Trimeric Complex

Work on the structure of Type I and III REases was severely frustrated by the inability to generate crystals at that time. Undeterred, David Dryden and co-workers used 3D structures and folds of Type II MTases to build a picture of the structural domains of the trimeric M·EcoKI (M_2S_1) (Fig. 10).

This DNA recognition/modification unit of the pentameric EcoKI complex required SAM for binding and distinguished hemimethylated (m6A-modified) DNA from unmodified DNA (Dryden et al. 1993; Powell and Murray 1995). Mutational analysis plus sequence comparisons and tertiary structure modeling indicated six motifs in the HsdM subunit common to the gamma class of Type II MTases (Cooper and Dryden 1994; Willcock et al. 1994; Dryden et al. 1995; Sturrock and Dryden 1997). This suggested a primordial MTase gene for Type I and Type II enzymes. In the model, two HsdM subunits (linked by the HsdS subunit) clamped the DNA, resembling two Type II MTases stacked together. Partial proteolysis indicated interaction of the carboxyl terminus of HsdR with HsdS (Davies et al. 1999b).

Mutational analysis, combined with alignment of 51 Type I HsdS TRD sequences in the database, and secondary structure predictions led to a tentative tertiary structure resembling that of M·HhaI (O'Neill et al. 1998). Each TRD would fit into the major groove and recognize the DNA, with the HsdM subunits arranged on either side of HsdS, allowing them to encircle the DNA and methylate the target adenines. Did this indicate that the MTases derived from a common ancestor with one monomeric TRD and a separate catalytic subunit (like some Type II MTases, e.g., AquI [Pinarbasi et al. 2003; Roberts et al. 2003b])? Two additional HsdR subunits would be responsible for bidirectional translocation and cleavage although complexes with a single HsdR could translocate DNA (Dryden et al. 1997; Janscak et al. 1998; Firman and Szczelkun

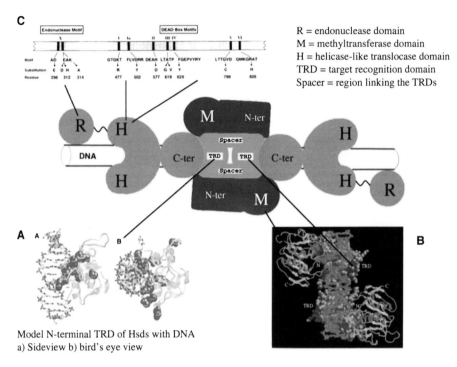

R = endonuclease domain
M = methyltransferase domain
H = helicase-like translocase domain
TRD = target recognition domain
Spacer = region linking the TRDs

Model N-terminal TRD of Hsds with DNA
a) Sideview b) bird's eye view

FIGURE 10. Models of EcoKI and of the structure of the EcoKI restriction complex. The two TRDs of the specificity subunit HsdS (green) recognize the two halves of the DNA recognition site AAC(N6)GTGC. The TRDs are linked by conserved sequence regions that function as subunit interfaces and also define the length of the nonspecific DNA sequence in the middle of the recognition site. Two HsdM modification subunits (blue) bind to the conserved regions of HsdS via their amino- and carboxy-terminal domains. They wrap around the DNA helix on the opposite side of HsdS, allowing access of the methyltransferase domain of HsdM to DNA, presumably using base flipping as described for Type II MTases. Two HsdR subunits (orange) associate with HsdM and HsdS via the carboxyl terminus. The central part of the protein is involved in translocation and contains "DEAD box" motifs, characteristic of helicases (H). These motifs probably fold into two domains (IA and 2A) to form a cleft through which the DNA would pass (resembling a "RecA-like" structure that may be common to all helicases/translocases). EcoKI belongs to helicase superfamily 2 (SF2), whose members are believed to guide the DNA via regions outside the IA and IIA domains toward the cleft involving interactions with the DNA backbone (and not the bases), in line with the function of EcoKI as a DNA translocase rather than helicase. In the amino terminus of HsdR is a motif "X" characteristic of endonucleases (R), which is the PD···(D/E)XK common core (Figs. 6 and 7). The enzyme binds the target site via HsdM and HsdS using SAM as cofactor for binding and distinguishing between hemimethylated and unmodified DNA. If unmodified, the enzyme undergoes a large conformational change and translocates the DNA past itself, while remaining bound to the recognition site, creating large loops visible by EM and AFM, concomitant with ATP hydrolysis. The model rests on extensive genetic, biochemical, and biophysical evidence (see text for further details and references).

(*Legend continued on following page.*)

2000). This model was supported by other extensive experimental data, including DNA footprinting, fluorescence anisotropy, gel retardation, protein–DNA cross-linking, and measurements of the hydrodynamic shape of wild-type protein and mutants (Dryden et al. 1993, 1995; Powell et al. 1993, 1998a,b, 2003; Cooper and Dryden 1994; Willcock et al. 1994; Chen et al. 1995; Powell and Murray 1995; Sturrock and Dryden 1997; O'Neill et al. 1998, 2001).

The Molecular Motors of Type I and Type III REases

In addition to the above studies, which generated support for common ancestry of Type I and II DNA recognition domains and MTase functions, progress was made with respect to the curious translocation and restriction properties of the Type I and III REases. Multiple sequence alignments and structure predictions indicated an amino-terminal conserved motif "X" in the HsdR subunit of EcoKI, which was also present in other Type I and Type III enzymes and resembled the PD $\cdots$ (D/E)XK motif in Type II REases, described previously (Titheradge et al. 1996). This "X" motif is the active site, as "X" mutants could no longer restrict DNA but retained the ability to hydrolyze ATP and translocate DNA in vivo, thus uncoupling translocation and restriction (Davies et al. 1999a,b; Janscak et al. 1999b, 2001; Wang et al. 2000; Chang and Julin 2001). Similarly, "X" mutations in the carboxyl terminus of the Res subunit of EcoP15I also abolished DNA cleavage without affecting ATP hydrolysis (Janscak et al. 2001).

The putative translocation domain of HsdR of EcoKI and Res of EcoP15I shared so-called "DEAD box" (or helicase) motifs with proteins involved in replication, recombination, transcription, and repair that could unwind DNA, backtrack disrupted replication forks, remodel chromatin, or remove stalled RNA polymerase duplexes (West 1996; Park et al. 2002; Maluf et al. 2003; Whitehouse et al. 2003). This included DNA helicases and

---

FIGURE 10. (*Continued.*) (*Inset A*) A model of amino acids 43–157 from the amino-terminal TRD of EcoKI interaction with DNA (Sturrock and Dryden 1997). (*Inset B*) A front view from a partial model of a Type I MTase bound to DNA constructed using two copies of the structure of Type II MTases bound to DNA. The TRD regions are based on the structure of the TRD from M·HhaI and the methyltransferase domains in the catalytic domains of M·TaqI. Space filling shows sites of mutations resulting in loss of specificity and activity. (*Inset C*) Section of the HsdR subunit showing mutational analysis of conserved endonuclease and "DEAD box" (helicase-like) motifs. (Reprinted from Loenen 2003; originally adapted from Davies et al. 1999, with permission from Elsevier; *A*, from O'Neill et al. 1998, reproduced with permission from EMBO; also see Sturrock and Dryden 1997; *B*, reprinted from Dryden et al. 1995, with permission from Springer Nature; *C*, reprinted from Davies et al. 1999a,b, with permission from Elsevier.)

related AAA$^+$ ATPases found in a wide variety of proteins from bacteria to humans. Did Type I and III REases share this functionality (West 1996)? The "DEAD box" motifs folded into a so-called RecA-like fold with a large cleft, through which the DNA could be either pushed or pulled (Gorbalenya and Koonin 1991; Murray et al. 1993; Titheradge et al. 1996; Aravind et al. 1999, 2000; Davies et al. 1999a,b; Caruthers and McKay 2002; Singleton and Wigley 2002, 2003). Sequence alignments and secondary structure predictions indicated that Type I REases belonged to the superfamily 2 (SF2) of the helicases (Gorbalenya and Koonin 1991; Murray et al. 1993; Titheradge et al. 1996; Hall and Matson 1999). This would be in line with mounting evidence that SF2 members often translocated or remodeled DNA without opening up the double helix (unlike many known members of the SF1, SF3, and SF4 superfamilies that did unwind DNA [Singleton and Wigley 2002]). It was likely that translocation by Type I REases would proceed via DNA–backbone interactions without strand separation or recognition of specific bases.

The crystal structures of several helicases revealed an ATP pocket consisting of the so-called "Walker" A and B boxes (helicase motifs I and II) first identified in ATP synthase (Walker et al. 1982) and an additional component Motif VI (Yao et al. 1997; Theis et al. 1999; Caruthers et al. 2000; Putnam et al. 2001; Caruthers and McKay 2002; Singleton and Wigley 2002; McClelland and Szczelkun 2004). These three motifs were strongly conserved in HsdR and Res (McClelland and Szczelkun 2004). Mutations in the "DEAD box" motifs in HsdR of EcoKI and Res of EcoP1I confirmed their importance in ATPase and endonuclease activity (Gorbalenya and Koonin 1991; Webb et al. 1996; Saha and Rao 1997; Davies et al. 1998, 1999a; Saha et al. 1998; Hall and Matson 1999; Singleton and Wigley 2002, 2003). Mutations in Walker A and B of EcoKI affected ATP binding and ATP hydrolysis, as expected; those in the other motifs provided the first formal proof for their involvement in translocation. Evidence for coupling of ATP hydrolysis to translocation was obtained using purified EcoKI mutant proteins: It could not linearize supercoiled DNA, had negligible ATPase activity in vitro, and also failed to translocate DNA in vivo. These latter results deserve particular mention, because of the novel use of EcoKI-mediated transfer of T7 DNA from the phage head into the cell (Fig. 11; Davies et al. 1999a; Garcia and Molineux 1999). EcoKI could pull the entire T7 chromosome (~39 kb) into *E. coli* at ~100–200 bp/sec (Davies et al. 1999a; Garcia and Molineux 1999), similar to the rates obtained in vitro with EcoKI and EcoR124I (Studier and Bandyopadhyay 1988; Firman and Szczelkun 2000).

Mark Szczelkun compared nearly 200 characterized and putative open reading frames (ORFs) encoding HsdR and Res subunits in Type I and III systems, respectively, which confirmed the conservation of the "X" and "DEAD

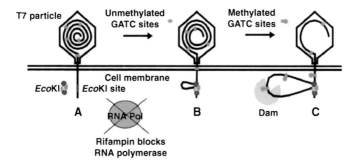

FIGURE 11. In vivo translocation assay of EcoKI. A single target for EcoKI provides the means of bringing the T7 genome into an EcoKI-restricting cell (Davies et al. 1999a). (*A*) Infection of the cell commences with insertion of the first 1000 bp of the T7 genome, which carries one unmodified recognition site for EcoKI. (*B*) Normal entry of T7 DNA is mediated by RNA polymerases, both *E. coli* and T7, which can be blocked by rifampicin and chloramphenicol. DNA translocation by EcoKI bound to its unmodified target site substitutes for RNA polymerase and pulls in the DNA from the phage head. (*C*) *E. coli* Dam methylates GATC sites that enter the cell. The fraction of DNA that has entered the cell can thus be estimated by comparing digests with the methylation-sensitive DpnI and the methylation-insensitive Sau3A. The entire T7 chromosome (~39 kb) can be pulled into the cell by EcoKI and the rate of entry was calculated to be 100–200 bp/sec (Davies et al. 1999a; Garcia and Molineux 1999), a figure similar to those (200–400 bp/sec) obtained from in vitro experiments with EcoKI and EcoR124I (Studier and Bandyopadhyay 1988; Firman and Szczelkun 2000). (Note that for technical reasons a mutant T7 0.3⁻ phage was used, which will be explained elsewhere.) (Reprinted with permission of Microbiology Society from Murray 2002.)

box" motifs (Gorbalenya and Koonin 1991; Murray et al. 1993; Titheradge et al. 1996; Davies et al. 1999b; Janscak et al. 1999b, 2001; McClelland and Szczelkun 2004). In HsdR, "X" was always ahead of the motor, in Res, behind it, in line with the long-held assumption that HsdR cut distant to the recognition site, and Res proximal. The comparison revealed a new putative helicase motif, the Q-tip helix (McClelland and Szczelkun 2004), also found in PcrA, rep, UvrB, RecG, and RuvB, and implicated in the activity of DEAD-box RNA helicases, RuvB, and BLM (Subramanya et al. 1996; Korolev et al. 1997; Bähr et al. 1998; Theis et al. 1999; Velankar et al. 1999; Iwasaki et al. 2000; Putnam et al. 2001; Singleton et al. 2001; Tanner et al. 2003). The importance of the Q-tip in Type I and III REases remained to be established. Outside these regions, homology was low and polypeptides sometimes lacked amino or carboxyl termini. The 39 Type III Res sequences split into two groups, IIIA and IIIB. The IIIA sequences were closely related, whereas the IIIB sequences were less well conserved, and, intriguingly, resembled HsdR more than IIIA (Titheradge et al. 1996; Davies et al. 1998, 1999a). However, both IIIA and IIIB REases cleaved DNA typical of Type III enzymes, and not Type I, perhaps indicative of a gradual transition between Type I, IIIA, and IIIB (McClelland and Szczelkun 2004).

HsdR or Res had no activity on their own, like many other superfamily members (Cooper and Dryden 1994; Dryden et al. 1997, 2001; Sturrock and Dryden 1997; O'Neill et al. 1998; Murray 2000; Rao et al. 2000; Szczelkun 2000; Bourniquel and Bickle 2002; Delagoutte and von Hippel 2003). Were one or two HsdR subunits used or needed? This reflected the debate over whether superfamily members act as monomers or dimers in their respective complexes (Dryden et al. 1997; Janscak et al. 1998; Firman and Szczelkun 2000; Nanduri et al. 2002; Maluf et al. 2003). EcoKI probably could translocate bidirectionally in vivo as well as in vitro, but EcoBI apparently could not, at least not in vitro (Powell et al. 1998b). Were the HsdR subunits acting independently of each other? This was a puzzle, because translocation studies with a single HsdR suggested that when a subunit released DNA during movement, it could bind and translocate the DNA on the opposite side of the complex (Firman and Szczelkun 2000; McClelland and Szczelkun 2004). Did perhaps nonspecific DNA wrap around the complex and stimulate translocation and/or cleavage (Mernagh and Kneale 1996; Szczelkun et al. 1996; McClelland and Szczelkun 2004)? It was too early to say (Szczelkun et al. 1996; Janscak et al. 1999a). Translocation by HsdR might involve contacting nonspecific DNA adjacent to the recognition site in a cleft, which would close and reopen, a process governed by ATP, Mg^{2+}, and probably SAM, to fuel and control the conformational changes. This would be in line with observations of substantial movements and rearrangements of different domains after ATP hydrolysis, which would allow helicase or translocase activity, respectively (Korolev et al. 1997; Hall and Matson 1999; Velankar et al. 1999; Singleton et al. 2001; Singleton and Wigley 2002; Mahdi et al. 2003).

The debate on the choice of the cut site with respect to the recognition site had started in the 1980s with Bill Studier's collision model (Chapter 6; Studier and Bandyopadhyay 1988). Evidence for cooperation between sites was supported by additional in vivo experiments and in vitro by atomic force microscopy (AFM) (Webb et al. 1996; O'Neill et al. 1998; Ellis et al. 1999). This was good news, as AFM, in contrast to the harsh treatment used in the EM, allowed gentle sample preparation via noncovalent attachment of protein–DNA complexes to a mica surface in aqueous solution. These data showed more efficient cleavage of linear DNA with two sites than with one site (Ellis et al. 1999). Two EcoKI complexes apparently bound their respective recognition sites on opposite sides of the target plasmid, dimerized, looped the DNA, and cut it ~7 min after addition of ATP, because of stalling of the complex upon excessive DNA supercoiling or maximal contraction of the DNA loop between the two bound EcoKI molecules. Interestingly, translocase mutants were still capable of dimerization, which might occur between any two occupied (not only adjacent) EcoKI sites, although interaction between adjacent sites was most probable (Ellis et al. 1999; Berge et al. 2000). This led to a variant of the collision model in which two EcoKI complexes could dimerize before translocation (Ellis et al. 1999). Collision with another protein or structure (e.g., a Holliday junction) would also stop translocation and result in DNA cleavage (Studier and Bandyopadhyay 1988; Janscak et al. 1999a; Murray 2000; Dryden et al. 2001).

Many questions remained: Would HsdR touch one strand or both strands of the dsDNA via backbone contacts; and what about the step size or amount of DNA transported per physical step? SF1 helicases stepped anything from 1–2 bp, 3–5 bp, or even 23 bp (Roman and Kowalczykowski 1989; Ali and Lohman 1997; Bianco and Kowalczykowski 2000; Dillingham et al. 2000; Kim et al. 2002). And why would no cleavage occur during initial translocation? Was the translocation rate too high or was the "X" site in the wrong conformation to contact the DNA? Easier to answer was the question about the nature of the DNA ends at the cut site: sticky or blunt end? This proved to depend on the REase: EcoKI (Type IA), EcoAI (Type IB), and EcoR124I (Type IC) cut randomly without preference for particular sequences, with $5'$ and $3'$ overhangs of varying length (Jindrova et al. 2005). The final conclusion was that two REases were needed for DSBs, each one providing one catalytic center for cleavage of one strand (Jindrova et al. 2005).

As mentioned in the Introduction, research on the Type III REases was limited to the enzymes from phage P1 and plasmid P15. It was unclear why EcoP1I and EcoP15I needed a second Mod subunit in the MTase, as methylation occurred on only one strand of the recognition sequence (Humbelin et al. 1988; Ahmad et al. 1995). The Mod subunits dictated specific recognition and methylation and the Res subunits translocation and cutting. In the case of EcoP15I, Res_2Mod_2 (again, why two of each subunit?) acted as MTase in the absence of ATP and as either MTase or REase in the presence of ATP, depending on the methylation state of the recognition site (Janscak et al. 2001). Did the second Mod subunit perhaps stabilize the complex via nonspecific DNA binding, similar to the Type IIS REase BspMI (Gormley et al. 2002)?

Although DNA translocation had been proven unambiguously for Type I enzymes, there was no convincing evidence for a similar mechanism for Type III enzymes (Murray 2000; Rao et al. 2000; Szczelkun 2000; Dryden et al. 2001; Bourniquel and Bickle 2002). Based on the Type I model, cleavage would occur after DNA tracking using ATP as the energy source and collision by two Res_2Mod_2 complexes, which remained attached to their two head-to-head recognition sites via the MTase part of Res_2Mod_2 (Fig. 12; Krüger et al. 1995; Meisel et al. 1995; Saha and Rao 1995; Rao et al. 2000; Dryden

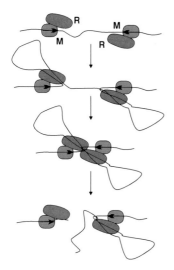

FIGURE 12. DNA tracking collision model for Type III REases (one Mod and Res subunit are shown for clarity [Meisel et al. 1995]). A pair of head-to-head–oriented recognition sites (→) is occupied by one enzyme molecule each. Mod is shown in blue and Res in red. Both enzyme-site complexes use ATP to translocate DNA (Meisel et al. 1995; Saha and Rao 1995), shown by a loop of increasing size, and convergently track until they collide. The resulting collision complex elicits a conformational change and results in cleavage of the DNA 25–27 bp downstream from one of the two recognition sites. This destabilizes the complex, preventing restriction of the second site (Janscak et al. 2001). (Adapted from Meisel et al. 1995, with permission from EMBO.)

et al. 2001; Bourniquel and Bickle 2002). Collision would result in a conformational change and cleavage 25–27 bp downstream from one of the two recognition sites. This would destabilize the complex, preventing restriction of the second site (Janscak et al. 2001). Uncertainty remained as to whether translocation occurred in one or both directions and whether perhaps two Res subunits cooperated to translocate DNA unidirectionally (McClelland and Szczelkun 2004).

Restriction Alleviation of Type I REases by ClpXP

Both early gene transfer experiments and later studies supported the notion that Type I genes easily replaced alleles with different recognition sites or, alternatively, mutant genes encoding nonmodifying proteins (Arber 1965; Arber and Linn 1969; Prakash-Cheng and Ryu 1993; Prakash-Cheng et al. 1993; Naito et al. 1995; Kobayashi 1996, 1998, 2001; O'Neill et al. 1997; Nakayama and Kobayashi 1998; Hurst and Werren 2001). This indicated that the incoming REase did not destroy the host DNA, despite the absence of cognate methylation on the chromosome. What was the mechanism behind this restriction alleviation (RA)? Did it perhaps involve subunit (dis)assembly (Dryden et al. 1997; Janscak et al. 1998)? Or was there perhaps another, general, protective mechanism to protect unmodified sites on the host DNA upon DNA damage, DNA repair, transfer of *hsd* genes, and/or recombination (Bertani and Weigle 1953; Makovets et al. 1998, 1999; Murray 2000, 2002; Doronina and Murray 2001)? Both in vivo and in vitro experiments helped to resolve this important issue. Complementation between lambda *hsdK* phages and host mutants, as well as western blots led to the discovery of intricate posttranslational control of the HsdR subunit of EcoKI by the ATP-dependent ClpXP protease (Makovets et al. 1998, 1999; Doronina and Murray 2001). Surprisingly, use of HsdR and HsdM mutants showed that the HsdR subunit was degraded by ClpXP when modification was impaired but only after assembly of a specific DNA/EcoKI-translocation-proficient complex. In other words, ClpXP would degrade HsdR *during* translocation, but not before (Fig. 13). In line with this, "DEAD box" mutants were resistant to degradation, whereas "X" mutants would be degraded like wild-type enzyme. Western blots showed this ClpXP-dependent degradation to be solely aimed at HsdR, whereas HsdM remained intact.

This extraordinary control of restriction to prevent chromosome degradation, if modification became even temporarily insufficient, was also active against EcoAI but not EcoR124I (Makovets et al. 1998, 1999; Doronina and Murray 2001). Although foreign (phage) DNA was destroyed by a reconstituted restriction-proficient methyltransfer-deficient EcoKI complex, all (600-odd) chromosomal target sequences remained unharmed, which led to the concept of "self" and "non-self" (Murray 2002). This was in contrast with Type II REases, which did cut the host DNA when modification of the host DNA became inadequate. These REases maintained themselves by cleaving non-self DNA, and therefore Ichizo Kobayashi called these enzymes "selfish" (Kobayashi 2001). The experiments with EcoKI and other Type I systems by Noreen Murray and colleagues clearly indicated otherwise (O'Neill

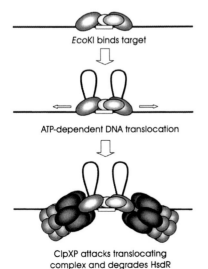

EcoKI binds target

ATP-dependent DNA translocation

ClpXP attacks translocating
complex and degrades HsdR

FIGURE 13. Model for the mechanism of ClpXP-dependent proteolytic control of restriction by EcoKI. ATP-dependent translocation of DNA by EcoKI occurs after the enzyme binds unmodified recognition sequences on the host chromosome. However, ClpXP recognizes HsdR during translocation and destroys the restriction subunits of the EcoKI complex (but leaves the trimeric methylase intact), thereby preventing further translocation and cutting of the chromosome. (Reprinted from Murray 2000, with permission from American Society for Microbiology; see also Loenen 2003.)

et al. 1997; Murray 2002). The "pro selfish" camp stated that the REase would not only destroy incoming foreign DNA but also enhance the frequency of horizontal transfer of R-M systems into other genomes by generating recombinogenic free DNA ends in the cell (Jeltsch and Pingoud 1996; Kobayashi 2001), somewhat similar to homing endonucleases (Gimble 2000). The proposed role for selfish and non-selfish REases need not be contradictive. DNA translocation of R-M systems could aid incorporation of unmodified DNA into the recipient chromosome after conjugation or transduction of large chunks of incoming DNA. Random cleavage by Type I enzymes into smaller fragments might generate suitable DNA substrates for the host's major recombination complex RecBC at Chi sites, and benefit populations as well as individual cells (Price and Bickle 1986; Barcus and Murray 1995; Murray 2002). In other words, although not essential, R-M systems might influence the stability of chromosomes at the population level, which must be neither too static nor too fluid, allowing influx of foreign DNA to enhance survival of the population when conditions change, leading to mosaic sequences and slow evolution toward new species (Price and Bickle 1986; Wilkins 2000; Murray 2002; Arber 2003). The role of translocation by Type I enzymes in this level of maintenance of chromosome integrity and evolution and dissection of RA pathways by bacterial hosts and their enemies would await further exploration (Murray 2002).

REFERENCES

The Seventh NEB Meeting on DNA Restriction and Modification August 24–29, 2015. Uniwersytet Gdański ulica, Kladki 24, Gdańsk, Poland.

Adzuma K, Mizuuchi K. 1989. Interaction of proteins located at a distance along DNA: mechanism of target immunity in the Mu DNA strand-transfer reaction. *Cell* **57**: 41–47.

Aggarwal AK. 1995. Structure and function of restriction endonucleases. *Curr Opin Struct Biol* **5**: 11–19.

Ahmad I, Krishnamurthy V, Rao DN. 1995. DNA recognition by the *Eco*P15I and *Eco*PI modification methyltransferases. *Gene* **157**: 143–147.

Ali JA, Lohman TM. 1997. Kinetic measurement of the step size of DNA unwinding by *Escherichia coli* UvrD helicase. *Science* **275**: 377–380.

Alves J, Vennekohl P. 2004. Protein engineering of restriction enzymes. In *Restriction endonucleases* (ed. Pingoud A), pp. 393–411. Springer, Berlin.

Alves J, Rüter T, Geiger R, Fliess A, Maass G, Pingoud A. 1989. Changing the hydrogen-bonding potential in the DNA binding site of EcoRI by site-directed mutagenesis drastically reduces the enzymatic activity, not, however, the preference of this restriction endonuclease for cleavage within the site-GAATTC. *Biochemistry* **28**: 2678–2684.

Anderson JE. 1993. Restriction endonucleases and modification methylases. *Curr Opin Struct Biol* **3**: 24–30.

Anton BP, Heiter DF, Benner JS, Hess EJ, Greenough L, Moran LS, Slatko BE, Brooks JE. 1997. Cloning and characterization of the *Bgl*II restriction-modification system reveals a possible evolutionary footprint. *Gene* **187**: 19–27.

Aravind L, Makarova KS, Koonin EV. 2000. SURVEY AND SUMMARY: Holliday junction resolvases and related nucleases: identification of new families, phyletic distribution and evolutionary trajectories. *Nucleic Acids Res* **28**: 3417–3432.

Aravind L, Walker DR, Koonin EV. 1999. Conserved domains in DNA repair proteins and evolution of repair systems. *Nucleic Acids Res* **27**: 1223–1242.

Arber W. 1965. Host-controlled modification of bacteriophage. *Ann Rev Microbiol* **19**: 365–378.

Arber W. 1966. Host specificity of DNA produced by *Escherichia coli*. 9. Host-controlled modification of bacteriophage fd. *J Mol Biol* **20**: 483–496.

Arber W. 1979. Promotion and limitation of genetic exchange. *Science* **205**: 361–365.

Arber W. 2000. Genetic variation: molecular mechanisms and impact on microbial evolution. *FEMS Microbiol Rev* **24**: 1–7.

Arber W. 2003. Elements for a theory of molecular evolution. *Gene* **317**: 3–11.

Arber W, Linn S. 1969. DNA modification and restriction. *Ann Rev Biochem* **38**: 467–500.

Arber W, Morse ML. 1965. Host specificity of DNA produced by *Escherichia coli*. VI. Effects on bacterial conjugation. *Genetics* **51**: 137–148.

Athanasiadis A, Vlassi M, Kotsifaki D, Tucker PA, Wilson KS, Kokkinidis M. 1994. Crystal structure of PvuII endonuclease reveals extensive structural homologies to EcoRV. *Nat Struct Biol* **1**: 469–475.

Bähr A, De Graeve F, Kedinger C, Chatton B. 1998. Point mutations causing Bloom's syndrome abolish ATPase and DNA helicase activities of the BLM protein. *Oncogene* **17**: 2565–2571.

Baldwin GS, Vipond IB, Halford SE. 1995. Rapid reaction analysis of the catalytic cycle of the *Eco*RV restriction endonuclease. *Biochemistry* **34**: 705–714.

Ban C, Yang W. 1998. Structural basis for MutH activation in *E. coli* mismatch repair and relationship of MutH to restriction endonucleases. *EMBO J* **17**: 1526–1534.

Bannister D, Glover SW. 1968. Restriction and modification of bacteriophages by R⁺ strains of *Escherichia coli* K12. *Biochem Biophys Res Commun* **30**: 735–738.

Bannister D, Glover SW. 1970. The isolation and properties of non-restricting mutants of two different host specificities associated with drug resistance factors. *J Gen Microbiol* **61**: 63–71.

Barcus VA, Murray NE. 1995. Barriers to recombination: restriction. *Population genetics of bacteria* (ed. Baumberg S, Young JPW, Saunders SR, Wellington EMH), pp. 31–58. Society for General Microbiology, Cambridge University Press, Cambridge.

Bath AJ, Milsom SE, Gormley NA, Halford SE. 2002. Many type IIs restriction endonucleases interact with two recognition sites before cleaving DNA. *J Biol Chem* **277**: 4024–4033.

Baxter BK, Topal MD. 1993. Formation of a cleavasome: enhancer DNA-2 stabilizes an active conformation of *Nae*I dimer. *Biochemistry* **32**: 8291–8298.

Beerli RR, Barbas CF. 2002. Engineering polydactyl zinc-finger transcription factors. *Nat Biotech* **20**: 135–141.

Beese LS, Steitz TA. 1991. Structural basis for the 3′–5′ exonuclease activity of *Escherichia coli* DNA polymerase I: a two metal ion mechanism. *EMBO J* **10**: 25–33.

Bennett SP, Halford SE. 1989. Recognition of DNA by type II restriction enzymes. *Curr Top Cell Regul* **30**: 57–104.

Berge T, Ellis DJ, Dryden DT, Edwardson JM, Henderson RM. 2000. Translocation-independent dimerization of the *Eco*KI endonuclease visualized by atomic force microscopy. *Biophys J* **79**: 479–484.

Bertani G, Weigle JJ. 1953. Host controlled variation in bacterial viruses. *J Bacteriol* **65**: 113–121.

Bianco PR, Kowalczykowski SC. 2000. Translocation step size and mechanism of the RecBC DNA helicase. *Nature* **405**: 368–372.

Bibikova M, Carroll D, Segal DJ, Trautman JK, Smith J, Kim YG, Chandrasegaran S. 2001. Stimulation of homologous recombination through targeted cleavage by chimeric nucleases. *Mol Cell Biol* **21:** 289–297.

Bibikova M, Golic M, Golic KG, Carroll D. 2002. Targeted chromosomal cleavage and mutagenesis in *Drosophila* using zinc-finger nucleases. *Genetics* **161:** 1169–1175.

Bickle TA. 1993. The ATP-dependent restriction enzymes. In *Nucleases*, 2nd ed. (ed. Linn SM, Lloyd SR, Roberts RJ), pp. 88–109. Cold Spring Harbor Laboratory Press, Cold Spring Harbor, NY.

Bickle TA, Krüger DH. 1993. Biology of DNA restriction. *Microbiol Rev* **57:** 434–450.

Bigger CH, Murray K, Murray NE. 1973. Recognition sequence of a restriction enzyme. *Nature: New Biol* **244:** 7–10.

Bilcock DT, Halford SE. 1999. DNA restriction dependent on two recognition sites: activities of the *Sfi*I restriction-modification system in *Escherichia coli*. *Mol Microbiol* **31:** 1243–1254.

Bilcock DT, Daniels LE, Bath AJ, Halford SE. 1999. Reactions of type II restriction endonucleases with 8-base pair recognition sites. *J Biol Chem* **274:** 36379–36386.

Bitinaite J, Schildkraut I. 2002. Self-generated DNA termini relax the specificity of *Sgr*AI restriction endonuclease. *Proc Natl Acad Sci* **99:** 1164–1169.

Bitinaite J, Wah DA, Aggarwal AK, Schildkraut I. 1998. *Fok*I dimerization is required for DNA cleavage. *Proc Natl Acad Sci* **95:** 10570–10575.

Bitinaite J, Mitkaite G, Dauksaite V, Jakubauskas A, Timinskas A, Vaisvila R, Lubys A, Janulaitis A. 2002. Evolutionary relationship of *Alw*26I, *Eco*31I and *Esp*3I, restriction endonucleases that recognise overlapping sequences. *Mol Genet Genomics* **267:** 664–672.

Black CB, Cowan JA. 1994. *Inorg Chem* **33:** 5805–5808.

Blattler MO, Wenz C., Pingoud A, Benner SA. 1998. Distorting duplex DNA by dimethylenesulfone substitution: a new class of 'transition state analog' inhibitors of restriction enzymes. *J Am Chem Soc* **120:** 2674–2675.

Blumenthal RM, Cheng X. 2001. A Taq attack displaces bases. *Nat Struct Biol* **8:** 101–103.

Bochtler M, Szczepanowski RH, Tamulaitis G, Grazulis S, Czapinska H, Manakova E, Siksnys V. 2006. Nucleotide flips determine the specificity of the Ecl18kI restriction endonuclease. *EMBO J* **25:** 2219–2229.

Bourniquel AA, Bickle TA. 2002. Complex restriction enzymes: NTP-driven molecular motors. *Biochimie* **84:** 1047–1059.

Boyer HW, Chow LT, Dugaiczyk A, Hedgpeth J, Goodman HM. 1973. DNA substrate site for the EcoRII restriction endonuclease and modification methylase. *Nature: New Biol* **244:** 40–43.

Bozic D, Grazulis S, Siksnys V, Huber R. 1996. Crystal structure of *Citrobacter freundii* restriction endonuclease Cfr10I at 2.15 A resolution. *J Mol Biol* **255:** 176–186.

Brooks JE, Nathan PD, Landry D, Sznyter LA, Waite-Rees P, Ives CL, Moran LS, Slatko BE, Benner JS. 1991. Characterization of the cloned *Bam*HI restriction modification system: its nucleotide sequence, properties of the methylase, and expression in heterologous hosts. *Nucleic Acids Res* **19:** 841–850.

Bujnicki JM. 2000a. Homology modelling of the DNA 5mC methyltransferase M.BssHII. Is permutation of functional subdomains common to all subfamilies of DNA methyltransferases? *Int J Biol Macromol* **27:** 195–204.

Bujnicki JM. 2000b. Phylogeny of the restriction endonuclease-like superfamily inferred from comparison of protein structures. *J Mol Evol* **50:** 39–44.

Bujnicki JM. 2001a. A model of structure and action of Sau3AI restriction endonuclease that comprises two MutH-like endonuclease domains within a single polypeptide. *Acta Microbiol Pol* **50:** 219–231.

Bujnicki JM. 2001b. Understanding the evolution of restriction-modification systems: clues from sequence and structure comparisons. *Acta Biochim Pol* **48**: 935–967.

Bujnicki JM. 2003. Crystallographic and bioinformatic studies on restriction endonucleases: inference of evolutionary relationships in the "midnight zone" of homology. *Curr Protein Pept Sci* **4**: 327–337.

Bujnicki JM. 2004. Molecular phylogenetics of restriction endonucleases. In *Restriction endonucleases* (ed. Pingoud A), pp. 63–93. Springer, Berlin.

Bujnicki JM, Rychlewski L. 2001a. Grouping together highly diverged PD-(D/E)XK nucleases and identification of novel superfamily members using structure-guided alignment of sequence profiles. *J Mol Microbiol Biotechnol* **3**: 69–72.

Bujnicki JM, Rychlewski L. 2001b. The herpesvirus alkaline exonuclease belongs to the restriction endonuclease PD-(D/E)XK superfamily: insight from molecular modeling and phylogenetic analysis. *Virus Genes* **22**: 219–230.

Bujnicki JM, Radlinska M, Rychlewski L. 2001. Polyphyletic evolution of type II restriction enzymes revisited: two independent sources of second-hand folds revealed. *Trends Biochem Sci* **26**: 9–11.

Carlson K, Kosturko LD. 1998. Endonuclease II of coliphage T4: a recombinase disguised as a restriction endonuclease? *Mol Microbiol* **27**: 671–676.

Carrick KL, Topal MD. 2003. Amino acid substitutions at position 43 of *Nae*I endonuclease. Evidence for changes in *Nae*I structure. *J Biol Chem* **278**: 9733–9739.

Caruthers JM, McKay DB. 2002. Helicase structure and mechanism. *Curr Opin Struct Biol* **12**: 123–133.

Caruthers JM, Johnson ER, McKay DB. 2000. Crystal structure of yeast initiation factor 4A, a DEAD-box RNA helicase. *Proc Natl Acad Sci* **97**: 13080–13085.

Cech D, Pein CD, Kubareva EA, Gromova ES, Oretskaya ES, Shabarova ZA. 1988. Influence of modifications on the cleavage of oligonucleotide duplexes by EcoRII and MvaI endonucleases. *Nucleosides Nucleotides* **7**: 585–588.

Chang HW, Julin DA. 2001. Structure and function of the *Escherichia coli* RecE protein, a member of the RecB nuclease domain family. *J Biol Chem* **276**: 46004–46010.

Chen A, Powell LM, Dryden DT, Murray NE, Brown T. 1995. Tyrosine 27 of the specificity polypeptide of EcoKI can be UV crosslinked to a bromodeoxyuridine-substituted DNA target sequence. *Nucleic Acids Res* **23**: 1177–1183.

Cheng X, Blumenthal RM. 2002. Cytosines do it, thymines do it, even pseudouridines do it—base flipping by an enzyme that acts on RNA. *Structure* **10**: 127–129.

Cheng X, Roberts RJ. 2001. AdoMet-dependent methylation, DNA methyltransferases and base flipping. *Nucleic Acids Res* **29**: 3784–3795.

Cheng X, Kumar S, Klimasauskas S, Roberts RJ. 1993a. Crystal structure of the HhaI DNA methyltransferase. *Cold Spring Harbor Symp Quant Biol* **58**: 331–338.

Cheng X, Kumar S, Posfai J, Pflugrath JW, Roberts RJ. 1993b. Crystal structure of the HhaI DNA methyltransferase complexed with *S*-adenosyl-L-methionine. *Cell* **74**: 299–307.

Cheng X, Balendiran K, Schildkraut I, Anderson JE. 1994. Structure of *Pvu*II endonuclease with cognate DNA. *EMBO J* **13**: 3927–3935.

Chevalier BS, Monnat RJ Jr, Stoddard BL. 2001. The homing endonuclease I-*Cre*I uses three metals, one of which is shared between the two active sites. *Nat Struct Biol* **8**: 312–316.

Chinen A, Uchiyama I, Kobayashi I. 2000. Comparison between *Pyrococcus horikoshii* and *Pyrococcus abyssi* genome sequences reveals linkage of restriction-modification genes with large genome polymorphisms. *Gene* **259**: 109–121.

Choi J. 1994. "Crystal structure analysis of site-directed mutants of EcoRI endonuclease complexed to DNA." PhD thesis, University of Pittsburgh.

Colandene JD, Topal MD. 1998. The domain organization of *Nae*I endonuclease: separation of binding and catalysis. *Proc Natl Acad Sci* **95**: 3531–3536.

Conlan LH, Dupureur CM. 2002a. Dissecting the metal ion dependence of DNA binding by *Pvu*II endonuclease. *Biochemistry* **41**: 1335–1342.

Conlan LH, Dupureur CM. 2002b. Multiple metal ions drive DNA association by *Pvu*II endonuclease. *Biochemistry* **41**: 14848–14855.

Conlan LH, Jose TJ, Thornton KC, Dupureur CM. 1999. Modulating restriction endonuclease activities and specificities using neutral detergents. *BioTechniques* **27**: 955–960.

Connolly BA, Eckstein F, Pingoud A. 1984. The stereochemical course of the restriction endonuclease *Eco*RI-catalyzed reaction. *J Biol Chem* **259**: 10760–10763.

Connolly BA, Liu HH, Parry D, Engler LE, Kurpiewski MR, Jen-Jacobson L. 2001. Assay of restriction endonucleases using oligonucleotides. *Methods Mol Biol* **148**: 465–490.

Conrad M, Topal MD. 1989. DNA and spermidine provide a switch mechanism to regulate the activity of restriction enzyme *Nae*I. *Proc Natl Acad Sci* **86**: 9707–9711.

Conrad M, Topal MD. 1992. Modified DNA fragments activate *Nae*I cleavage of refractory DNA sites. *Nucleic Acids Res* **20**: 5127–5130.

Cooper LP, Dryden DT. 1994. The domains of a type I DNA methyltransferase. Interactions and role in recognition of DNA methylation. *J Mol Biol* **236**: 1011–1021.

Cowan JA. 1998. Metal activation of enzymes in nucleic acid biochemistry. *Chem Rev* **98**: 1067–1088.

Cowan JA. 2004. Role of metal ions in promoting DNA binding and cleavage by restriction endonucleases. In *Restriction endonucleases* (ed. Pingoud A), pp. 339–360. Springer, Berlin.

Davies GP, Powell LM, Webb JL, Cooper LP, Murray NE. 1998. *Eco*KI with an amino acid substitution in any one of seven DEAD-box motifs has impaired ATPase and endonuclease activities. *Nucleic Acids Res* **26**: 4828–4836.

Davies GP, Kemp P, Molineux IJ, Murray NE. 1999a. The DNA translocation and ATPase activities of restriction-deficient mutants of *Eco*KI. *J Mol Biol* **292**: 787–796.

Davies GP, Martin I, Sturrock SS, Cronshaw A, Murray NE, Dryden DT. 1999b. On the structure and operation of type I DNA restriction enzymes. *J Mol Biol* **290**: 565–579.

Deibert M, Grazulis S, Janulaitis A, Siksnys V, Huber R. 1999. Crystal structure of *Mun*I restriction endonuclease in complex with cognate DNA at 1.7 Å resolution. *EMBO J* **18**: 5805–5816.

Deibert M, Grazulis S, Sasnauskas G, Siksnys V, Huber R. 2000. Structure of the tetrameric restriction endonuclease NgoMIV in complex with cleaved DNA. *Nat Struct Biol* **7**: 792–799.

Delagoutte E, von Hippel PH. 2003. Helicase mechanisms and the coupling of helicases within macromolecular machines. Part II: integration of helicases into cellular processes. *Q Rev Biophys* **36**: 1–69.

Demple B, Johnson A, Fung D. 1986. Exonuclease III and endonuclease IV remove 3' blocks from DNA synthesis primers in H_2O_2-damaged *Escherichia coli*. *Proc Natl Acad Sci* **83**: 7731–7735.

Dillingham MS, Wigley DB, Webb MR. 2000. Demonstration of unidirectional single-stranded DNA translocation by PcrA helicase: measurement of step size and translocation speed. *Biochemistry* **39**: 205–212.

Dominguez MA Jr, Thornton KC, Melendez MG, Dupureur CM. 2001. Differential effects of isomeric incorporation of fluorophenylalanines into PvuII endonuclease. *Proteins* **45:** 55–61.

Dorner LF, Schildkraut I. 1994. Direct selection of binding proficient/catalytic deficient variants of *Bam*HI endonuclease. *Nucleic Acids Res* **22:** 1068–1074.

Doronina VA, Murray NE. 2001. The proteolytic control of restriction activity in *Escherichia coli* K-12. *Mol Microbiol* **39:** 416–428.

Dryden DT. 2013. The architecture of restriction enzymes. *Structure* **21:** 1720–1721.

Dryden DT, Cooper LP, Murray NE. 1993. Purification and characterization of the methyltransferase from the type 1 restriction and modification system of *Escherichia coli* K12. *J Biol Chem* **268:** 13228–13236.

Dryden DT, Sturrock SS, Winter M. 1995. Structural modelling of a type I DNA methyltransferase. *Nat Struct Biol* **2:** 632–635.

Dryden DT, Cooper LP, Thorpe PH, Byron O. 1997. The *in vitro* assembly of the *Eco*KI type I DNA restriction/modification enzyme and its *in vivo* implications. *Biochemistry* **36:** 1065–1076.

Dryden DT, Murray NE, Rao DN. 2001. Nucleoside triphosphate-dependent restriction enzymes. *Nucleic Acids Res* **29:** 3728–3741.

Dunn TM, Hahn S, Ogden S, Schleif RF. 1984. An operator at –280 base pairs that is required for repression of araBAD operon promoter: addition of DNA helical turns between the operator and promoter cyclically hinders repression. *Proc Natl Acad Sci* **81:** 5017–5020.

Dupureur CM, Conlan LH. 2000. A catalytically deficient active site variant of PvuII endonuclease binds Mg(II) ions. *Biochemistry* **39:** 10921–10927.

Dupureur CM, Dominguez MA Jr. 2001. The PD···(D/E)XK motif in restriction enzymes: a link between function and conformation. *Biochemistry* **40:** 387–394.

Dupureur CM, Hallman LM. 1999. Effects of divalent metal ions on the activity and conformation of native and 3-fluorotyrosine-*Pvu*II endonucleases. *Eur J Biochem/FEBS* **261:** 261–268.

Durai S, Mani M, Kandavelou K, Wu J, Porteus MH, Chandrasegaran S. 2005. Zinc finger nucleases: custom-designed molecular scissors for genome engineering of plant and mammalian cells. *Nucleic Acids Res* **33:** 5978–5990.

Ellis DJ, Dryden DT, Berge T, Edwardson JM, Henderson RM. 1999. Direct observation of DNA translocation and cleavage by the *Eco*KI endonuclease using atomic force microscopy. *Nat Struct Biol* **6:** 15–17.

Embleton ML, Siksnys V, Halford SE. 2001. DNA cleavage reactions by type II restriction enzymes that require two copies of their recognition sites. *J Mol Biol* **311:** 503–514.

Engler LE. 1998. "Specificity determinants in the BamHI-endonuclease-DNA interaction." PhD thesis, University of Pittsburgh.

Engler LE, Welch KK, Jen-Jacobson L. 1997. Specific binding by *Eco*RV endonuclease to its DNA recognition site GATATC. *J Mol Biol* **269:** 82–101.

Engler LE, Sapienza P, Dorner LF, Kucera R, Schildkraut I, Jen-Jacobson L. 2001. The energetics of the interaction of *Bam*HI endonuclease with its recognition site GGATCC. *J Mol Biol* **307:** 619–636.

Erskine SG, Halford SE. 1998. Reactions of the eco RV restriction endonuclease with fluorescent oligodeoxynucleotides: identical equilibrium constants for binding to specific and non-specific DNA. *J Mol Biol* **275:** 759–772.

Firman K, Szczelkun MD. 2000. Measuring motion on DNA by the type I restriction endonuclease *Eco*R124I using triplex displacement. *EMBO J* **19:** 2094–2102.

Fliess A, Wolfes H, Seela F, Pingoud A. 1988. Analysis of the recognition mechanism involved in the EcoRV catalyzed cleavage of DNA using modified oligodeoxynucleotides. *Nucleic Acids Res* **16:** 11781–11793.

Flores H, Osuna J, Heitman J, Soberon X. 1995. Saturation mutagenesis of His[114] of *Eco*RI reveals relaxed-specificity mutants. *Gene* **157:** 295–301.

Friedhoff P, Kolmes B, Gimadutdinow O, Wende W, Krause KL, Pingoud A. 1996. Analysis of the mechanism of the *Serratia* nuclease using site-directed mutagenesis. *Nucleic Acids Res* **24:** 2632–2639.

Friedhoff P, Lurz R, Luder G, Pingoud A. 2001. Sau3AI, a monomeric type II restriction endonuclease that dimerizes on the DNA and thereby induces DNA loops. *J Biol Chem* **276:** 23581–23588.

Fritz A, Kuster W, Alves J. 1998. Asn[141] is essential for DNA recognition by *Eco*RI restriction endonuclease. *FEBS Lett* **438:** 66–70.

Fuxreiter M, Osman R. 2001. Probing the general base catalysis in the first step of BamHI action by computer simulations. *Biochemistry* **40:** 15017–15023.

Gabbara S, Bhagwat AS. 1992. Interaction of *Eco*RII endonuclease with DNA substrates containing single recognition sites. *J Biol Chem* **267:** 18623–18630.

Garcia LR, Molineux IJ. 1999. Translocation and specific cleavage of bacteriophage T7 DNA in vivo by *Eco*KI. *Proc Natl Acad Sci* **96:** 12430–12435.

Garcia RA, Bustamante CJ, Reich NO. 1996. Sequence-specific recognition of cytosine C5 and adenine N6 DNA methyltransferases requires different deformations of DNA. *Proc Natl Acad Sci* **93:** 7618–7622.

Geiger R, Rüter T, Alves J, Fliess A, Wolfes H, Pingoud V, Urbanke C, Maass G, Pingoud A, Dusterhoft A, et al. 1989. Genetic engineering of EcoRI mutants with altered amino acid residues in the DNA binding site: physicochemical investigations give evidence for an altered monomer/dimer equilibrium for the Gln144Lys145 and Gln144Lys145Lys200 mutants. *Biochemistry* **28:** 2667–2677.

Ghosh A, Passaris I, Tesfazgi Mebrhatu M, Rocha S, Vanoirbeek K, Hofkens J, Aertsen A. 2014. Cellular localization and dynamics of the Mrr type IV restriction endonuclease of *Escherichia coli*. *Nucleic Acids Res* **42:** 3908–3918.

Gimble FS. 2000. Invasion of a multitude of genetic niches by mobile endonuclease genes. *FEMS Microbiol Lett* **185:** 99–107.

Gorbalenya AE, Koonin EV. 1991. Endonuclease (R) subunits of type-I and type-III restriction-modification enzymes contain a helicase-like domain. *FEBS Lett* **291:** 277–281.

Gormley NA, Hillberg AL, Halford SE. 2002. The type IIs restriction endonuclease *Bsp*MI is a tetramer that acts concertedly at two copies of an asymmetric DNA sequence. *J Biol Chem* **277:** 4034–4041.

Grable J, Frederick CA, Samudzi C, Jen-Jacobson L, Lesser D, Greene P, Boyer HW, Itakura K, Rosenberg JM. 1984. Two-fold symmetry of crystalline DNA-EcoRI endonuclease recognition complexes. *J Biomol Struct Dyn* **1:** 1149–1160.

Grabowski G, Jeltsch A, Wolfes H, Maass G, Alves J. 1995. Site-directed mutagenesis in the catalytic center of the restriction endonuclease EcoRI. *Gene* **157:** 113–118.

Grasby JA, Connolly BA. 1992. Stereochemical outcome of the hydrolysis reaction catalyzed by the *Eco*RV restriction endonuclease. *Biochemistry* **31:** 7855–7861.

Grazulis S, Deibert M, Rimseliene R, Skirgaila R, Sasnauskas G, Lagunavicius A, Repin V, Urbanke C, Huber R, Siksnys V. 2002. Crystal structure of the *Bse*634I restriction endonuclease: comparison of two enzymes recognizing the same DNA sequence. *Nucleic Acids Res* **30:** 876–885.

Grigorescu A, Horvath M, Wilkosz PA, Chandrasekhar K, Rosenberg JM. 2004. The integration of recognition and cleavage: x-ray structures of pre-transition state complex, post-reactive complex, and the DNA-free endonuclease. In *Restriction endonucleases* (ed. Pingoud A), pp. 137–177. Springer, Berlin, Heidelberg.

Hager PW, Reich NO, Day JP, Coche TG, Boyer HW, Rosenberg JM, Greene PJ. 1990. Probing the role of glutamic acid 144 in the EcoRI endonuclease using aspartic acid and glutamine replacements. *J Biol Chem* **265**: 21520–21526.

Halford SE. 2001. Hopping, jumping and looping by restriction enzymes. *Biochem Soc Trans* **29**: 363–374.

Halford SE, Marko JF. 2004. How do site-specific DNA-binding proteins find their targets? *Nucleic Acids Res* **32**: 3040–3052.

Halford SE, Bilcock DT, Stanford NP, Williams SA, Milsom SE, Gormley NA, Watson MA, Bath AJ, Embleton ML, Gowers DM, et al. 1999. Restriction endonuclease reactions requiring two recognition sites. *Biochem Soc Trans* **27**: 696–699.

Hall MC, Matson SW. 1999. Helicase motifs: the engine that powers DNA unwinding. *Mol Microbiol* **34**: 867–877.

Hashimoto H, Horton JR, Zhang X, Bostick M, Jacobsen SE, Cheng X. 2008. The SRA domain of UHRF1 flips 5-methylcytosine out of the DNA helix. *Nature* **455**: 826–829.

Hattman S. 1973. Plasmid-controlled variation in the content of methylated bases in single-stranded DNA bacteriophages M13 and fd. *J Mol Biol* **74**: 749–752.

Hedgpeth J, Goodman HM, Boyer HW. 1972. DNA nucleotide sequence restricted by the RI endonuclease. *Proc Natl Acad Sci* **69**: 3448–3452.

Heitman J. 1993. On the origins, structures and functions of restriction-modification enzymes. *Genet Eng* **15**: 57–108.

Heitman J, Model P. 1990. Mutants of the *Eco*RI endonuclease with promiscuous substrate specificity implicate residues involved in substrate recognition. *EMBO J* **9**: 3369–3378.

Hickman AB, Li Y, Mathew SV, May EW, Craig NL, Dyda F. 2000. Unexpected structural diversity in DNA recombination: the restriction endonuclease connection. *Mol Cell* **5**: 1025–1034.

Horton NC, Perona JJ. 1998. Role of protein-induced bending in the specificity of DNA recognition: crystal structure of EcoRV endonuclease complexed with d(AAAGAT)+d(ATCTT). *J Mol Biol* **277**: 779–787.

Horton NC, Perona JJ. 2000. Crystallographic snapshots along a protein-induced DNA-bending pathway. *Proc Natl Acad Sci* **97**: 5729–5734.

Horton JR, Nastri HG, Riggs PD, Cheng X. 1998a. Asp34 of *Pvu*II endonuclease is directly involved in DNA minor groove recognition and indirectly involved in catalysis. *J Mol Biol* **284**: 1491–1504.

Horton NC, Newberry KJ, Perona JJ. 1998b. Metal ion-mediated substrate-assisted catalysis in type II restriction endonucleases. *Proc Natl Acad Sci* **95**: 13489–13494.

Horton NC, Dorner LF, Perona JJ. 2002. Sequence selectivity and degeneracy of a restriction endonuclease mediated by DNA intercalation. *Nat Struct Biol* **9**: 42–47.

Horton JR, Blumenthal RM, Cheng X. 2004a. Restriction endonucleases: structure of the conserved catalytic core and the role of metal ions in DNA cleavage. In *Restriction endonucleases* (ed. Pingoud A), pp. 361–392. Springer, Berlin.

Horton JR, Ratner G, Banavali NK, Huang N, Choi Y, Maier MA, Marquez VE, MacKerell AD Jr, Cheng X. 2004b. Caught in the act: visualization of an intermediate in the DNA base-flipping pathway induced by HhaI methyltransferase. *Nucleic Acids Res* **32**: 3877–3886.

Horton JR, Zhang X, Maunus R, Yang Z, Wilson GG, Roberts RJ, Cheng X. 2006. DNA nicking by HinP1I endonuclease: bending, base flipping and minor groove expansion. *Nucleic Acids Res* **34:** 939–948.

Horton JR, Wang H, Mabuchi MY, Zhang X, Roberts RJ, Zheng Y, Wilson GG, Cheng X. 2014. Modification-dependent restriction endonuclease, MspJI, flips 5-methylcytosine out of the DNA helix. *Nucleic Acids Res* **42:** 12092–12101.

Hsieh PC, Xiao JP, O'Loane D, Xu SY. 2000. Cloning, expression, and purification of a thermostable nonhomodimeric restriction enzyme, BslI. *J Bacteriol* **182:** 949–955.

Huai Q, Colandene JD, Chen Y, Luo F, Zhao Y, Topal MD, Ke H. 2000. Crystal structure of NaeI—an evolutionary bridge between DNA endonuclease and topoisomerase. *EMBO J* **19:** 3110–3118.

Huai Q, Colandene JD, Topal MD, Ke H. 2001. Structure of NaeI-DNA complex reveals dual-mode DNA recognition and complete dimer rearrangement. *Nat Struct Biol* **8:** 665–669.

Humbelin M, Suri B, Rao DN, Hornby DP, Eberle H, Pripfl T, Kenel S, Bickle TA. 1988. Type III DNA restriction and modification systems *Eco*P1 and *Eco*P15. Nucleotide sequence of the *Eco*P1 operon, the *Eco*P15 *mod* gene and some *Eco*P1 *mod* mutants. *J Mol Biol* **200:** 23–29.

Hurst GD, Werren JH. 2001. The role of selfish genetic elements in eukaryotic evolution. *Nat Rev Genet* **2:** 597–606.

Ivanenko T, Heitman J, Kiss A. 1998. Mutational analysis of the function of Met137 and Ile197, two amino acids implicated in sequence-specific DNA recognition by the *Eco*RI endonuclease. *Biol Chem* **379:** 459–465.

Iwasaki H, Han YW, Okamoto T, Ohnishi T, Yoshikawa M, Yamada K, Toh H, Daiyasu H, Ogura T, Shinagawa H. 2000. Mutational analysis of the functional motifs of RuvB, an AAA⁺ class helicase and motor protein for holliday junction branch migration. *Mol Microbiol* **36:** 528–538.

Janscak P, Dryden DT, Firman K. 1998. Analysis of the subunit assembly of the typeIC restriction-modification enzyme *Eco*R124I. *Nucleic Acids Res* **26:** 4439–4445.

Janscak P, MacWilliams MP, Sandmeier U, Nagaraja V, Bickle TA. 1999a. DNA translocation blockage, a general mechanism of cleavage site selection by type I restriction enzymes. *EMBO J* **18:** 2638–2647.

Janscak P, Sandmeier U, Bickle TA. 1999b. Single amino acid substitutions in the HsdR subunit of the type IB restriction enzyme EcoAI uncouple the DNA translocation and DNA cleavage activities of the enzyme. *Nucleic Acids Res* **27:** 2638–2643.

Janscak P, Sandmeier U, Szczelkun MD, Bickle TA. 2001. Subunit assembly and mode of DNA cleavage of the type III restriction endonucleases *Eco*P1I and *Eco*P15I. *J Mol Biol* **306:** 417–431.

Janulaitis A, Stakenas P, Berlin Yu A. 1983. A new site-specific endodeoxyribonuclease from *Citrobacter freundii*. *FEBS Lett* **161:** 210–212.

Janulaitis A, Vaisvila R, Timinskas A, Klimasauskas S, Butkus V. 1992. Cloning and sequence analysis of the genes coding for *Eco*57I type IV restriction-modification enzymes. *Nucleic Acids Res* **20:** 6051–6056.

Jeltsch A. 1999. Circular permutations in the molecular evolution of DNA methyltransferases. *J Mol Evol* **49:** 161–164.

Jeltsch A, Pingoud A. 1996. Horizontal gene transfer contributes to the wide distribution and evolution of type II restriction-modification systems. *J Mol Evol* **42:** 91–96.

Jeltsch A, Urbanke C. 2004. Sliding or hopping? How restriction enzymes find their way on DNA. In *Restriction endonucleases* (ed. Pingoud A), pp. 95–110. Springer, Berlin.

Jeltsch A, Alves J, Maass G, Pingoud A. 1992. On the catalytic mechanism of *Eco*RI and *Eco*RV. A detailed proposal based on biochemical results, structural data and molecular modelling. *FEBS Lett* **304:** 4–8.

Jeltsch A, Alves J, Oelgeschläger T, Wolfes H, Maass G, Pingoud A. 1993a. Mutational analysis of the function of Gln115 in the *Eco*RI restriction endonuclease, a critical amino acid for recognition of the inner thymidine residue in the sequence -GAATTC- and for coupling specific DNA binding to catalysis. *J Mol Biol* **229:** 221–234.

Jeltsch A, Alves J, Wolfes H, Maass G, Pingoud A. 1993b. Substrate-assisted catalysis in the cleavage of DNA by the *Eco*RI and *Eco*RV restriction enzymes. *Proc Natl Acad Sci* **90:** 8499–8503.

Jeltsch A, Kroger M, Pingoud A. 1995a. Evidence for an evolutionary relationship among type-II restriction endonucleases. *Gene* **160:** 7–16.

Jeltsch A, Maschke H, Selent U, Wenz C, Köhler E, Connolly BA, Thorogood H, Pingoud A. 1995b. DNA binding specificity of the *Eco*RV restriction endonuclease is increased by Mg^{2+} binding to a metal ion binding site distinct from the catalytic center of the enzyme. *Biochemistry* **34:** 6239–6246.

Jen-Jacobson L. 1997. Protein-DNA recognition complexes: conservation of structure and binding energy in the transition state. *Biopolymers* **44:** 153–180.

Jen-Jacobson L, Engler LE, Lesser DR, Kurpiewski MR, Yee C, McVerry B. 1996. Structural adaptations in the interaction of *Eco*RI endonuclease with methylated GAATTC sites. *EMBO J* **15:** 2870–2882.

Jen-Jacobson L, Lesser D.R., Kurpiewski M. 1991. DNA sequence discrimination by *Eco*RI endonuclease. In *Nucleic acids and molecular biology* (ed. Eckstein F, Lilley DMJ), pp. 142–170. Springer, Heidelberg.

Jen-Jacobson L, Engler LE, Jacobson LA. 2000. Structural and thermodynamic strategies for site-specific DNA binding proteins. *Structure* **8:** 1015–1023.

Jindrova E, Schmid-Nuoffer S, Hamburger F, Janscak P, Bickle TA. 2005. On the DNA cleavage mechanism of Type I restriction enzymes. *Nucl Acids Res* **33:** 1760–1766.

Jo K, Topal MD. 1995. DNA topoisomerase and recombinase activities in Nae I restriction endonuclease. *Science* **267:** 1817–1820.

Johnson MS, Sutcliffe MJ, Blundell TL. 1990. Molecular anatomy: phyletic relationships derived from three-dimensional structures of proteins. *J Mol Evol* **30:** 43–59.

Jones S, Daley DT, Luscombe NM, Berman HM, Thornton JM. 2001. Protein–RNA interactions: a structural analysis. *Nucleic Acids Res* **29:** 943–954.

Jose TJ, Conlan LH, Dupureur CM. 1999. Quantitative evaluation of metal ion binding to PvuII restriction endonuclease. *J Biol Inorg Chem* **4:** 814–823.

Kandavelou K, Chandrasegaran S. 2009. Custom-designed molecular scissors for site-specific manipulation of the plant and mammalian genomes. *Methods Mol Biol* **544:** 617–636.

Kandavelou K, Mani M, Durai S, Chandrasegaran S. 2005. "Magic" scissors for genome surgery. *Nat Biotechnol* **23:** 686–687.

Kandavelou K, Mani M, Durai S, Chandrasegaran S. 2004. Engineering and applications of chemeric nucleases. In *Restriction endonucleases* (ed. Pingoud A), pp. 413–434. Springer, Berlin.

Kandavelou K, Ramalingam S, London V, Mani M, Wu J, Alexeev V, Civin CI, Chandrasegaran S. 2009. Targeted manipulation of mammalian genomes using designed zinc finger nucleases. *Biochem Biophys Res Commun* **388:** 56–61.

Kim YC, Grable JC, Love R, Greene PJ, Rosenberg JM. 1990. Refinement of Eco RI endonuclease crystal structure: a revised protein chain tracing. *Science* **249:** 1307–1309.

Kim Y, Choi J, Grable JC, Greene P, Hager P, Rosenberg JM. 1994. In *Structural biology: the state of the art* (ed. Sarma RH, Sarma MH), pp. 225–246. Adenine Press, Schenectady, NY.

Kim DE, Narayan M, Patel SS. 2002. T7 DNA helicase: a molecular motor that processively and unidirectionally translocates along single-stranded DNA. *J Mol Biol* **321:** 807–819.

King G, Murray NE. 1994. Restriction enzymes in cells, not eppendorfs. *Trends Microbiol* **2:** 465–469.

King K, Benkovic SJ, Modrich P. 1989. Glu-111 is required for activation of the DNA cleavage center of *Eco*RI endonuclease. *J Biol Chem* **264:** 11807–11815.

Kirsanova OV, Baskunov VB, Gromova ES. 2004. Type IIE and IIF restriction endonucleases interacting with two recognition sites on DNA. *Mol Biol* **38:** 886–900.

Kita K, Kotani H, Sugisaki H, Takanami M. 1989. The *Fok*I restriction-modification system. I. Organization and nucleotide sequences of the restriction and modification genes. *J Biol Chem* **264:** 5751–5756.

Klimasauskas S, Kumar S, Roberts RJ, Cheng X. 1994. HhaI methyltransferase flips its target base out of the DNA helix. *Cell* **76:** 357–369.

Kobayashi I. 1996. DNA modification and restriction: selfish behavior of an epigenetic system. In *Epigenetic mechanisms of gene regulation* (ed. Russo V, Martienssen R, Riggs A), pp. 155–172. Cold Spring Harbor Laboratory Press, Cold Spring Harbor, NY.

Kobayashi I. 1998. Selfishness and death: raison d'etre of restriction, recombination and mitochondria. *Trends Genet* **14:** 368–374.

Kobayashi I. 2001. Behavior of restriction-modification systems as selfish mobile elements and their impact on genome evolution. *Nucleic Acids Res* **29:** 3742–3756.

Kobayashi I. 2004. Restriction-modification systems as minimal forms of life. In *Restriction endonucleases* (ed. Pingoud A), pp. 19–62. Springer, Berlin.

Kong H. 1998. Analyzing the functional organization of a novel restriction modification system, the BcgI system. *J Mol Biol* **279:** 823–832.

Kong H, Smith CL. 1997. Substrate DNA and cofactor regulate the activities of a multifunctional restriction-modification enzyme, *Bcg*I. *Nucleic Acids Res* **25:** 3687–3692.

Korolev S, Hsieh J, Gauss GH, Lohman TM, Waksman G. 1997. Major domain swiveling revealed by the crystal structures of complexes of *E. coli* Rep helicase bound to single-stranded DNA and ADP. *Cell* **90:** 635–647.

Kostrewa D, Winkler FK. 1995. Mg^{2+} binding to the active site of *Eco*RV endonuclease: a crystallographic study of complexes with substrate and product DNA at 2 Å resolution. *Biochemistry* **34:** 683–696.

Kosykh VG, Puntezhis SA, Bur'ianov Ia I, Baev AA. 1982. [Isolation, purification and properties of restriction endonuclease EcoRII]. *Biokhimiia (Moscow, Russia)* **47:** 619–625.

Kovall RA, Matthews BW. 1998. Structural, functional, and evolutionary relationships between λ-exonuclease and the type II restriction endonucleases. *Proc Natl Acad Sci* **95:** 7893–7897.

Kovall RA, Matthews BW. 1999. Type II restriction endonucleases: structural, functional, and evolutionary relationships. *Curr Opin Chem Biol* **3:** 578–583.

Krüger DH, Bickle TA. 1983. Bacteriophage survival: multiple mechanisms for avoiding the deoxyribonucleic acid restriction systems of their hosts. *Microbiol Rev* **47:** 345–360.

Krüger DH, Barcak GJ, Reuter M, Smith HO. 1988. *Eco*RII can be activated to cleave refractory DNA recognition sites. *Nucleic Acids Res* **16:** 3997–4008.

Krüger DH, Schroeder C, Reuter M, Bogdarina IG, Buryanov YI, Bickle TA. 1985. DNA methylation of bacterial viruses T3 and T7 by different DNA methylases in *Escherichia coli* K12 cells. *EUr J Biochem/FEBS* **150:** 323–330.

Krüger DH, Kupper D, Meisel A, Reuter M, Schroeder C. 1995. The significance of distance and orientation of restriction endonuclease recognition sites in viral DNA genomes. *FEMS Microbiol Rev* **17:** 177–184.

Kupper D, Reuter M, Mackeldanz P, Meisel A, Alves J, Schroeder C, Krüger DH. 1995. Hyperexpressed *Eco*RII renatured from inclusion bodies and native enzyme both exhibit essential cooperativity with two DNA sites. *Protein Expression Purif* **6:** 1–9.

Kuster W. 1998. "Bedeutung hydrophober kontakte fur die sequenzspezifische DNA-erkennung der restriktionsendonuklease EcoRI." Doctoral thesis, University of Hanover.

Kuz'min NP, Loseva SP, Beliaeva R, Kravets AN, Solonin AS. 1984. [EcoRV restrictase: physical and catalytic properties of homogenous enzyme]. *Mol Biol (Mosc)* **18:** 197–204.

Lacks S, Greenberg B. 1975. A deoxyribonuclease of *Diplococcus pneumoniae* specific for methylated DNA. *J Biol Chem* **250:** 4060–4066.

Lesser DR, Kurpiewski MR, Jen-Jacobson L. 1990. The energetic basis of specificity in the EcoRI endonuclease–DNA interaction. *Science* **250:** 776–786.

Lesser DR, Kurpiewski MR, Waters T, Connolly BA, Jen-Jacobson L. 1993. Facilitated distortion of the DNA site enhances *Eco*RI endonuclease-DNA recognition. *Proc Natl Acad Sci* **90:** 7548–7552.

Loenen WA, Dryden DT, Raleigh EA, Wilson GG. 2014a. Type I restriction enzymes and their relatives. *Nucleic Acids Res* **42:** 20–44.

Loenen WA, Dryden DT, Raleigh EA, Wilson GG, Murray NE. 2014b. Highlights of the DNA cutters: a short history of the restriction enzymes. *Nucleic Acids Res* **42:** 3–19.

Loenen WAM. 2003. Tracking EcoKI and DNA fifty years on: a golden story full of surprises. *Nucleic Acids Res* **31:** 7059–7069.

Looney MC, Moran LS, Jack WE, Feehery GR, Benner JS, Slatko BE, Wilson GG. 1989. Nucleotide sequence of the *Fok*I restriction-modification system: separate strand-specificity domains in the methyltransferase. *Gene* **80:** 193–208.

Lukacs CM, Aggarwal AK. 2001. *Bgl*II and *Mun*I: what a difference a base makes. *Curr Opin Struct Biol* **11:** 14–18.

Lukacs CM, Kucera R, Schildkraut I, Aggarwal AK. 2000. Understanding the immutability of restriction enzymes: crystal structure of *Bgl*II and its DNA substrate at 1.5 Å resolution. *Nat Struct Biol* **7:** 134–140.

Luscombe NM, Laskowski RA, Thornton JM. 2001. Amino acid–base interactions: a three-dimensional analysis of protein–DNA interactions at an atomic level. *Nucleic Acids Res* **29:** 2860–2874.

Mahdi AA, Briggs GS, Sharples GJ, Wen Q, Lloyd RG. 2003. A model for dsDNA translocation revealed by a structural motif common to RecG and Mfd proteins. *EMBO J* **22:** 724–734.

Makovets S, Titheradge AJ, Murray NE. 1998. ClpX and ClpP are essential for the efficient acquisition of genes specifying type IA and IB restriction systems. *Mol Microbiol* **28:** 25–35.

Makovets S, Doronina VA, Murray NE. 1999. Regulation of endonuclease activity by proteolysis prevents breakage of unmodified bacterial chromosomes by type I restriction enzymes. *Proc Natl Acad Sci* **96:** 9757–9762.

Maluf NK, Fischer CJ, Lohman TM. 2003. A dimer of *Escherichia coli* UvrD is the active form of the helicase in vitro. *J Mol Biol* **325:** 913–935.

Mani M, Kandavelou K, Dy FJ, Durai S, Chandrasegaran S. 2005a. Design, engineering, and characterization of zinc finger nucleases. *Biochem Biophys Res Commun* **335:** 447–457.

Mani M, Smith J, Kandavelou K, Berg JM, Chandrasegaran S. 2005b. Binding of two zinc finger nuclease monomers to two specific sites is required for effective double-strand DNA cleavage. *Biochem Biophys Res Commun* **334:** 1191–1197.

Martin AM, Sam MD, Reich NO, Perona JJ. 1999. Structural and energetic origins of indirect readout in site-specific DNA cleavage by a restriction endonuclease. *Nat Struct Biol* **6:** 269–277.

McClelland SE, Szczelkun MD. 2004. The Type I and III restriction endonucleases: structural elements in molecular motors that process DNA. In *Restriction endonucleases* (ed. Pingoud A), pp. 111–135. Springer, Berlin.

McKane M, Milkman R. 1995. Transduction, restriction and recombination patterns in *Escherichia coli*. *Genetics* **139:** 35–43.

Meisel A, Mackeldanz P, Bickle TA, Krüger DH, Schroeder C. 1995. Type III restriction endonucleases translocate DNA in a reaction driven by recognition site-specific ATP hydrolysis. *EMBO J* **14:** 2958–2966.

Mernagh DR, Kneale GG. 1996. High resolution footprinting of a type I methyltransferase reveals a large structural distortion within the DNA recognition site. *Nucleic Acids Res* **24:** 4853–4858.

Mernagh DR, Taylor IA, Kneale GG. 1998. Interaction of the type I methyltransferase M. *Eco*R124I with modified DNA substrates: sequence discrimination and base flipping. *Biochem J* **336** (Pt 3): 719–725.

Milsom SE, Halford SE, Embleton ML, Szczelkun MD. 2001. Analysis of DNA looping interactions by type II restriction enzymes that require two copies of their recognition sites. *J Mol Biol* **311:** 515–527.

Mizuuchi K, Nobbs TJ, Halford SE, Adzuma K, Qin J. 1999. A new method for determining the stereochemistry of DNA cleavage reactions: application to the *Sfi*I and *Hpa*II restriction endonucleases and to the MuA transposase. *Biochemistry* **38:** 4640–4648.

Mordasini T, Curioni A, Andreoni W. 2003. Why do divalent metal ions either promote or inhibit enzymatic reactions? The case of *Bam*HI restriction endonuclease from combined quantum-classical simulations. *J Biol Chem* **278:** 4381–4384.

Mucke M, Lurz R, Mackeldanz P, Behlke J, Krüger DH, Reuter M. 2000. Imaging DNA loops induced by restriction endonuclease EcoRII. A single amino acid substitution uncouples target recognition from cooperative DNA interaction and cleavage. *J Biol Chem* **275:** 30631–30637.

Mucke M, Grelle G, Behlke J, Kraft R, Krüger DH, Reuter M. 2002. *Eco*RII: a restriction enzyme evolving recombination functions? *EMBO J* **21:** 5262–5268.

Mucke M, Krüger DH, Reuter M. 2003. Diversity of type II restriction endonucleases that require two DNA recognition sites. *Nucleic Acids Res* **31:** 6079–6084.

Muir RS, Flores H, Zinder ND, Model P, Soberon X, Heitman J. 1997. Temperature-sensitive mutants of the EcoRI endonuclease. *J Mol Biol* **274:** 722–737.

Murray NE. 2000. Type I restriction systems: sophisticated molecular machines (a legacy of Bertani and Weigle). *Microbiol Mol Biol Rev* **64:** 412–434.

Murray NE. 2002. 2001 Fred Griffith review lecture. Immigration control of DNA in bacteria: self versus non-self. *Microbiology* **148:** 3–20.

Murray NE, Daniel AS, Cowan GM, Sharp PM. 1993. Conservation of motifs within the unusually variable polypeptide sequences of type I restriction and modification enzymes. *Mol Microbiol* **9:** 133–143.

Naito T, Kusano K, Kobayashi I. 1995. Selfish behavior of restriction-modification systems. *Science* **267:** 897–899.

Nakayama Y, Kobayashi I. 1998. Restriction-modification gene complexes as selfish gene entities: roles of a regulatory system in their establishment, maintenance, and apoptotic mutual exclusion. *Proc Natl Acad Sci* **95:** 6442–6447.

Nanduri B, Byrd AK, Eoff RL, Tackett AJ, Raney KD. 2002. Pre-steady-state DNA unwinding by bacteriophage T4 Dda helicase reveals a monomeric molecular motor. *Proc Natl Acad Sci* **99:** 14722–14727.

Needels MC, Fried SR, Love R, Rosenberg JM, Boyer HW, Greene PJ. 1989. Determinants of *Eco*RI endonuclease sequence discrimination. *Proc Natl Acad Sci.* **86:** 3579–3583.

Newman PC, Nwosu VU, Williams DM, Cosstick R, Seela F, Connolly BA. 1990a. Incorporation of a complete set of deoxyadenosine and thymidine analogues suitable for the study of protein nucleic acid interactions into oligodeoxynucleotides. Application to the EcoRV restriction endonuclease and modification methylase. *Biochemistry* **29:** 9891–9901.

Newman PC, Williams DM, Cosstick R, Seela F, Connolly BA. 1990b. Interaction of the EcoRV restriction endonuclease with the deoxyadenosine and thymidine bases in its recognition hexamer d(GATATC). *Biochemistry* **29:** 9902–9910.

Newman M, Strzelecka T, Dorner LF, Schildkraut I, Aggarwal AK. 1994a. Structure of restriction endonuclease *Bam*HI and its relationship to *Eco*RI. *Nature* **368:** 660–664.

Newman M, Strzelecka T, Dorner LF, Schildkraut I, Aggarwal AK. 1994b. Structure of restriction endonuclease *Bam*HI phased at 1.95 A resolution by MAD analysis. *Structure* **2:** 439–452.

Newman M, Strzelecka T, Dorner LF, Schildkraut I, Aggarwal AK. 1995. Structure of *Bam*HI endonuclease bound to DNA: partial folding and unfolding on DNA binding. *Science* **269:** 656–663.

Newman M, Lunnen K, Wilson G, Greci J, Schildkraut I, Phillips SE. 1998. Crystal structure of restriction endonuclease *Bgl*I bound to its interrupted DNA recognition sequence. *EMBO J* **17:** 5466–5476.

Nobbs TJ, Halford SE. 1995. DNA cleavage at two recognition sites by the *Sfi*I restriction endonuclease: salt dependence of *cis* and *trans* interactions between distant DNA sites. *J Mol Biol* **252:** 399–411.

Nobusato A, Uchiyama I, Kobayashi I. 2000a. Diversity of restriction-modification gene homologues in *Helicobacter pylori*. *Gene* **259:** 89–98.

Nobusato A, Uchiyama I, Ohashi S, Kobayashi I. 2000b. Insertion with long target duplication: a mechanism for gene mobility suggested from comparison of two related bacterial genomes. *Gene* **259:** 99–108.

Nunes-Duby SE, Kwon HJ, Tirumalai RS, Ellenberger T, Landy A. 1998. Similarities and differences among 105 members of the Int family of site-specific recombinases. *Nucleic Acids Res* **26:** 391–406.

O'Neill M, Chen A, Murray NE. 1997. The restriction-modification genes of *Escherichia coli* K-12 may not be selfish: they do not resist loss and are readily replaced by alleles conferring different specificities. *Proc Natl Acad Sci* **94:** 14596–14601.

O'Neill M, Dryden DT, Murray NE. 1998. Localization of a protein–DNA interface by random mutagenesis. *EMBO J* **17:** 7118–7127.

O'Neill M, Powell LM, Murray NE. 2001. Target recognition by EcoKI: the recognition domain is robust and restriction-deficiency commonly results from the proteolytic control of enzyme activity. *J Mol Biol* **307:** 951–963.

Oelgeschläger T, Geiger R, Rüter T, Alves J, Fliess A, Pingoud A. 1990. Probing the function of individual amino acid residues in the DNA binding site of the *Eco*RI restriction endonuclease by analysing the toxicity of genetically engineered mutants. *Gene* **89:** 19–27.

Oller AR, Vanden Broek W, Conrad M, Topal MD. 1991. Ability of DNA and spermidine to affect the activity of restriction endonucleases from several bacterial species. *Biochemistry* **30**: 2543–2549.

Pabo CO, Peisach E, Grant RA. 2001. Design and selection of novel Cys_2His_2 zinc finger proteins. *Ann Rev Biochem* **70**: 313–340.

Panne D, Raleigh EA, Bickle TA. 1999. The McrBC endonuclease translocates DNA in a reaction dependent on GTP hydrolysis. *J Mol Biol* **290**: 49–60.

Park JS, Marr MT, Roberts JW. 2002. *E. coli* transcription repair coupling factor (Mfd protein) rescues arrested complexes by promoting forward translocation. *Cell* **109**: 757–767.

Pein CD, Reuter M, Cech D, Krüger DH. 1989. Oligonucleotide duplexes containing CC(A/T)GG stimulate cleavage of refractory DNA by restriction endonuclease *Eco*RII. *FEBS Lett* **245**: 141–144.

Pein CD, Reuter M, Meisel A, Cech D, Krüger DH. 1991. Activation of restriction endonuclease EcoRII does not depend on the cleavage of stimulator DNA. *Nucleic Acids Res* **19**: 5139–5142.

Petrauskene OV, Karpova EA, Gromova ES, Guschlbauer W. 1994. Two subunits of *Eco*RII restriction endonuclease interact with two DNA recognition sites. *Biochem Biophys Res Commun* **198**: 885–890.

Petrauskene OV, Babkina OV, Tashlitsky VN, Kazankov GM, Gromova ES. 1998. *Eco*RII endonuclease has two identical DNA-binding sites and cleaves one of two co-ordinated recognition sites in one catalytic event. *FEBS Lett* **425**: 29–34.

Piatrauskene OV, Tashlitskii VN, Brevnov MG, Bakman I, Gromova ES. 1996. [Kinetic modeling of the mechanism of allosteric interactions of restriction endonuclease EcoRII with two DNA segments]. *Biokhimiia (Moscow, Russia)* **61**: 1257–1269.

Piekarowicz A, Golaszewska M, Sunday AO, Siwinska M, Stein DC. 1999. The *Hae*IV restriction modification system of *Haemophilus aegyptius* is encoded by a single polypeptide. *J Mol Biol* **293**: 1055–1065.

Pieper U, Pingoud A. 2002. A mutational analysis of the PD···D/EXK motif suggests that McrC harbors the catalytic center for DNA cleavage by the GTP-dependent restriction enzyme McrBC from *Escherichia coli*. *Biochemistry* **41**: 5236–5244.

Pieper U, Schweitzer T, Groll DH, Pingoud A. 1999. Defining the location and function of domains of McrB by deletion mutagenesis. *Biol Chem* **380**: 1225–1230.

Pinarbasi H, Pinarbasi E, Hornby DP. 2003. The small subunit of M · *Aqu*I is responsible for sequence-specific DNA recognition and binding in the absence of the catalytic domain. *J Bacteriol* **185**: 1284–1288.

Pingoud A. 2004. *Restriction endonucleases*. Springer, Berlin.

Pingoud A, Jeltsch A. 1997. Recognition and cleavage of DNA by type-II restriction endonucleases. *Eur J Biochem/FEBS* **246**: 1–22.

Pingoud A, Jeltsch A. 2001. Structure and function of type II restriction endonucleases. *Nucleic Acids Res* **29**: 3705–3727.

Pingoud V, Kubareva E, Stengel G, Friedhoff P, Bujnicki JM, Urbanke C, Sudina A, Pingoud A. 2002. Evolutionary relationship between different subgroups of restriction endonucleases. *J Biol Chem* **277**: 14306–14314.

Pingoud A, Fuxreiter M, Pingoud V, Wende W. 2005. Type II restriction endonucleases: structure and mechanism. *Cell Mol Life Sci* **62**: 685–707.

Ponting CP, Russell RR. 2002. The natural history of protein domains. *Ann Rev Biophys Biomol Struct* **31**: 45–71.

Porteus MH, Baltimore D. 2003. Chimeric nucleases stimulate gene targeting in human cells. *Science* **300:** 763.

Powell LM, Murray NE. 1995. S-adenosyl methionine alters the DNA contacts of the *Eco*KI methyltransferase. *Nucleic Acids Res* **23:** 967–974.

Powell LM, Dryden DT, Willcock DF, Pain RH, Murray NE. 1993. DNA recognition by the *Eco*K methyltransferase. The influence of DNA methylation and the cofactor *S*-adenosyl-L-methionine. *J Mol Biol* **234:** 60–71.

Powell LM, Connolly BA, Dryden DT. 1998a. The DNA binding characteristics of the trimeric *Eco*KI methyltransferase and its partially assembled dimeric form determined by fluorescence polarisation and DNA footprinting. *J Mol Biol* **283:** 947–961.

Powell LM, Dryden DT, Murray NE. 1998b. Sequence-specific DNA binding by EcoKI, a type IA DNA restriction enzyme. *J Mol Biol* **283:** 963–976.

Powell LM, Lejeune E, Hussain FS, Cronshaw AD, Kelly SM, Price NC, Dryden DT. 2003. Assembly of *Eco*KI DNA methyltransferase requires the C-terminal region of the HsdM modification subunit. *Biophys Chem* **103:** 129–137.

Prakash-Cheng A, Ryu J. 1993. Delayed expression of in vivo restriction activity following conjugal transfer of *Escherichia coli hsdK* (restriction-modification) genes. *J Bacteriol* **175:** 4905–4906.

Prakash-Cheng A, Chung SS, Ryu J. 1993. The expression and regulation of *hsdK* genes after conjugative transfer. *Mol Gen Genet* **241:** 491–496.

Price C, Bickle TA. 1986. A possible role for DNA restriction in bacterial evolution. *Microbiol Sci* **3:** 296–299.

Putnam CD, Clancy SB, Tsuruta H, Gonzalez S, Wetmur JG, Tainer JA. 2001. Structure and mechanism of the RuvB Holliday junction branch migration motor. *J Mol Biol* **311:** 297–310.

Raleigh EA, Wilson G. 1986. *Escherichia coli* K-12 restricts DNA containing 5-methylcytosine. *Proc Natl Acad Sci* **83:** 9070–9074.

Raleigh EA, Murray NE, Revel H, Blumenthal RM, Westaway D, Reith AD, Rigby PW, Elhai J, Hanahan D. 1988. McrA and McrB restriction phenotypes of some *E. coli* strains and implications for gene cloning. *Nucleic Acids Res* **16:** 1563–1575.

Ramalingam S, Kandavelou K, Rajenderan R, Chandrasegaran S. 2011. Creating designed zinc-finger nucleases with minimal cytotoxicity. *J Mol Biol* **405:** 630–641.

Rao DN, Saha S, Krishnamurthy V. 2000. ATP-dependent restriction enzymes. *Prog Nucleic Acid Res Mol Biol* **64:** 1–63.

Reid SL, Parry D, Liu HH, Connolly BA. 2001. Binding and recognition of GATATC target sequences by the EcoRV restriction endonuclease: a study using fluorescent oligonucleotides and fluorescence polarization. *Biochemistry* **40:** 2484–2494.

Repin VE, Lebedev LR, Puchkova L, Serov GD, Tereschenko T, Chizikov VE, Andreeva I. 1995. New restriction endonucleases from thermophilic soil bacteria. *Gene* **157:** 321–322.

Reuter M, Kupper D, Meisel A, Schroeder C, Krüger DH. 1998. Cooperative binding properties of restriction endonuclease EcoRII with DNA recognition sites. *J Biol Chem* **273:** 8294–8300.

Reuter M, Schneider-Mergener J, Kupper D, Meisel A, Mackeldanz P, Krüger DH, Schroeder C. 1999. Regions of endonuclease EcoRII involved in DNA target recognition identified by membrane-bound peptide repertoires. *J Biol Chem* **274:** 5213–5221.

Reuter M, Mucke M, Krüger DH. 2004. Structure and function of Type IIE restriction endonucleases—or: from a plasmid that restricts phage replication to a new molecular DNA recognition mechanism. In *Restriction endonucleases* (ed. Pingoud A), pp. 261–295. Springer, Berlin.

Rigden DJ, Setlow P, Setlow B, Bagyan I, Stein RA, Jedrzejas MJ. 2002. PrfA protein of *Bacillus* species: prediction and demonstration of endonuclease activity on DNA. *Protein Sci* **11:** 2370–2381.

Rippe K, von Hippel PH, Langowski J. 1995. Action at a distance: DNA-looping and initiation of transcription. *Trends Biochem Sci* **20:** 500–506.

Roberts RJ, Cheng X. 1998. Base flipping. *Ann Rev Bochem* **67:** 181–198.

Roberts RJ, Halford S.E. 1993. Type II restriction enzymes. In *Nucleases*, 2nd ed. (ed. Linn SM, Lloyd RS, Roberts RJ), pp. 35–88. Cold Spring Harbor Laboratory Press, Cold Spring Harbor, NY.

Roberts RJ, Macelis D. 1991. Restriction enzymes and their isoschizomers. *Nucleic Acids Res* **19** (Suppl): 2077–2109.

Roberts RJ, Macelis D. 1993. REBASE—restriction enzymes and methylases. *Nucleic Acids Res* **21:** 3125–3137.

Roberts RJ, Belfort M, Bestor T, Bhagwat AS, Bickle TA, Bitinaite J, Blumenthal RM, Degtyarev S, Dryden DT, Dybvig K, et al. 2003a. A nomenclature for restriction enzymes, DNA methyltransferases, homing endonucleases and their genes. *Nucleic Acids Res* **31:** 1805–1812.

Roberts RJ, Vincze T, Posfai J, Macelis D. 2003b. REBASE: restriction enzymes and methyltransferases. *Nucleic Acids Res* **31:** 418–420.

Roberts RJ, Macelis D, Vincze T, Pósfai J. 2004. The genomics of restriction and modification. In *5th NEB Meeting on Restriction and Modification (Bristol UK, 2004).* http://rebase.neb.com/cgi-bin/statlist.

Roman LJ, Kowalczykowski SC. 1989. Characterization of the adenosinetriphosphatase activity of the *Escherichia coli* RecBCD enzyme: relationship of ATP hydrolysis to the unwinding of duplex DNA. *Biochemistry* **28:** 2873–2881.

Rosati O. 1999. "Untersuchung und design von DNA-kontakten der restriktionsendonuklease EcoRI inner- und ausserhalb der erkennungssequenz." Doctoral thesis, University of Hanover.

Rosenberg JM. 1991. Structure and function of restriction endonucleases. *Curr Opin Struct Biol* **1:** 104–113.

Saha S, Rao DN. 1995. ATP hydrolysis is required for DNA cleavage by *Eco*PI restriction enzyme. *J Mol Biol* **247:** 559–567.

Saha S, Rao DN. 1997. Mutations in the Res subunit of the *Eco*PI restriction enzyme that affect ATP-dependent reactions. *J Mol Biol* **269:** 342–354.

Saha S, Ahmad I, Reddy YV, Krishnamurthy V, Rao DN. 1998. Functional analysis of conserved motifs in type III restriction-modification enzymes. *Biol Chem* **379:** 511–517.

Samudzi CT. 1990. "Use of the molecular replacement method in structural studies of *Eco*RI endonuclease." PhD thesis, University of Pittsburgh.

Sapranauskas R, Sasnauskas G, Lagunavicius A, Vilkaitis G, Lubys A, Siksnys V. 2000. Novel subtype of type IIs restriction enzymes. *Bfi*I endonuclease exhibits similarities to the EDTA-resistant nuclease Nuc of *Salmonella typhimurium*. *J Biol Chem* **275:** 30878–30885.

Scheuring Vanamee E, Viadiu H, Lukacs CM, Aggarwal AK. 2004. Two of a kind: BamHI and BglII. In *Restriction endonucleases* (ed. Pingoud A), pp. 215–236. Springer, Berlin.

Schildkraut I, Banner CD, Rhodes CS, Parekh S. 1984. The cleavage site for the restriction endonuclease *Eco*RV is 5′-GAT/ATC-3′. *Gene* **27:** 327–329.

Schleif R. 1992. DNA looping. *Ann Rev Biochem* **61:** 199–223.

Schulze C, Jeltsch A, Franke I, Urbanke C, Pingoud A. 1998. Crosslinking the *Eco*RV restriction endonuclease across the DNA-binding site reveals transient intermediates and conformational changes of the enzyme during DNA binding and catalytic turnover. *EMBO J* **17:** 6757–6766.

Selent U, Rüter T, Köhler E, Liedtke M, Thielking V, Alves J, Oelgeschläger T, Wolfes H, Peters F, Pingoud A. 1992. A site-directed mutagenesis study to identify amino acid residues involved in the catalytic function of the restriction endonuclease *Eco*RV. *Biochemistry* **31:** 4808–4815.

Senesac JH, Allen JR. 1995. Oligonucleotide activation of the type IIe restriction enzyme NaeI for digestion of refractory sites. *BioTechniques* **19:** 990–993.

Sharp PM, Kelleher JE, Daniel AS, Cowan GM, Murray NE. 1992. Roles of selection and recombination in the evolution of type I restriction-modification systems in enterobacteria. *Proc Natl Acad Sci* **89:** 9836–9840.

Sidorova NY, Rau DC. 1996. Differences in water release for the binding of *Eco*RI to specific and nonspecific DNA sequences. *Proc Natl Acad Sci* **93:** 12272–12277.

Sidorova N, Rau D.C. 2004. The role of water in the EcoRI-DNA binding. In *Restriction endonucleases* (ed. Pingoud A), pp. 319–337. Springer, Berlin.

Šikšnys V, Zareckaja N, Vaisvila R, Timinskas A, Stakenas P, Butkus V, Janulaitis A. 1994. CAATTG-specific restriction-modification *mun*I genes from *Mycoplasma*: sequence similarities between R·*Mun*I and R·*Eco*RI. *Gene* **142:** 1–8.

Šikšnys V, Timinskas A, Klimasauskas S, Butkus V, Janulaitis A. 1995. Sequence similarity among type-II restriction endonucleases, related by their recognized 6-bp target and tetranucleotide-overhang cleavage. *Gene* **157:** 311–314.

Šikšnys V, Skirgaila R, Sasnauskas G, Urbanke C, Cherny D, Grazulis S, Huber R. 1999. The Cfr10I restriction enzyme is functional as a tetramer. *J Mol Biol* **291:** 1105–1118.

Šikšnys V, Grazulis S., Huber R. 2004. Structure and function of the tetrameric restriction enzymes. In *Restriction endonucleases* (ed. Pingoud A), pp. 237–259. Springer, Berlin.

Singleton MR, Wigley DB. 2002. Modularity and specialization in superfamily 1 and 2 helicases. *J Bacteriol* **184:** 1819–1826.

Singleton MR, Wigley DB. 2003. Multiple roles for ATP hydrolysis in nucleic acid modifying enzymes. *EMBO J* **22:** 4579–4583.

Singleton MR, Scaife S, Wigley DB. 2001. Structural analysis of DNA replication fork reversal by RecG. *Cell* **107:** 79–89.

Skirgaila R, Grazulis S, Bozic D, Huber R, Siksnys V. 1998. Structure-based redesign of the catalytic/metal binding site of Cfr10I restriction endonuclease reveals importance of spatial rather than sequence conservation of active centre residues. *J Mol Biol* **279:** 473–481.

Stahl F, Wende W, Jeltsch A, Pingoud A. 1996. Introduction of asymmetry in the naturally symmetric restriction endonuclease EcoRV to investigate intersubunit communication in the homodimeric protein. *Proc Natl Acad Sci* **93:** 6175–6180.

Stahl F, Wende W, Jeltsch A, Pingoud A. 1998a. The mechanism of DNA cleavage by the type II restriction enzyme EcoRV: Asp36 is not directly involved in DNA cleavage but serves to couple indirect readout to catalysis. *Biol Chem* **379:** 467–473.

Stahl F, Wende W, Wenz C, Jeltsch A, Pingoud A. 1998b. Intra- vs intersubunit communication in the homodimeric restriction enzyme *Eco*RV: Thr 37 and Lys 38 involved in indirect readout are only important for the catalytic activity of their own subunit. *Biochemistry* **37:** 5682–5688.

Stanford NP, Halford SE, Baldwin GS. 1999. DNA cleavage by the *Eco*RV restriction endonuclease: pH dependence and proton transfers in catalysis. *J Mol Biol* **288:** 105–116.

Stankevicius K, Lubys A, Timinskas A, Vaitkevicius D, Janulaitis A. 1998. Cloning and analysis of the four genes coding for *Bpu*10I restriction-modification enzymes. *Nucleic Acids Res* **26:** 1084–1091.

Stasiak A. 1980a. [Restriction enzymes. I. Mechanisms of action of type II restriction-modification systems (author's transl)]. *Postępy Biochem* **26:** 343–367.

Stasiak A. 1980b. [Restriction enzymes. II. Mechanisms of action of type I and III restriction-modification systems (author's transl)]. *Postępy Biochem* **26:** 369–387.

Stein DC, Chien R, Seifert HS. 1992. Construction of a *Neisseria gonorrhoeae* MS11 derivative deficient in NgoMI restriction and modification. *J Bacteriol* **174:** 4899–4906.

Stewart FJ, Raleigh EA. 1998. Dependence of McrBC cleavage on distance between recognition elements. *Biol Chem* **379:** 611–616.

Stewart FJ, Panne D, Bickle TA, Raleigh EA. 2000. Methyl-specific DNA binding by McrBC, a modification-dependent restriction enzyme. *J Mol Biol* **298:** 611–622.

Stover T, Köhler E, Fagin U, Wende W, Wolfes H, Pingoud A. 1993. Determination of the DNA bend angle induced by the restriction endonuclease EcoRV in the presence of Mg^{2+}. *J Biol Chem* **268:** 8645–8650.

Strzelecka T, Newman M, Dorner LF, Knott R, Schildkraut I, Aggarwal AK. 1994. Crystallization and preliminary X-ray analysis of restriction endonuclease *Bam*HI-DNA complex. *J Mol Biol* **239:** 430–432.

Stuckey JA, Dixon JE. 1999. Crystal structure of a phospholipase D family member. *Nat Struct Biol* **6:** 278–284.

Studier FW, Bandyopadhyay PK. 1988. Model for how type I restriction enzymes select cleavage sites in DNA. *Proc Natl Acad Sci* **85:** 4677–4681.

Sturrock SS, Dryden DT. 1997. A prediction of the amino acids and structures involved in DNA recognition by type I DNA restriction and modification enzymes. *Nucleic Acids Res* **25:** 3408–3414.

Su TJ, Tock MR, Egelhaaf SU, Poon WC, Dryden DT. 2005. DNA bending by M·EcoKI methyltransferase is coupled to nucleotide flipping. *Nucleic Acids Res* **33:** 3235–3244.

Subramanya HS, Bird LE, Brannigan JA, Wigley DB. 1996. Crystal structure of a DExx box DNA helicase. *Nature* **384:** 379–383.

Sugisaki H, Kanazawa S. 1981. New restriction endonucleases from *Flavobacterium okeanokoites* (*Fok*I) and *Micrococcus luteus* (*Mlu*I). *Gene* **16:** 73–78.

Sun J, Viadiu H, Aggarwal AK, Weinstein H. 2003. Energetic and structural considerations for the mechanism of protein sliding along DNA in the nonspecific BamHI-DNA complex. *Biophys J* **84:** 3317–3325.

Szczelkun MD. 2000. How do proteins move along DNA? Lessons from type-I and type-III restriction endonucleases. *Essays in biochemistry: motor proteins* (ed. Banting G, Higgins SJ), Vol. 35, pp. 131–143. Portland Press, London.

Szczelkun MD, Connolly BA. 1995. Sequence-specific binding of DNA by the EcoRV restriction and modification enzymes with nucleic acid and cofactor analogues. *Biochemistry* **34:** 10724–10733.

Szczelkun MD, Halford SE. 1996. Recombination by resolvase to analyse DNA communications by the *Sfi*I restriction endonuclease. *EMBO J* **15:** 1460–1469.

Szczelkun MD, Dillingham MS, Janscak P, Firman K, Halford SE. 1996. Repercussions of DNA tracking by the type IC restriction endonuclease *Eco*R124I on linear, circular and catenated substrates. *EMBO J* **15:** 6335–6347.

Szybalski W, Kim SC, Hasan N, Podhajska AJ. 1991. Class-IIS restriction enzymes—a review. *Gene* **100**: 13–26.

Takano T, Watanabe T, Fukasawa T. 1968. Mechanism of host-controlled restriction of bacteriophage λ by R factors in *Escherichia coli* K12. *Virology* **34**: 290–302.

Tamulaitis G, Solonin AS, Siksnys V. 2002. Alternative arrangements of catalytic residues at the active sites of restriction enzymes. *FEBS Lett* **518**: 17–22.

Tanner NK, Cordin O, Banroques J, Doere M, Linder P. 2003. The Q motif: a newly identified motif in DEAD box helicases may regulate ATP binding and hydrolysis. *Mol Cell* **11**: 127–138.

Taylor JD, Halford SE. 1989. Discrimination between DNA sequences by the EcoRV restriction endonuclease. *Biochemistry* **28**: 6198–6207.

Taylor JD, Badcoe IG, Clarke AR, Halford SE. 1991. *Eco*RV restriction endonuclease binds all DNA sequences with equal affinity. *Biochemistry* **30**: 8743–8753.

Theis K, Chen PJ, Skorvaga M, Van Houten B, Kisker C. 1999. Crystal structure of UvrB, a DNA helicase adapted for nucleotide excision repair. *EMBO J* **18**: 6899–6907.

Thielking V, Alves J, Fliess A, Maass G, Pingoud A. 1990. Accuracy of the *Eco*RI restriction endonuclease: binding and cleavage studies with oligodeoxynucleotide substrates containing degenerate recognition sequences. *Biochemistry* **29**: 4682–4691.

Thielking V, Selent U, Köhler E, Wolfes H, Pieper U, Geiger R, Urbanke C, Winkler FK, Pingoud A. 1991. Site-directed mutagenesis studies with EcoRV restriction endonuclease to identify regions involved in recognition and catalysis. *Biochemistry* **30**: 6416–6422.

Thielking V, Selent U, Köhler E, Landgraf A, Wolfes H, Alves J, Pingoud A. 1992. Mg^{2+} confers DNA binding specificity to the *Eco*RV restriction endonuclease. *Biochemistry* **31**: 3727–3732.

Thomas MP, Brady RL, Halford SE, Sessions RB, Baldwin GS. 1999. Structural analysis of a mutational hot-spot in the *Eco*RV restriction endonuclease: a catalytic role for a main chain carbonyl group. *Nucleic Acids Res* **27**: 3438–3445.

Thorogood H, Grasby JA, Connolly BA. 1996. Influence of the phosphate backbone on the recognition and hydrolysis of DNA by the *Eco*RV restriction endonuclease. A study using oligodeoxynucleotide phosphorothioates. *J Biol Chem* **271**: 8855–8862.

Titheradge AJ, Ternent D, Murray NE. 1996. A third family of allelic *hsd* genes in *Salmonella enterica*: sequence comparisons with related proteins identify conserved regions implicated in restriction of DNA. *Mol Microbiol* **22**: 437–447.

Titheradge AJ, King J, Ryu J, Murray NE. 2001. Families of restriction enzymes: an analysis prompted by molecular and genetic data for type ID restriction and modification systems. *Nucleic Acids Res* **29**: 4195–4205.

Todd AE, Orengo CA, Thornton JM. 2002. Plasticity of enzyme active sites. *Trends Biochem Sci* **27**: 419–426.

Topal MD, Conrad M. 1993. Changing endonuclease *Eco*RII Tyr308 to Phe abolishes cleavage but not recognition: possible homology with the Int-family of recombinases. *Nucleic Acids Res* **21**: 2599–2603.

Topal MD, Thresher RJ, Conrad M, Griffith J. 1991. NaeI endonuclease binding to pBR322 DNA induces looping. *Biochemistry* **30**: 2006–2010.

Tsutakawa SE, Jingami H, Morikawa K. 1999a. Recognition of a TG mismatch: the crystal structure of very short patch repair endonuclease in complex with a DNA duplex. *Cell* **99**: 615–623.

Tsutakawa SE, Muto T, Kawate T, Jingami H, Kunishima N, Ariyoshi M, Kohda D, Nakagawa M, Morikawa K. 1999b. Crystallographic and functional studies of very short patch repair endonuclease. *Mol Cell* **3**: 621–628.

van der Woerd MJ, Pelletier JJ, Xu S, Friedman AM. 2001. Restriction enzyme BsoBI-DNA complex: a tunnel for recognition of degenerate DNA sequences and potential histidine catalysis. *Structure* **9**: 133–144.

Van Heuverswyn H, Fiers W. 1980. Recognition sequence for the restriction endonuclease *Bgl*I from *Bacillus globigii*. *Gene* **9**: 195–203.

Vanamee ES, Viadiu H, Kucera R, Dorner L, Picone S, Schildkraut I, Aggarwal AK. 2005. A view of consecutive binding events from structures of tetrameric endonuclease *Sfi*I bound to DNA. *EMBO J* **24**: 4198–4208.

Velankar SS, Soultanas P, Dillingham MS, Subramanya HS, Wigley DB. 1999. Crystal structures of complexes of PcrA DNA helicase with a DNA substrate indicate an inchworm mechanism. *Cell* **97**: 75–84.

Venclovas C, Timinskas A, Siksnys V. 1994. Five-stranded β-sheet sandwiched with two α-helices: a structural link between restriction endonucleases *Eco*RI and *Eco*RV. *Proteins* **20**: 279–282.

Vermote CL, Halford SE. 1992. *Eco*RV restriction endonuclease: communication between catalytic metal ions and DNA recognition. *Biochemistry* **31**: 6082–6089.

Viadiu H, Aggarwal AK. 1998. The role of metals in catalysis by the restriction endonuclease *Bam*HI. *Nat Struct Biol* **5**: 910–916.

Viadiu H, Aggarwal AK. 2000. Structure of BamHI bound to nonspecific DNA: a model for DNA sliding. *Mol Cell* **5**: 889–895.

Viadiu H, Vanamee ES, Jacobson EM, Schildkraut I, Aggarwal AK. 2003. Crystallization of restriction endonuclease *Sfi*I in complex with DNA. *Acta Crystallogr, Sect D: Biol Crystallogr* **59**: 1493–1495.

Vilkaitis G, Lubys A, Merkiene E, Timinskas A, Janulaitis A, Klimasauskas S. 2002. Circular permutation of DNA cytosine-N4 methyltransferases: *in vivo* coexistence in the BcnI system and *in vitro* probing by hybrid formation. *Nucleic Acids Res* **30**: 1547–1557.

Vinogradova MV, Gromova ES, Kosykh VG, Bur'ianov Ia I, Shabarova ZA. 1990. [Interaction of *Eco*RII restriction and modification enzymes with synthetic DNA fragments. Determination of the size of *Eco*RII binding site]. *Molekuliarnaia Biologiia* **24**: 847–850.

Vipond IB, Halford SE. 1993. Structure–function correlation for the *Eco*RV restriction enzyme: from non-specific binding to specific DNA cleavage. *Molr Microbiol* **9**: 225–231.

Vipond IB, Halford SE. 1995. Specific DNA recognition by *Eco*RV restriction endonuclease induced by calcium ions. *Biochemistry* **34**: 1113–1119.

Vipond IB, Baldwin GS, Halford SE. 1995. Divalent metal ions at the active sites of the EcoRV and *Eco*RI restriction endonucleases. *Biochemistry* **34**: 697–704.

Vitkute J, Maneliene Z, Petrusyte M, Janulaitis A. 1997. *Bpl*I, a new *Bcg*I-like restriction endonuclease, which recognizes a symmetric sequence. *Nucleic Acids Res* **25**: 4444–4446.

Vovis GF, Horiuchi K, Zinder ND. 1975. Endonuclease R-*Eco*RII restriction of bacteriophage f1 DNA in vitro: ordering of genes V and VII, location of an RNA promotor for gene VIII. *J Virol* **16**: 674–684.

Wah DA, Hirsch JA, Dorner LF, Schildkraut I, Aggarwal AK. 1997. Structure of the multimodular endonuclease FokI bound to DNA. *Nature* **388**: 97–100.

Wah DA, Bitinaite J, Schildkraut I, Aggarwal AK. 1998. Structure of FokI has implications for DNA cleavage. *Proc Natl Scad Sci* **95**: 10564–10569.

Walker JE, Saraste M, Runswick MJ, Gay NJ. 1982. Distantly related sequences in the α- and β-subunits of ATP synthase, myosin, kinases and other ATP-requiring enzymes and a common nucleotide binding fold. *EMBO J* **1**: 945–951.

Wang J, Chen R, Julin DA. 2000. A single nuclease active site of the *Escherichia coli* RecBCD enzyme catalyzes single-stranded DNA degradation in both directions. *J Biol Chem* **275:** 507–513.

Waters TR, Connolly BA. 1994. Interaction of the restriction endonuclease *Eco*RV with the deoxyguanosine and deoxycytidine bases in its recognition sequence. *Biochemistry* **33:** 1812–1819.

Webb JL, King G, Ternent D, Titheradge AJ, Murray NE. 1996. Restriction by *Eco*KI is enhanced by co-operative interactions between target sequences and is dependent on DEAD box motifs. *EMBO J* **15:** 2003–2009.

Welsh AJ, Halford SE, Scott DJ. 2004. Analysis of Type II restriction endonucleases that interact with two recognition sites. In *Restriction endonucleases* (ed. Pingoud A), pp. 297–317. Springer, Berlin.

Wentzell LM, Halford SE. 1998. DNA looping by the *Sfi*I restriction endonuclease. *J Mol Biol* **281:** 433–444.

Wentzell LM, Nobbs TJ, Halford SE. 1995. The *Sfi*I restriction endonuclease makes a four-strand DNA break at two copies of its recognition sequence. *J Mol Biol* **248:** 581–595.

Wenz C, Jeltsch A, Pingoud A. 1996. Probing the indirect readout of the restriction enzyme *Eco*RV. Mutational analysis of contacts to the DNA backbone. *J Biol Chem* **271:** 5565–5573.

West SC. 1996. DNA helicases: new breeds of translocating motors and molecular pumps. *Cell* **86:** 177–180.

Whitehouse I, Stockdale C, Flaus A, Szczelkun MD, Owen-Hughes T. 2003. Evidence for DNA translocation by the ISWI chromatin-remodeling enzyme. *Mol Cell Biol* **23:** 1935–1945.

Wilkins BM. 2000. Plasmid promiscuity: meeting the challenge of DNA immigration. *Env Microbiol* **4:** 495–500.

Wilkosz PA, Chandrasekhar WK, Rosenberg JM. 1995. Preliminary characterization of *Eco*RI-DNA co-crystals: incomplete factorial design of oligonucleotide sequences. *Acta Crystallogr, Sect D: Biol Crystallogr* **51:** 938–945.

Willcock DF, Dryden DT, Murray NE. 1994. A mutational analysis of the two motifs common to adenine methyltransferases. *EMBO J* **13:** 3902–3908.

Wilson GG. 1991. Organization of restriction-modification systems. *Nucleic Acids Res* **19:** 2539–2566.

Windolph S, Alves J. 1997. Influence of divalent cations on inner-arm mutants of restriction endonuclease *Eco*RI. *Eur J Biochem/FEBS* **244:** 134–139.

Winkler FK. 1992. Structure and function of restriction endonucleases. *Curr Opin Struct Biol* **2:** 93–99.

Winkler FK. 1994. DNA totally flipped-out by methylase. *Structure* **2:** 79–83.

Winkler FK, Prota AE. 2004. Structure and function of EcoRV endonuclease. In *Restriction endonucleases* (ed. Pingoud A), pp. 179–214. Springer, Berlin.

Winkler FK, Banner DW, Oefner C, Tsernoglou D, Brown RS, Heathman SP, Bryan RK, Martin PD, Petratos K, Wilson KS. 1993. The crystal structure of *Eco*RV endonuclease and of its complexes with cognate and non-cognate DNA fragments. *EMBO J* **12:** 1781–1795.

Wolfes H, Alves J, Fliess A, Geiger R, Pingoud A. 1986. Site directed mutagenesis experiments suggest that Glu 111, Glu 144 and Arg 145 are essential for endonucleolytic activity of EcoRI. *Nucleic Acids Res* **14:** 9063–9080.

Wright DJ, King K, Modrich P. 1989. The negative charge of Glu-111 is required to activate the cleavage center of EcoRI endonuclease. *J Biol Chem* **264:** 11816–11821.

Wu J, Kandavelou K, Chandrasegaran S. 2007. Custom-designed zinc finger nucleases: what is next? *Cell Mol Life Sci* **64:** 2933–2944.

Yang CC, Topal MD. 1992. Nonidentical DNA-binding sites of endonuclease *Nae*I recognize different families of sequences flanking the recognition site. *Biochemistry* **31:** 9657–9664.

Yao N, Hesson T, Cable M, Hong Z, Kwong AD, Le HV, Weber PC. 1997. Structure of the hepatitis C virus RNA helicase domain. *Nat Struct Biol* **4:** 463–467.

Yolov AA, Gromova ES, Kubareva EA, Potapov VK, Shabarova ZA. 1985. Interaction of EcoRII restriction and modification enzymes with synthetic DNA fragments. V. Study of single-strand cleavages. *Nucleic Acids Res* **13:** 8969–8981.

Yoshimori R, Roulland-Dussoix D, Boyer HW. 1972. R factor-controlled restriction and modification of deoxyribonucleic acid: restriction mutants. *J Bacteriol* **112:** 1275–1279.

Zaremba M, Urbanke C, Halford SE, Siksnys V. 2004. Generation of the *Bfi*I restriction endonuclease from the fusion of a DNA recognition domain to a non-specific nuclease from the phospholipase D superfamily. *J Mol Biol* **336:** 81–92.

Zaremba M, Sasnauskas G, Urbanke C, Siksnys V. 2005. Conversion of the tetrameric restriction endonuclease Bse634I into a dimer: oligomeric structure-stability-function correlations. *J Mol Biol* **348:** 459–478.

Zaremba M, Sasnauskas G, Urbanke C, Siksnys V. 2006. Allosteric communication network in the tetrameric restriction endonuclease Bse634I. *J Mol Biol* **363:** 800–812.

Zaremba M, Sasnauskas G, Siksnys V. 2012. The link between restriction endonuclease fidelity and oligomeric state: a study with Bse634I. *FEBS Lett* **586:** 3324–3329.

Zhou XE, Wang Y, Reuter M, Mucke M, Krüger DH, Meehan EJ, Chen L. 2004. Crystal structure of type IIE restriction endonuclease *Eco*RII reveals an autoinhibition mechanism by a novel effector-binding fold. *J Mol Biol* **335:** 307–319.

WWW RESOURCE

http://rebase.neb.com/rebase/rebase.html The Restriction Enzyme Database.

CHAPTER 8

Improved Detection Methods, Single-Molecule Studies, and Whole-Genome Analyses Result in Novel Insights on Structures, Functions, and Applications of Type I, II, III, and IV Restriction Enzymes: ~2004–2016

INTRODUCTION

As mentioned in Chapter 7, with the arrival of whole-genome sequencing projects it has become clear that the Type I subclasses IA–IE and Type III R-M systems are common in bacteria and archaea (http://rebase.neb.com/rebase/rebase.html). The subdivision of Type II REases in 2003 into 11 subtypes based on behavior and cleavage properties (Roberts et al. 2003; Chapter 7) was helpful, but sometimes puzzling: Some REases would fit in more than one category or in none properly; unrelated proteins could be assigned to one or more of these subtypes; and some REases restricted DNA/RNA hybrids (a useful property to study small regulatory RNAs!) (Roberts et al. 2003; Murray et al. 2010; Loenen et al. 2014b). Although this subdivision remains useful, enzyme structures and/or domain organizations can be used as an alternative for classification (Niv et al. 2007; Pingoud et al. 2014). The notion that REases are evolutionarily related despite the lack of sequence similarity has grown more and more compelling, especially because of the increase in crystal structures with the PD···(D/E)XK fold (called the "PD fold"). As discussed in Chapter 7 (and depicted in Figs. 6 and 7 in that chapter), the structural studies by the group of Virginjuis Šikšnys provided formal evidence for a conserved sequence motif in the active site of REases, called the PD (D/E)XK site. The active site residues of EcoRI (G↓AATTC), NgoMIV (G↓CCGGC), and

---

Chapter doi:10.1101/restrictionenzymes_8

Cfr10I (R↓CCGGY) appeared to be common to 11 other REases belonging to Type IIP, IIE, and IIF (Kovall and Matthews 1999; Pingoud et al. 2002; Šikšnys et al. 2004). The almost simultaneous appearance of the structures of BamHI, PvuII, and EcoRV elicited much excitement (Winkler et al. 1993; Cheng et al. 1994; Newman et al. 1994), and both the papers on BamHI (Newman et al. 1994) and PvuII (Cheng et al. 1994) discussed the core structural motifs identified in the paper by Venclovas et al. (1994) when the structures of EcoRI and EcoRV were compared.

Early attempts to change specificity had not been very successful (Wolfes et al. 1986; Jeltsch et al. 1996; Lukacs et al. 2000; Pingoud et al. 2014). Substitutions usually resulted in a decrease in activity, but without exception failed to produce substantial changes in specificity. These findings led to the important lesson that recognition did not simply involve amino acids in direct contact with the bases and the backbone but also required water molecules and a complex network of other interactions (Pingoud et al. 2014). Sequence-specific DNA recognition by REases often involved binding to B-DNA in the major groove, with or without DNA distortion, similar to many regulatory proteins (see e.g., Rohs et al. 2010; Pingoud et al. 2014). In contrast, recognition by base flipping was used by enzymes that do chemistry: MTases, DNA repair enzymes (Roberts and Cheng 1998; Cheng and Roberts 2001), but also some REases (Bochtler et al. 2006; Horton et al. 2006; Tamulaitis et al. 2007; Szczepanowski et al. 2008; Miyazono et al. 2014; Manakova et al. 2015).

The new class of Type IV REases, defined in 2003 (Roberts et al. 2003), are modification-dependent enzymes that recognize modified Cs and As (http://rebase.neb.com/rebase/rebase.html; Roberts et al. 2003). Early findings in the history of modified DNA date back quite a while, long before its importance became known. Modified DNA containing m5C was discovered in 1925 (Johnson and Coghill 1925), followed by m6A (Dunn and Smith 1955a,b), and hm5C in the 1950s (Hershey et al. 1953; Wyatt and Cohen 1953). The analysis of T* mutant phages in 1952 (Chapter 1; Luria and Human 1952) led to the discovery of enzymes that glycosylate hm5C (ghm5C) and of host genes that enable restriction of nonglucosylated phage DNA: *rglA* and *rglB* (restricts glucose-less DNA; Luria and Human 1952; Revel and Luria 1970). The *rglA* and *rglB* genes were renamed *mcrA* and *mcrBC* (modified cytosine restriction) (Noyer-Weidner et al. 1986; Raleigh and Wilson 1986) and were the first designated Type IV REases (Roberts et al. 2003).

Current interest in modified bases is high, as research into the dynamics of DNA modifications ("epigenetic phenomena") have become of paramount importance for research into all kingdoms (Loenen and Raleigh 2014). Interestingly, hm5C was already discovered in eukaryotic (rat) brain and liver in 1972 (Penn et al. 1972) but did not receive much attention until the discovery of the role of Tet (ten-eleven translocation) proteins, a topic outside the scope of this book (Tet proteins are involved in m5C conversion and hence in control of normal and malignant cell differentiation) (see e.g., Veron and Peters 2011; Pastor et al. 2013; Baumann 2014; Stower 2014; Lu et al. 2015; Hendrickson and Cairns 2016; Jeschke et al. 2016). By 1980, some eight types of modified bases in phage DNA had been described (Warren 1980). A little later the m4C modification was found in *Bacillus* (Janulaitis et al. 1983), which is often present in thermophilic and mesophilic bacteria (Ehrlich et al. 1985, 1987). For an extensive description of the techniques used to detect and analyze such modified bases, see Weigele and Raleigh (2016), whose review discusses the initial harsh chemical treatments, physiological methods, paper chromatography, anion exchange columns, high-performance liquid chromatography (HPLC), mass spectrometry (MS), and SMRT. SMRT technology analyzes fluorescently labeled nucleotides that are incorporated slightly slower when encountering modified bases in the template strand than unmodified template bases during the sequencing procedure. This method thus allows the analysis of the "methylome" (i.e., the distribution of methylated bases in the DNA of different organisms).

This chapter uses the reviews that appeared in 2014 in *Nucleic Acids Research* as starting material, plus selected talks and posters presented at the 7th NEB meeting in Gdansk in 2015 (Loenen and Raleigh 2014; Loenen et al. 2014a,b; Mruk and Kobayashi 2014; Pingoud et al. 2014; Rao et al. 2014). Groups in Atlanta (Cheng), Bangalore (Rao and Nagaraja), Baltimore (Chandrasegaran), Berlin (Reuter and Kruger), Bristol (Halford and Szczelkun), Delft (Dekker), Edinburgh (Dryden), Gdansk (Mruk and Skowron), Giessen (Pingoud), Moscow (Zavil'gel'skii), New York (Aggarwal), Piscataway (Bogdanova), Pittsburgh (Jen-Jacobson), Portsmouth (Kneale), Seattle (Stoddard), Tokyo (Kobayashi), Tucson (Horton), Vilnius (Lubys and Šikšnys), and Warsaw (Piekarowicz, Bujnicki, and Bochtler) and at NEB (Roberts, Raleigh, Morgan, and Wilson) made important contributions to the field, as discussed throughout this chapter. Control (C) proteins of Type II enzymes were studied by the groups of Bob Blumenthal in Toledo and Geoff Kneale in Portsmouth. Data about the four types from approximately 2004 onward will be discussed, including structures of some Type II REases, as well as the very first structures of the other types, which together reveal many new, unexpected, and amazing details about the mechanisms employed to prevent indiscriminate restriction by the REase (subunit). Other types of control of restriction were elucidated, via transcription regulation, DNA mimics, C proteins, or the cognate MTase. R-M genes and lone MTase genes in pathogenic organisms also became of great interest because they are linked to virulence via "phase variation" (Piekarowicz 2013). Yet another breakthrough was the discovery of the family of the Type II REase MmeI, which would finally allow the generation of the new specificities so long hoped for. As in the previous chapters, this final chapter starts with the Type II REases (Part A), followed by the ATP-dependent Type I (Part B) and III (Part C) R-M systems, and the modification-dependent Type IV REases (Part D). The final section discusses the phenomenon of phase variation used by pathogenic bacteria to combat phage and evade host immunity (Part E).

PART A: TYPE II ENZYMES

Introduction

By 2014, approximately 4000 REases had been identified belonging to more than 350 different prototype Type II REases (i.e., biochemically different) (Roberts et al. 2010; Pingoud et al. 2014). The majority of these prototypes had characterized or putative relatives in sequenced genomes, resulting in more than 8000 publications (http://rebase.neb.com/rebase/rebase.html; Roberts et al. 2010; Pingoud et al. 2014). Most Type II REases shared little amino acid sequence similarity, with the exception of, for example, EcoRI and RsrI, an early example of "neutral drift": EcoRI and RsrI (recognition site G↓AATTC) are identical in places with 50% overall identity (Aiken et al. 1986; Stephenson et al. 1989), allowing the construction of active hybrids (Chuluunbaatar et al. 2007). Also cases of mosaicism occur—for example, EcoRI, MunI (C↓AATTG), and MluCI (↓AATT) (Pingoud et al. 2014). In this large family, "compelling examples" (Pingoud et al. 2014) could be found of convergent (e.g., HaeIII [GG↓CC] and BsuRI [GG↓CC] [Wilson and Murray 1991]) and of divergent evolution (e.g., Bsu36I [CC↓TNAGG], BlpI [GC↓TNAGC], Bpu10I [CCTNAGC], and BbvCI [CCTCAGC] [Heiter et al. 2005]). In addition to the divalent cations Mg^{2+} and Mn^{2+}, discussed earlier, some Type II REases use Zn^{2+} (BslI [CCNNNNN↓NNGG], PacI [TTAAT↓TAA] [Vanamee et al. 2003; Shen et al. 2010; Horton 2015]) and Co^{2+}, Ni^{2+}, and Cu^{2+} (Pingoud et al. 2014). In the case of the well-known 8-bp cutter NotI (GC↓GGCCGC), the enzyme is dependent on Fe^{2+}, but this Fe^{2+} is incorporated in a structural Cys4 cluster (Lambert et al. 2008; Pingoud et al. 2014). Despite the high specificity of all enzymes, star activity on noncognate sites does occur, which can be partially inhibited by, for example, spermidine, hydrostatic pressure, mutation, or lowering enzyme concentrations (Pingoud et al. 2014).

The number of crystal structures rose from 16 in 2004 (Chapter 7; summarized in Horton et al. 2004) to 35 by 2014 (Pingoud et al. 2014) and to more than 50 new "de novo" (i.e., the first structure of a particular enzyme) enzyme structures in 2017 (Horton 2015). Figure 1A shows the REase structures in the Protein Data Base (PDB) by February 2, 2018 (http://rebase.neb

A

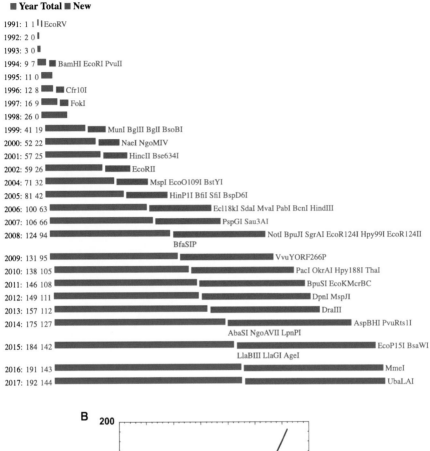

B

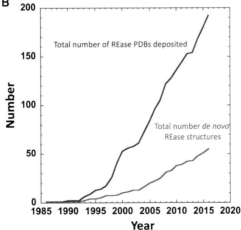

FIGURE 1. (*A*) The REase structures in the Protein Data Base (PDB) by February 2, 2008 (http://rebase.neb.com/cgi-bin/cryyearbar 2017). (*B*) Comparison of the total number of REase structures deposited in the PDB with de novo REase structures. (Courtesy of Nancy Horton.)

.com/cgi-bin/cryyearbar 2017). Figure 1B shows a graph displaying the difference between the total number of REase structures and the de novo structures in the PDB (due to follow-up structures with ligands and/or mutations of a particular enzyme that are also deposited in the PDB by April 2017 [Horton 2015]). Most of these enzymes carry the PD fold, but, in addition, structures of REases with PLD, GIY-YIG, HNH, and half-pipe folds have been elucidated (see page 191). The REases that were the first of their (sub)type (adapted from Horton 2015) are indicated in bold in Appendix 1, which lists selected REases studied from approximately 2004 to 2017, including some earlier references, where appropriate. Although these structures would greatly aid modeling studies, even for well-characterized REases the properties that determine specificity and selectivity remain difficult to predict, because the enzyme is fixed in the crystal and changes conformation during the catalysis, and the additional interactions involved in this "transition" state are not evident in the crystal structure (Lanio et al. 2000; Pingoud et al. 2014).

The picture emerging from all these publications is that (similar to other protein families) the various domains involved in DNA binding, specific recognition, restriction, ATP binding and hydrolysis, and methylation have been fused or separated in all sorts of ways during the course of evolution. As a result, enzymes may have one or two catalytic sites and cleave DNA in one or two steps, with or without sliding and detaching from their DNA and with or without looping (Embleton et al. 2004; Halford and Marko 2004; Halford et al. 2004; Pingoud et al. 2014). Several studies addressed the question of the contribution of 1D and 3D movements of the REases along the DNA in order to find their recognition site (Gowers and Halford 2003; Gowers et al. 2005). Often multimers would bind to two sites rather than acquiring a second catalytic domain, which would be evolutionarily simple. One interesting study by the Bristol group concerns the reaction mechanism of seven REases that recognize GGCGCC and cut at different positions (Gowers et al. 2004). Using plasmids with one or two copies of this sequence revealed five distinct mechanisms, much larger than generally thought at the time (Gowers et al. 2004). Another example includes enzymes specific for the CCNGG sequences (Fig. 1 in Sasnauskas et al. 2015a, adapted in Fig. 3). Nearly 70% of all Type II REases belong to three families; the rest remain "mysteries": They may be fringe members, or examples of new folds and DNA degradation mechanisms (see Pingoud et al. 2014 for further discussion).

Catalytic Domains of Type II REases

The PD···(D/E)XK Structural Fold

The PD···(D/E)XK fold (called the "PD" domain in this chapter) is present with variations in almost all Type II REases whose structures have been determined and is classified in the SCOP (Structural Classification of Proteins) database (http://scop.mrc-lmb.cam.ac.uk) as the REase-like fold (Niv et al. 2007; Steczkiewicz et al. 2012). The PD motif is often not easy to identify without information from a crystal structure, as the motif may vary and the amino acids involved are often in different locations along the polypeptide chain (Pingoud et al. 2014). Among 289 characterized Type II enzymes, 69% belonged to the PD superfamily (Orlowski and Bujnicki 2008) that includes the four nucleases mentioned in Chapter 7 (lambda exonuclease, MutH, VSR, and TnsA), but also, for example, RecB, *Sulfolobus solfataricus* Holliday-junction resolvase, and T7 endo I. The mechanism of catalysis continues to be the subject of study and debate. For example, the number of Mg^{2+} ions needed during catalysis remains uncertain (see Pingoud et al. 2014 for details and discussion).

The HNH and GIY-YIG Structural Domains

Other endonucleolytic motifs have been identified, including HNH and GIY-YIG motifs, found in homing endonucleases (HEases), Holliday-junction resolvases, exonucleases, nonspecific *Serratia* nuclease, and colicins (Friedhoff et al. 1999; Galburt et al. 1999; Jurica and Stoddard 1999; Pingoud et al. 2005a; Stoddard 2005; Kleinstiver et al. 2011, 2013). HNH examples are, for example, KpnI (GGTAC↓C) (Saravanan et al. 2004, 2007b; Vasu et al. 2013), Hpy99I (CGWCG↓), and PacI (TTAAT↓TAA), whereas two GIY-YIG REases, Eco29kI (CCGC↓GG) and Hpy188I (TCN↓GA), have been crystallized (Pertzev et al. 1997; Xu et al. 2000b; Bujnicki et al. 2001; Bujnicki 2004; Ibryashkina et al. 2007; Gasiunas et al. 2008; Kaminska et al. 2008; Orlowski and Bujnicki 2008; Mak et al. 2010; Mokrishcheva et al. 2011; Sokolowska et al. 2011). HNH motifs are often difficult to recognize because of the weak connection between the HNH and the residues that form the active site (Sokolowska et al. 2009). HNH enzymes use Mg^{2+} or Mn^{2+}, but also other ions (Ni^{2+}, Co^{2+}, Zn^{2+}, or Ca^{2+}), sometimes with $Cys4$-Zn^{2+} binding elements (called $\beta\beta\alpha$-metal fold), although many $Cys4$-Zn^{2+} motifs are not associated with catalytic sites but perform structural roles (Saravanan et al. 2004; Orlowski and Bujnicki 2008; Sokolowska et al. 2009; Shen et al. 2010; Pingoud et al. 2014).

Other Endonuclease Structural Domains

Thought unusual at the time, BfiI (ACTGGG [5/4]) was the first REase found that did not belong to the PD family: It carries the PLD nuclease domain and does not require Mg^{2+} for restriction (Sapranauskas et al. 2000). BfiI is a homodimer with a carboxy-terminal "B3-like" DNA-binding domain (DBD), which

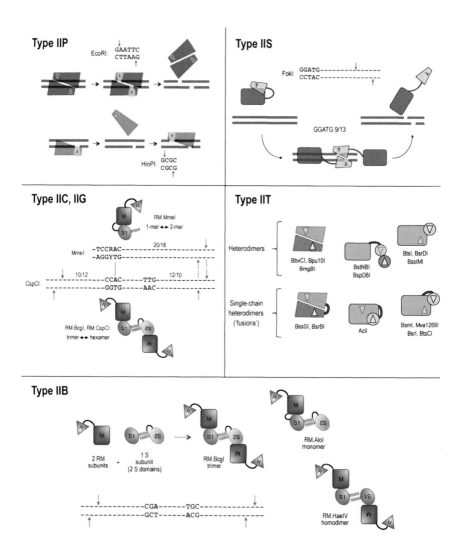

FIGURE 2. Subunit composition and cleavage mechanism of selected subtypes of Type II REases. Type IIP enzymes act mainly as homodimers (*top*) and cleave both DNA strands at once. Some act as dimers of dimers (homotetramers) instead and do the same. Still others act as monomers (*bottom*) and cleave the DNA strands separately, one after the other. Bright triangles represent catalytic sites. Type IIS enzymes generally bind as monomers but cleave as "transient" homodimers. Type IIB enzymes cleave on both sides of their bipartite recognition sequences. Their subunit/domain stoichiometry and polypeptide chain continuity varies. Three examples of primary forms are shown: BcgI, AloI, and HaeIV. These forms assemble in higher-order oligomers for cleavage. Type IIB enzymes display bilateral symmetry with respect to their methylation and cleavage positions. It is not clear whether they cleave to the left or to the right of the half-sequence bound. Type IIG enzymes (e.g., BcgI) might cleave upstream (*left*) of their bound recognition half-site.

(Legend continued on following page.)

resembles B3 domains of some plant transcription factors. The catalytic site is formed at the interface of the two amino-terminal domains (similar to that of Nuc endonuclease from *S. typhimurium*), and although it binds to two sites at once, it cleaves only one strand at a time via an unusual covalent enzyme–DNA intermediate. BfiI appears to swivel the catalytic site by 180° and the same residues perform the same reaction on both DNA strands (Lagunavicius et al. 2003; Sasnauskas et al. 2003, 2007, 2010; Gražulis et al. 2005; Golovenko et al. 2014; Pingoud et al. 2014). Using the classification into 11 subtypes, this enzyme may be assigned to six or more of these subtypes (Marshall and Halford 2010; Pingoud et al. 2014). The enzyme AspCNI (GCCGC [9/5]) has a PLD-like domain and cleaves poorly at high concentrations (Heiter et al. 2015). PLD REases are not as rare as previously thought (Sapranauskas et al. 2000), as REBASE BLAST identified more than 40 other putatives (Pingoud et al. 2014). Some ATP-dependent enzymes (e.g., NgoAVII and CglI) contain a B3-like DNA recognition domain and a PLD catalytic domain (Tamulaitienė et al. 2014).

Type II REase Subtypes

This Type II section gives an overview and update with examples of the 11 subtypes, using two reviews (Roberts et al. 2003; Pingoud et al. 2014) as starting material. Figure 2 shows the subunit composition and cleavage mechanism of selected subtypes of Type II REases. Note that Pingoud et al. (2014) do not always follow the REBASE classification (http://rebase.neb.com/rebase/rebase.html). The reason for this is that different subtypes do not necessarily group with the different branches of the REase evolutionary tree, as exemplified by, for example, members of the EcoRII "CCGG family" studied by the Vilnius group (Table 1), which all cut at the same site (in contrast to the site studied by the Bristol group mentioned above [Gowers et al. 2004]): SsoII ($\downarrow$CCNGG, Type IIP), EcoRII ($\downarrow$CCWGG, Type IIE), and NgoMIV (G$\downarrow$CCGGC, Type IIF) have similar DNA-binding sites and catalytic centers (Pingoud et al. 2002; Niv et al. 2007). Specificities for partly related, and even unrelated, sequences can nevertheless depend on the same structural framework:

FIGURE 2. (*Continued.*) All other Type IIG enzymes (e.g., MmeI) cleave downstream from the site, often with the same geometry. These proteins have very similar amino acid sequences, however, suggesting that somehow the reactions are the same. Type IIT enzymes cleave within or close to asymmetric sequences. Composition varies; they have two different catalytic sites: top-strand-specific and bottom-strand-specific. In some, both subunits/domains interact with the recognition sequence (*left* cartoons). In others, only the larger subunit/domain recognizes the DNA. (Reprinted from Pingoud et al. 2014.)

TABLE 1. The CCGG family studied by the Vilnius group of Virgis Šikšnys

Enzyme	Recognition site	Type	Structure	PDB ID[a]	Reference(s)
Ecl18kI	↓CCNGG	IIF	Dimer/ tetramer	2FQZ, 2GB7	Bochtler et al. 2006
EcoRII	↓CCWGG	IIE	Dimer	3HQF, 3HQG	Zhou et al. 2004; Golovenko et al. 2009
PspGI	↓CCWGG	IIP	Dimer	3BM3	Szczepanowski et al. 2008
PfoI	T↓CCNGGA	IIP	Dimer	Ms. in prep.	Manakova et al. 2015
Kpn2I	T↓CCGGA	IIP	Dimer	Ms. in prep.	Manakova et al. 2015
Cfr10I	R↓CCGGY	IIF	Tetramer	1CFR	Bozic et al. 1996
Bse634I	R↓CCGGY	IIF	Tetramer	3V1Z, 3V20, 3V21, 1KNV	Gražulis et al. 2002; Manakova et al. 2012
NgoMIV	G↓CCGGC	IIF	Tetramer	4ABT (cited in Manakova et al. 2012)	Deibert et al. 2000
BsaWI	W↓CCGGW	IIF	Dimer/ tetramer/ oligomer	4ZSF	Tamulaitis et al. 2015
AgeI	A↓CCGGT	IIP	Monomer/ dimer	5DWA, 5DWB, 5DWC	Manakova et al. 2015; Tamulaitienė et al. 2017
SgrAI	CR↓CCGGYG	IIF	Dimer/ tetramer/ oligomer	4C3G cryoEM 3MQY, 3N78, 3N7B, 3MQ6, 3DVO, 3DW9, 3DPG	Lyumkis et al. 2013; Little et al. 2011; Park et al. 2010; Dunten et al. 2008
UbaLAI	CC↓WGG	IIE	Monomer	5O63	Sasnauskas et al. 2015, 2017

Updated by Elena Manak, Gintautas Tamulaitis, and Giedrius Sasnauskas (November, 2017).

[a] PDB ID is the identification number in the Protein Data Bank.

↓CCNGG (SsoII), ↓CCWGG (PspGI/EcoRII), G↓CCGGC (NgoMIV), R↓CCGGY (Cfr10I), and MboI (↓GATC) (Pingoud et al. 2005c).

Type IIA

Type IIA enzymes usually have separate R and S domains, recognize asymmetric sequences, and cleave within or at a defined position in or close to this site (Roberts et al. 2003; Pingoud et al. 2014). Many have two MTases each modifying one strand of the recognition sequence, rather than a single MTase. Others are combined R-M enzymes, some with separate MTases. Kinetic studies indicate that Type IIA enzymes transiently dimerize for cooperative cleavage. Examples are BbvCI, which uses two different catalytic sites from different subunits (Bellamy et al. 2005; Heiter et al. 2005), and Mva1269I (GAATGC [1/−1], IIA/IIS), which uses two sites from different domains within the same protein (Armalyte et al. 2005).

Type IIB

Type IIB enzymes cleave on both sides of a bipartite site releasing ~34 bp (http://rebase.neb.com/rebase/rebase.html; Roberts et al. 2003; Marshall et al. 2007; Pingoud et al. 2014). Some enzymes comprise a large single RMS polypeptide with features in common with Type I enzymes. Sometimes SAM acts as the cofactor for R as well as for S. Most IIB enzymes can only restrict when bound to two sites, preferably in *cis*, or in *trans* on concatenates. The first IIB R-M system, BcgI (cloned in 1994 [Kong et al. 1994] and extensively studied by the Halford group), concertedly cleaves two double-strand bonds [(10/12) CGANNNNNNTGC (12/10)] (Kong et al. 1993; Marshall and Halford 2010; Sasnauskas et al. 2010; Marshall et al. 2011; Smith et al. 2013a,b; Pingoud et al. 2014). It can be considered IIB/G/H/S (Kong and Smith 1998; Jurenaite-Urbanaviciene et al. 2007; Marshall and Halford 2010; Marshall et al. 2011; Smith et al. 2013a,b; Pingoud et al. 2014), like some other Type IIB enzymes (http://rebase.neb.com/rebase/rebase.html). BcgI comprises two subunits, RM and S, for cleavage and methylation with a stoichiometry of $(RM)_2S_1$, comparable to the Type I pentamer $R_2M_2S_1$, but cutting at fixed positions (Kong et al. 1994; Kong and Smith 1997; Kong 1998). Other enzymes include, for example, BaeI [(10/15) ACNNNNGTAYC (12/7)], BsaXI [(9/12) ACNNNNCTCC (10/7)], and NgoAVIII [(12/14) GACNNNNNTGA (13/11)] (Sears et al. 1996; Marshall and Halford 2010). The BcgI-like enzymes modify both strands of their recognition sequences without additional MTases, and cleavage requires multiple $(RM)_2S_1$ complexes for double-strand cleavage on both sides of the recognition site (Marshall et al. 2007, 2011). The exact mechanism requires further investigation (see Marshall and Halford 2010 for discussion). Other IIB enzymes are, for example, AloI ([7/12] GAACNNNNNNTC [12/7]), PpiI ([7/12] GAACNNNNNCTC [13/8]), CjeI ([8/14] CCANNNNNNNGT [15/9]), and TstI ([8/13] CACNNNNNNTCC [12/7]) (Jurenaite-Urbanaviciene et al. 2007; Smith et al. 2014). Domain-swapping experiments suggest that, like Type I enzymes, TRD swapping may also be used to generate hybrid specificities of Type II enzymes (Jurenaite-Urbanaviciene et al. 2007). Domain swapping and circular permutation of subdomains of BsaXI ([9/12] ACNNNNNCTCC [10/7]), or deletion, resulted in either active protein with altered specificity, poor protein yields, or inactive enzymes, which allowed mapping of critical amino acids for the interaction between the RM subunit and the TRD of the S subunit (Xu et al. 2015).

Type IIC

Type IIC are combined RM enzymes (http://rebase.neb.com/rebase/rebase .html; Roberts et al. 2003; Pingoud et al. 2014). Most IIC bind as monomers to continuous and asymmetric sequences and cleave on one side of the recognition site at 1 turn, 1½ turn, or 2 turns away, whereas others cleave on both sides (i.e., IIB). Cleavage via transient dimerization is likely and is more efficient on DNA with multiple recognition sites or on addition of oligonucleotides. Examples are Eco57I (CTGAAG [16/14]), MmeI (TCCRAC [20/18]), and BpuSI (also called RM·BpuSI, GGGAC [10/14]), which are also considered to be IIC as well as IIE or IIG, respectively (see pages 202–203).

Type IIE

The prototype Type IIE enzymes are EcoRII and NaeI with separate domains for cleavage and allosteric activation (http://rebase.neb.com/rebase/rebase .html; Roberts et al. 2003; Pingoud et al. 2014), as discussed in Chapter 7 and, for example, Reuter et al. (2004). The Type IIE enzymes prove to be diverse in structure (Fig. 3). Figure 3 shows a comparison of the structures of NaeI (Fig. 3A), EcoRII (Fig. 3B), and a new Type IIE enzyme from an unknown bacterium, named UbaLAI (Sasnauskas et al. 2017) by the CCGG group in Vilnius (Fig. 4).

Nearly a dozen papers were published on EcoRII in collaborations between experts in the field of crystallography, AFM, and single-molecule studies (Zhou et al. 2002, 2003, 2004; Kruger and Reuter 2005; Tamulaitis et al. 2006a,b, 2008; Shlyakhtenko et al. 2007; Gilmore et al. 2009; Golovenko et al. 2009; Szczepek et al. 2009). A high-resolution crystal structure of the dimeric EcoRII was published in 2004, which revealed a hinge loop connecting the catalytic and allosteric activation domains (Zhou et al. 2002, 2003, 2004). The catalytic domain (comprised of two copies of the carboxy-terminal domain) had the PD fold, whereas the two amino-terminal regulatory/effector domains had a different DNA recognition fold with a large cleft. This fold was novel at the time, but is in fact the B3-like fold mentioned above and present in BfiI and NgoAVII (more specifically, it is a SCOP double-split β-barrel fold, of the DNA-binding pseudobarrel domain superfamily). The structure explained the mechanism of autoinhibition/activation of EcoRII, which was novel in REases, but similar to that described for various transcription factors (Zhou et al. 2004). This structure contained three possible DNA-binding regions, and in line with this, only

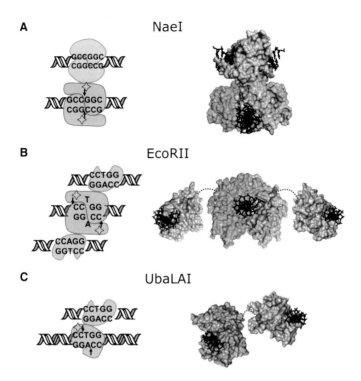

FIGURE 3. Diversity of Type IIE REases. In all panels PD···(D/E)XK subunits are colored in different shades of green, monomeric MvaI-like PD···(D/E)XK domains are red, catabolite activator protein (CAP)-like domains are orange and light brown, and B3-like domains are blue. Yellow diamonds in the cartoon representations denote the catalytic center(s) present in each enzyme. (*A*) NaeI (GCC↓GGC) is a Type II homodimer that simultaneously binds two recognition sites. One is cleaved by the EcoRV-like dimer of the catalytic N domains (Endo domains), whereas the second one bound to the CAP DNA-binding motif in the carboxy-terminal domain (Topo domain) stimulates cleavage of the first site (PDB ID: 1IAW) (Embleton et al. 2001; Huai et al. 2001). (*B*) EcoRII (↓CCWGG) is a Type IIE homodimer capable of simultaneous binding of three recognition sites. One is cleaved by the PspGI-like dimer of the catalytic C domains, whereas two others, one per EcoRII-N effector domain, stimulate cleavage of the first site (PDB IDs: 3hqf and 3hqg) (Tamulaitis et al. 2006a,b; Golovenko et al. 2009). (*C*) UbaLAI (CC↓WGG) is a novel monomeric REase consisting of an MvaI-like catalytic domain (red) and an EcoRII-N-like effector domain (blue; PDB ID to be published). UbaLAI requires two recognition sites for optimal activity, and, like NaeI and EcoRII, uses one copy of a recognition site to stimulate cleavage of a second copy. UbaLAI-N acts as a handle that tethers the monomeric UbaLAI-C domain to the DNA, thereby helping UbaLAI-C to perform two sequential DNA nicking reactions on the second recognition site during a single DNA-binding event (Sasnauskas et al. 2017). The structure of the UbaLAI-C domain is a model built using Modeller (Webb and Sali 2016). (Portions reprinted from Sasnauskas et al. 2017, courtesy of Gintautas Tamulaitis.)

A

B

FIGURE 4. (*A*) CCGG group photo. From *left* to *right*: Inga Songailiene, Gintautas Tamulaitis, Elena Mankova, Saulius Gražulis, Virgis Šikšnys, Giedrė Tamulaitienė, Giedrius Sasnauskas, and Mindaugas Zaremba. (*B*) Graduate students and postdoctoral colleagues from 1977 through 2011 at Steve Halford's retirement party (2011). His group, from *left* to *right*: Stuart Bellamy, Dave Scott, Rachel Smith, Kelly Sanders, Niall Gormley, Tim Nobbs, Mark Watson, Panos Soultanos, Geoff Baldwin, Steve Halford (in his DNA jumper), Darren Gowers, Mark Szczelkun, Neil Stanford, Jacqui Marshall, Barry Vipond, Yana Kovacheva, Katie Wood, Tony Maxwell, Isobel Kingston, John Taylor, Sophie Castell, Michelle Embleton, Christian Vermote, Alistair Jacklin, Alison Ackroyd, Fiona Preece, Susan Retter, Lucy Catto, Shelley Williams. Christian Parker and Denzil Bilcock were at the party but not in the photo. (Absent: Pete Luke, Paul Bennett, Samantha Hall, Lois Wenztell, Symon Erskine, Mark Oram, Abigail Bath, David Rusling, and Sumita Ganguly).

a plasmid with three recognition sites yielded linear DNA during a single turnover, whereas the same plasmid with only one or two sites did not (Tamulaitis et al. 2006b). AFM studies showed two-loop structures with an EcoRII dimer at the core of the three-site synaptosome (Shlyakhtenko et al. 2007). A variant of AFM (called high-speed AFM) allowed single-molecule imaging of the EcoRII protein (Gilmore et al. 2009). In this way, binding, translocation,

C

D

FIGURE 4. *(Continued.)* (*C*) Werner Arber, Noreen Murray, and D.N. Rao at the 6th NEB meeting in Bremen (2010). (*D*) Participants of the CSHL meeting in 2013: History of Restriction Enzymes. (*D*, Courtesy Cold Spring Harbor Laboratory Archives.)

and dissociation could be monitored, and they indicated that EcoRII can translocate along the DNA to search for a second binding site, after finding the first site. Dissociation from the loop structure resulted in either two monomers bound to the two sites or one dimer to one site (Gilmore et al. 2009). Further experiments showed the very different ways in which the enzyme interacted with the effector and substrate DNA. The carboxy-terminal domain flipped the central T:A base pair out, and interacted with the CC:GG half-sites, whereas the effector domain bound asymmetrically without pushing out the T:A base pair (Golovenko et al. 2009). Interestingly, the 7-bp cutter PfoI (T↓CCNGGA) also uses base flipping as part of its DNA recognition mechanism. But in this case the extrahelical bases are captured in binding pockets that are quite different from those in the related structurally characterized enzymes Ecl18kI, PspGI, and EcoRII-C (Manakova et al. 2015). PspGI (↓CCWGG) and Ecl18kI/SsoII (↓CCNGG) flip the central A and T (W) bases out of the helix, compressing the recognition sequence in effect to just CC-GG (Bochtler et al. 2006; Tamulaitis et al. 2007; Szczepanowski et al. 2008). Repression of catalysis by the amino-terminal domain was further analyzed by site-directed mutagenesis and addition of soluble peptides in *trans*, which revealed the structural elements essential for autoinhibition (Szczepek et al. 2009). The crystal structure of MvaI identified MvaI as a monomer that recognizes its pseudo-symmetric target sequence (CC↓WGG) asymmetrically (Kaus-Drobek et al. 2007). The enzyme has two lobes: a catalytic one that contacts the bases from the minor groove side, and the other that contacts those from the major groove. MvaI resembles BcnI (CC↓SGG), and also MutH, which nicks DNA rather than cutting both strands. The reason for this is clear: MvaI, BcnI, and MutH have a single catalytic site and just nick their substrates upon binding. Because the substrates of MvaI and BcnI are symmetric, these two enzymes can then bind in the opposite orientation and nick the other strand resulting in double-strand cleavage. The substrate of MutH (hemimethylated GATC) is asymmetric, and so MutH can only bind in one orientation and thus cannot cut the second strand. Different responses to slight substrate asymmetries, which could be altered by protein engineering, determine whether these monomeric REases make single-strand nicks or double-strand breaks (Sokolowska et al. 2007a; see Kaus-Drobek et al. 2007 for further details). For some other studies on the EcoRII and CCGG family, see Kubareva et al. (1992, 2000); Šikšnys et al. (2004); Pingoud et al. (2005b); Sud'ina et al. (2005); Zaremba et al. (2006); Fedotova et al. (2009); and Abrosimova et al. (2013), and the Type IIF section.

Type IIF

Type IIF bind two recognition sites and cleave all four strands at once as pairs of back-to-back dimers (http://rebase.neb.com/rebase/rebase.html; Roberts et al. 2003; Šikšnys et al. 2004; Zaremba et al. 2005; Pingoud et al. 2014). The structures of Cfr10I (R↓CCGGY), Bse634I (R↓CCGGY), and NgoMIV (G↓CCGGY) (Chapter 7) and the observed transient tetramerization of Ecl18kI (↓CCNGG) indicated that the boundaries between IIE and IIF are not strict (Šikšnys et al. 2004; Zaremba et al. 2005, 2010; Pingoud et al. 2014). Work on Bse634I continued in Vilnius (Zaremba et al. 2005, 2006, 2012; Manakova et al. 2012). The tetramer could be converted to a dimeric enzyme by mutation, and kinetic studies indicated two types of communication signals via the dimer–dimer interface in the tetramer: an inhibitory and an activating signal, which somehow control the catalytic and regulatory properties of the Bse634I and mutant proteins (Zaremba et al. 2005, 2006). Contrary to expectation, dimeric enzymes have the same fidelity toward their recognition site as the tetramer, because they act concertedly at two sites, thus providing a safety catch against cleavage at a single unmodified site (Zaremba et al. 2012). The structures of the SfiI (GGCCNNNN↓NGGCC) tetramer in complex with cognate DNA provided details on how SfiI recognized and cleaved its target DNA sites (Viadiu et al. 2003; Vanamee et al. 2005). Some other Type IIF enzymes are PluTI (GGCGC↓C) (Khan et al. 2010; Pingoud et al. 2014) and SgrAI (CR↓CCGGYG). SgrAI (Laue et al. 1990; Tautz et al. 1990; Capoluongo et al. 2000; Bitinaite and Schildkraut 2002; Daniels et al. 2003; Hingorani-Varma and Bitinaite 2003; Wood et al. 2005; Dunten et al. 2008, 2009; Park et al. 2010; Little et al. 2011; Lyumkis et al. 2013; Ma et al. 2013b; Horton 2015) is also a member of the CCGG family and preferentially cleaves concertedly at two sites. Interestingly, SgrAI assembles into homotetramers, and then other molecules join to generate helical structures with one DNA-bound homodimer after another. Adjacent homodimers are not back-to-back (i.e., 180°), but at ∼90°, and four homodimers form almost one turn of a left-hand spiral of 18 homodimers or perhaps even more. These SgrAI filaments have some star activity, probably as a result of asymmetry generated by the multimerization process (Fig. 5). Another interesting enzyme is the Type IIF homotetrameric GIY-YIG Cfr42I enzyme that is rather similar to the monomeric/dimeric Eco29kI enzyme, which supports the notion of convergent evolution of REases belonging to unrelated nuclease families toward homo-tetramers with a "safety catch" (Gasiunas et al. 2008).

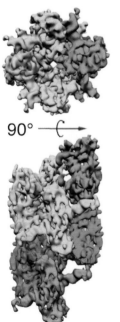

90°

FIGURE 5. CryoEM structure of SgrAI bound to DNA. Each SgrAI dimer is colored uniquely. This picture was made using PDB coordinates and surface rendering. (The original figure in the paper by Lyumkis et al. [2013] was the actual cryoEM envelope, carved up into different subunits.) (Adapted from Lyumkis et al. 2013, with permission from Elsevier.)

Type IIG

Type IIG are Type I-like combined RM systems, with an amino-terminal PD domain, and a γ-class MTase domain in a single protein (http://rebase.neb.com/rebase/rebase.html; Roberts et al. 2003; Niv et al. 2007; Loenen et al. 2014a; Pingoud et al. 2014). The S specificity subunit may be present as a separate subunit or as a domain attached to the carboxyl terminus of RM. IIG are stimulated by SAM or are SAM-dependent. This definition includes most IIB and IIC REases (Loenen et al. 2014a; Pingoud et al. 2014). Only one catalytic site is present in these domains, and cleavage of duplex DNA is thought to occur by the transient dimerization of neighboring enzyme molecules. Examples are Eco57I, MmeI, and BpuSI.

Eco57I was the first member of a new class of monomeric enzymes, initially called Type IV (like BspLU11III [GGGAC (10/14)] from *Bacillus* sp. LU11 [Lepikhov et al. 2001]), but renamed Type IIG enzymes (Janulaitis et al. 1992a,b), although it can also be considered IIE as it is accompanied by one additional MTase. It is a large RMS protein, cuts one and one-half turns away, and is useful for engineering (Janulaitis et al. 1992b; Rimseliene et al. 2003; Pingoud et al. 2014). It methylates the top strand of its asymmetric recognition site (CTGAAG [16/14]), whereas a separate MTase, M·Eco57I, methylates the adenine in the bottom strand (Janulaitis et al. 1992a). M·Eco57I can also methylate the same adenine in the top strand as Eco57I (Janulaitis et al. 1992a; Loenen et al. 2014a). Some other monomeric Type IIG have accompanying MTases that methylate m5C (BpuSI) or m4C (BseRI, GAGGAG [10/8]) (Loenen et al. 2014a).

MmeI is IIE/IIG/IIC and cuts two turns away (TCCRAC [20/18]). It was the first Type IIG enzyme to be purified and belongs to a large family of closely related enzymes with many different specificities (Boyd et al. 1986; Morgan et al. 2008; Loenen et al. 2014a). Based on in vitro studies, MmeI has also been named Type IIL, for lone-strand DNA modification (Morgan et al. 2009). As the enzyme does not require a head-to-head approach in vitro, there is disagreement on its mode of action: Does in vivo MmeI act on two inverted (head-to-head) recognition sequences like Type III enzymes (Dryden et al. 2011; Schwarz et al. 2011; Loenen et al. 2014a), using sliding or DNA looping between adjacent sites (Halford et al. 1999; Halford 2001), or perhaps bind DNA as a monomer and then form dimers or multimers before methylation or cleavage, similar to Type I enzymes (Loenen et al. 2014a)? But why would MmeI slide along the DNA, as the adenine that will eventually be methylated is likely to be flipped into the binding pocket on specific site recognition (Cooper et al. 2017; Bogdanove et al. 2018)? MmeI requires at least two bound specificity sites for cutting. Unlike FokI, adding excess enzyme in solution, without a specific site, does not stimulate cutting. Richard (Rick) Morgan suggests a model that includes the requirement for enzyme bound at two (or possibly four) sites to come together for cutting (Cooper et al. 2017; Bogdanove et al. 2018). As methylation is effective at single sites, this process does not require dimerization of the enzyme.

MmeI has been well characterized (Boyd et al. 1986; Tucholski et al. 1995, 1998; Nakonieczna et al. 2007, 2009; Morgan et al. 2008, 2009; Callahan et al. 2011, 2016), and rational engineering based on sequence alignments and mutational analysis led to altered specificities that could be predicted (Morgan and Luyten 2009; Morgan et al. 2009). Changes in the S domain alter the recognition site for both R and M (like Type I enzymes), and hence members of the MmeI family have been able to diverge widely in the course of evolution (Morgan et al. 2008, 2009; Morgan and Luyten 2009). Certain different pairs of amino acids are specific for alternative base pairs in the recognition sequence: for example, Glu^{806}...Arg^{808} in MmeI (TCCRAC) specifies the third C, whereas Lys^{806}...Asp^{808} specifies G at that position (TCCRAG). The crystal structure has been solved (Callahan et al. 2011, 2016). Together with the structure of MmeI in complex with DNA (and SAM-analog sinefungin) (Callahan et al. 2011), these data on the MmeI family allowed the construction of REases with novel predictable DNA recognition and restriction properties, which had "long been a goal of modern biology" (Callahan et al. 2016) and previously denied for EcoRI and EcoRV. With this in mind, Geoff Wilson pondered whether one could predict and design new specificities of other enzymes (e.g., Type I HsdS) or even predict those of putative HsdS subunits in REBASE based on sequence data alone (Loenen et al. 2014a). The answer to this is yes, as in recent times Rick Morgan has predicted and made specificity changes in Type I HsdS systems (R Morgan, in prep.).

BpuSI (GGGAC [10/14]) is IIG or IIS and has two MTases (Shen et al. 2011; Sarrade-Loucheur et al. 2013; Pingoud et al. 2014). The crystal structure indicates that it resembles the well-characterized carboxy-terminal cleavage domain of FokI (GGATG [9/13]) and produces 5′ sticky ends (Wah et al. 1997, 1998; Shen et al. 2011). This is unusual because most Type IIG enzymes create 3′ overhangs, indicating that their catalytic sites cleave across the minor groove of DNA rather than across the major groove. BpuSI was crystallized without DNA and evidently must undergo significant structural rearrangements to bind DNA and carry out catalysis (Shen et al. 2011). This means that the carboxy-terminal S domain must rotate with respect to the R and M domains and reorganize in order to bind DNA (also seen with other REases) (Shen et al. 2011; Sarrade-Loucheur et al. 2013; Pingoud et al. 2014).

Type IIH

Type IIH are hybrid Type IIP-like (e.g., GACNNN↓NNGTC) REases with an m6A MTase (Pingoud et al. 2014). M·AhdI is a tetramer of M and S subunits, suggestive of the ancestral form of Type I MTases. As such, they have been called "Type 1½" RM systems, and a "missing link" between Type I and II IIH enzymes, but as they have proved rather common, this distinction may no longer be relevant (Marks et al. 2003; Pingoud et al. 2014).

Type IIM

Type IIM enzymes recognize methylated DNA. The well-known DpnI (Lacks and Greenberg 1975; Pingoud et al. 2014) cuts Gm6A↓TC as a monomer, one strand at a time. The complementary specificities of DpnI and DpnII have been useful for site-directed mutagenesis, as DpnII cuts unmethylated ↓GATC) sites (Lacks and Greenberg 1977). DpnI has the amino-terminal PD domain and a carboxy-terminal winged-helix (wH) allosteric activator domain. Both domains bind methylated DNA with sequence specificity (Lacks and Greenberg 1975; Siwek et al. 2012; Mierzejewska et al. 2014; Pingoud et al. 2014). A new addition to this subtype is the BisI (Gm5CNGC) enzyme and its relatives (Xu et al. 2016). Some enzymes that recognize methylated DNA and are classified as Type IV enzymes would also fit into the IIM subtype, if they cut at specific sites (see the section Part D: Type IV Enzymes).

Type IIP

The best-known orthodox Type IIP palindromic REases are, of course, EcoRI and EcoRV. Type IIP cleave symmetric recognition sequences and have a single domain in which recognition and cleavage functions are integrated (Pingoud et al. 2014). They tend to have a single cognate MTase, although some have two MTases. The IIP REases can be monomeric but most are homodimers or homotetramers. Multimers usually cleave both DNA strands in one binding event, whereas monomers need to cleave sequentially first one strand, then the other, because of the opposite 5′ to 3′ polarity of the DNA strands (Gowers et al. 2004; Pingoud et al. 2014). In line with this prediction, the BcnI (CC↓SGG) monomer first localizes the recognition site by 1D and 3D diffusion, and nicks one DNA strand; it then diffuses from the nicked site, turns 180°, diffuses back, and cleaves the other (unnicked) strand (Sokolowska et al. 2007b; Kostiuk et al. 2011, 2015, 2017; Sasnauskas et al. 2011).

Single-molecule studies with EcoRV provided evidence for fast 3D sliding and jumping of EcoRV on nonspecific DNA following a slow initial 1D diffusion (Bonnet et al. 2008). Using optical tweezers with fluorescence tracking, it became clear that the enzyme stays in close contact with the DNA during sliding (Bonnet et al. 2008; Biebricher et al. 2009). Aneel Aggarwal and coworkers analyzed the structure of BstYI, a thermophilic REase that cleaves 5'-Pu/ GATCPy-3', a degenerate version of the BamHI (G↓GATCC) and BglII (A↓GATCT) recognition sites. A comparison of free BstYI with BamHI and BglII revealed a strong structural likeness between these enzymes, but in addition, BstYI also contained an extra "arm" domain possibly related to the thermostability of BstYI (Townson et al. 2004). The cocrystal structure with DNA revealed a mechanism of degenerate DNA recognition, which will stimulate thoughts about the possibilities and limitations in altering specificities of closely related REases (Townson et al. 2005). Interestingly, an isoschizomer of BamHI, OkrAI (G↓GATCC), is a much smaller version of BamHI, which recognizes the DNA in a similar manner, "a rare opportunity to compare two REases that work on exactly the same DNA substrate" (Vanamee et al. 2011).

The group of Ichizo Kobayashi studied regulation of the EcoRI operon (see page 214) (Liu and Kobayashi 2007; Liu et al. 2007), whereas the group of Linda Jen-Jacobsen in Pittsburgh continued studies on EcoRI with respect to the mechanism of coupling between DNA recognition specificity and catalysis (Kurpiewski et al. 2004), the inhibition by Cu^{2+} ions of Mg^{2+}-catalyzed DNA cleavage (Ji et al. 2014), and the relaxed specificity and structure of promiscuous mutants of EcoRI that cleave at EcoRI* sites (Sapienza et al. 2005, 2007, 2014). As EcoRI* sites are not protected by M·EcoRI, promiscuous mutants are deleterious to the host. They encountered "unanticipated and counterintuitive observations" that three EcoRI mutants with such relaxed specificity in vivo nevertheless bound more tightly than wild-type EcoRI to the cognate site (GAATTC) in vitro and even preferred that site to EcoRI* sites (Sapienza et al. 2005). How could this be? Using structural and thermodynamic analyses, this question was addressed further (Sapienza et al. 2007, 2014). The crystal structure of the promiscuous mutant A138T homodimer in complex with the cognate site was nearly identical to that of the wild-type complex, except that the threonine138 side chains interacted with bases 5' to the GAATTC site. This would enable A138T to form complexes with EcoRI* sites that structurally resembled the specific wild-type complex with GAATTC (Sapienza et al. 2007). The importance of these flanking bases was also confirmed by the finding that AAATTC sites with an adjacent 5'-purine-pyrimidine (5'-RY) were cleaved much faster (up to 170× faster!). This and further thermodynamic analyses supported the notion that specificity relied on a series of cooperative events that were "uniquely associated with specific recognition" (Sapienza et al. 2014).

SwaI (ATTT↓AAAT) (Dedkov and Degtyarev 1998) and PacI (TTAAT↓ TAA) both recognize AT-rich DNA sequences, but their protein structures are completely different (Shen et al. 2010). In the case of PacI, the normal base-pairing is completely disrupted in the bound structure: "two bases on each strand are unpaired, four are engaged in noncanonical A:A and T:T base pairs, and the remaining two bases are matched with new Watson–Crick partners." This suggests that PacI is an unusual REase that recognizes its target site via contacts not visualized in the DNA-bound cocrystal structure (Shen et al. 2010). Whereas PacI is elongated and follows the track of the DNA helix (Shen et al. 2010), SwaI is flattened and horseshoe-shaped (Shen et al. 2015). SwaI has an open conformation with the DNA-binding surface accessible from the outside. When bound to DNA, the enzyme is closed and completely encircles the DNA. Like PacI, SwaI profoundly distorts the DNA on binding, but in a different way (Shen et al. 2010). In SwaI, the central T:A and A:T bases are unpaired, and the two adenines switch positions and stack on each other in the reverse order. This is accompanied by a ∼50° bend in the helix and severe compression of the major groove, much as is seen in EcoRV (GAT↓ATC) (Winkler et al. 1993). The authors had no idea "how this surprising reversal in base order takes place" (Shen et al. 2015, 2017). Like EcoP15I, which has been used to count Huntington's disease CAG repeats, TseI (G↓CWGC) is also useful for the analysis of A:A and T:T mismatches in CAG and CTG repeats in this dreadful disease (Moncke-Buchner et al. 2002; Ma et al. 2013a).

Type IIS

Type IIS cut at a fixed distance from the recognition site (http://rebase.neb .com/rebase/rebase.html; Szybalski et al. 1991; Roberts et al. 2003; Welsh et al. 2004; Niv et al. 2007; Pingoud et al. 2014). The recognition and cleavage domains are separated by a linker region allowing fusion of the cleavage domain to other recognition modules, thus generating novel specificities. They usually have two MTases, each methylating one of the two strands (m6A or m5C). The name Type IIS (for "shifted") enzymes was first coined by Wacław Szybalski and coworkers at the University of Wisconsin, who devised "ingenious applications" in the 1980s (Hasan et al. 1986; Kim et al. 1988; Pingoud et al. 2014). All Type IIB, IIC, and IIG REases can be considered IIS (cut outside their recognition sites), but all share the integral γ-class MTase as described above.

FokI (GGATG [9/13]) is one of the earliest, most-studied, Type IIS enzymes with a DNA recognition domain and a separate cleavage domain, which has been used extensively for genome engineering (Sugisaki and Kanazawa 1981; Nwankwo and Wilson 1987; Mandecki and Bolling 1988; Kaczorowski et al. 1989; Kita et al. 1989a,b; Landry et al. 1989; Looney et al. 1989; Sugisaki et al. 1989; Goszczynski and McGhee 1991; Szybalski et al. 1991; Li et al. 1993; Skowron et al. 1993; Kim et al. 1994; Waugh and Sauer 1994; Yonezawa and

Sugiura 1994; Kim et al. 1996a, 1997, 1998; Skowron et al. 1996; Hirsch et al. 1997; Wah et al. 1997, 1998; Bitinaite et al. 1998; Leismann et al. 1998; Chandrasegaran and Smith 1999; Friedrich et al. 2000; Vanamee et al. 2001; Catto et al. 2006; Laurens et al. 2012; Pernstich and Halford 2012; Rusling et al. 2012; Guilinger et al. 2014b; Mino et al. 2014; Pingoud et al. 2014). The accompanying two MTases are fused into a single protein. Single-particle EM studies provided new insights into the activation mechanism of FokI and avoidance of aspecific cleavage (Vanamee et al. 2007). FokI crystals show the catalytic domain to be hidden behind the DNA recognition domain, which will require a substantial conformational change before cutting can take place after dimerization of two catalytic domains (Bitinaite et al. 1998; Pingoud et al. 2014). Details on the cleavage mechanism still need to be sorted out, but the need for two enzyme molecules for catalysis appears to be quite common (Embleton et al. 2001; Welsh et al. 2004; Catto et al. 2006, 2008; Gemmen et al. 2006; Sanders et al. 2009; Pingoud et al. 2014). Other enzymes also have a FokI-like domain—for example, StsI (GGATG [10/14]) (Kita et al. 1992a,b) and Mva1269I (GAATGC [1/−1]) (Armalyte et al. 2005).

Type IIT

Type IIT enzymes are heterodimers with two subunits (e.g., Bpu10I, BbvCI) or heterotetramers (e.g., BslI [CCNNNNN↓NNGG] [http://rebase.neb.com/rebase/rebase.html; Roberts et al. 2003; Pingoud et al. 2014]). IIT use two different catalytic sites for cleavage. Some enzymes are single chain (e.g., Mva1269I uses an EcoRI-like domain and a FokI-like domain) (Armalyte et al. 2005; Pingoud et al. 2014). Type IIT systems usually have two MTases (either separate proteins or fused as a single protein) that each modify one strand. They are useful after conversion to strand-specific nicking enzymes (see Chan et al. 2011 for a review, and page 209)—for example, BbvCI has two catalytic sites from different subunits, each cleaving its own strand (Bellamy et al. 2005; Heiter et al. 2005).

The "Half-Pipe"

PabI of *Pyrococcus abyssi* (GTA↓C) was thought to be a bona fide REase, as it was found near a MTase gene (Pingoud et al. 2014). However, it is a homodimeric DNA glycosylase with a unique structure and flips all four purines out of the helix, leaving the pyrimidines as intrahelical "orphans" (Ishikawa et al. 2005; Watanabe et al. 2006; Miyazono et al. 2007, 2014; Pingoud et al. 2014; Kojima and Kobayashi 2015). Close isoschizomers of PabI are ubiquitous in *Helicobacter pylori* strains. Whether PabI is involved in genetic rearrangements remains to be investigated (Pingoud et al. 2014).

Type II Enzymes as Tools for Gene Targeting

Fusions

As briefly discussed in Chapter 7, Srinivasan Chandrasegaran at Johns Hopkins School of Medicine pioneered what is now termed gene targeting by fusing the REase endonuclease domain of FokI to a zinc-finger protein to create a novel engineered zinc-finger nuclease (ZFN) (Li and Chandrasegaran 1993). ZFNs usually contain three to six Zn fingers (each ~30 aa) with a ββα fold that binds one Zn^{2+} via 2Cys + 2His (Miller et al. 1985; Klug 2010a,b). Each finger recognizes a 3-bp target sequence via four amino acids that project from the α-helix into the major DNA groove (Durai et al. 2005; Wu et al. 2007). Two different three-finger ZFNs will recognize an 18-bp sequence, sufficient to be unique in the human genome. Such constructs have been used with considerable success, although they tend to be less specific than expected (Urnov et al. 2005, 2010; Carroll 2011a,b; Gabriel et al. 2011; Handel and Cathomen 2011; Pattanayak et al. 2011; Perez-Pinera et al. 2012a; see also Carroll 2014; Carroll and Beumer 2014; Hendel et al. 2015).

ZNF-based engineered highly specific REases can be used for gene targeting by introducing a dsDNA break into a complex genome and thereby stimulating homologous recombination (Yanik et al. 2013; Carroll 2014). With the exception of engineered homing endonucleases ("meganucleases") with integrated DBD and catalytic domains (Galetto et al. 2009), the other engineered nucleases have distinct DBD and catalytic domains. ZFNs usually have the cleavage domain of FokI (Li et al. 1992, 1993; Li and Chandrasegaran 1993; Waugh and Sauer 1993; Kim et al. 1994, 1996b; Chandrasegaran and Smith 1999; Bibikova et al. 2002; Urnov et al. 2005; Miller et al. 2007; Szczepek et al. 2007; Mino et al. 2009; Mori et al. 2009; Carroll 2011a,b; Gabriel et al. 2011; Handel and Cathomen 2011; Pattanayak et al. 2011; Ramalingam et al. 2011, 2013; Handel et al. 2012; Bhakta et al. 2013; Pingoud et al. 2014), but PvuII (CAG↓CTG) has also been used for this purpose (Schierling et al. 2012).

Nonspecific ("off-target") cleavage can be reduced by mutations in the dimerization surface (Miller et al. 2007; Szczepek et al. 2007), but according to Steve Halford the off-target problem might well be due to dimerization between a specific and a nonspecific ZFN (Halford et al. 2011).

The fusion construct with PvuII (CAG↓CTG) was slightly better than that with FokI (Schierling et al. 2012; Pingoud et al. 2014), but fusions with the transcription activator-like effector (TALE) proteins, where one module recognizes one base (Fig. 6A,B; Pingoud et al. 2014), were an improvement on engineered nuclease constructs. These proteins contain many (up to 35) nearly identical repeats of ~34 aa. The 13th residue in each repeat recognizes the

DNA base. The repeats form a superhelix around the DNA, following the track of the major groove for several turns. The individual repeats are left-handed two-helix bundles that, one after the other, juxtapose the 13th amino acid of each repeat to adjacent bases in one strand of the DNA (Deng et al. 2012; Mak et al. 2012, 2013).

In the case of PvuII, the DBD of a TALE protein is fused via a linker of defined length to the homodimeric REase (Fig. 6B). Wild-type PvuII (wtPvuII) is shown on the left in Figure 6Ba, and a variant of PvuII as a TALE-linked monomer (scPvuII) on the right in Figure 6Ba. A model of a TALE-PvuII fusion protein was constructed using the structures PDB 1pvi (Cheng et al. 1994) and PDB 3ugm (Mak et al. 2012) on a DNA composed of the PvuII recognition site and two TALE target sites upstream of and downstream from the PvuII recognition site, separated by 6 bp (Fig. 6Bb). The fusion protein is a dimer of identical subunits, each composed of a PvuII subunit and a TALE protein.

TALE-based nucleases (engineered TALE nucleases [TALENs]), based on FokI and PvuII, proved much better tools for genome manipulations than did ZFNs (Miller et al. 2011; Perez-Pinera et al. 2012b; Joung and Sander 2013; Yanik et al. 2013), but they also have some off-target activity. Profiling of 30 different unique TALENs for the ability of potential off-target cleavage using in vitro selection and high-throughput sequencing resulted in 76 predicted off-target substrates in the human genome, 16 of which were accessible and modified by TALENs in human cells (Guilinger et al. 2014a). This analysis allowed the construction of a TALEN variant with ~10× lower off-target activity in human cells (Guilinger et al. 2014a).

In 2014, FokI was fused to Cas9, which cleaves dsDNA at a sequence programmed by a short single-stranded guide RNA (Guilinger et al. 2014b). Unfortunately, genome editing by Cas9 can also result in off-target DNA recognition. Fusions of catalytically inactive Cas9 and FokI nuclease (fCas9) modified target DNA sites with >140-fold higher specificity than wild-type Cas9 and with an efficiency similar to that of paired Cas9 "nickases" that cleave only one DNA strand each. The specificity of fCas9 was at least fourfold higher than that of paired nickases and may be a good strategy for highly specific genome-wide editing (Guilinger et al. 2014b). Use of very long (up to 10 kb) homologous flanking arms for break repair also improves targeting (Baker et al. 2017).

Nickases (Nicking Enzymes)

Another approach to gene targeting has been the use of nickases. Precise incisions in genomic DNA are required for (faithful) homologous recombination,

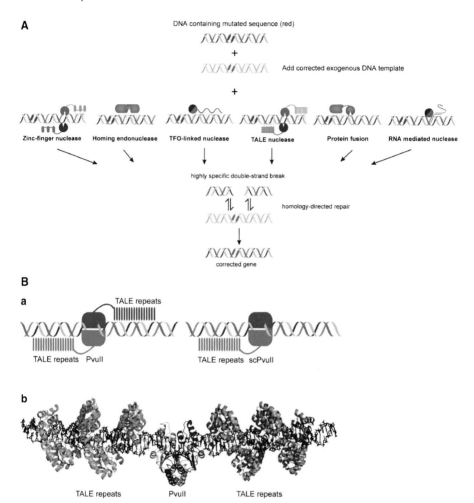

FIGURE 6. Engineering of Type II REases as tools for gene targeting. (*A*) Engineered highly specific endonucleases that can be used for gene targeting by introducing a double-strand break into a complex genome and thereby stimulating homologous recombination (Yanik et al. 2013). With the exception of engineered homing endonucleases ("meganucleases") in which the function of DNA binding and DNA cleavage is present in the same polypeptide chain (Galetto et al. 2009), the other engineered nucleases consist of separate DNA-binding (green) and DNA-cleavage (blue) modules. ZFNs and TALENs usually have the nonspecific cleavage domain of the restriction endonuclease FokI as DNA-cleavage module, but the restriction endonuclease PvuII can also be used for this purpose (Schierling et al. 2012; Yanik et al. 2013). PvuII has also been employed in triple-helix-forming oligonucleo-tide (TFO)-linked nucleases (Eisenschmidt et al. 2005) and in protein fusions (with catalyti-cally inactive I-SceI) (Fonfara et al. 2012) as DNA-cleavage module. ZFNs, TALENs, and TFO-linked nucleases are programmable, as are the RNA-mediated nucleases (Jinek et al. 2012) modified after Pingoud and Wende (2011). (Reprinted from Yanik et al. 2013.) (*B*) TALE-PvuII fusion proteins. (*a*) Scheme of the architecture of TALE–PvuII fusion proteins.

(*Legend continued on following page.*)

but dsDNA breaks would activate the error-prone, nonhomologous end-joining (NHEJ) pathway. This led to the idea that a nicking domain that would cut only one DNA strand might work better than a cleavage domain, and could be used for DNA repair studies and other DNA manipulations (e.g., terminal labeling, genome mapping, and DNA amplification) (Chan et al. 2011; Xiao et al. 2011). The large subunits of some heterodimeric REases (e.g., some Type IIT and IIS) can function as nicking enzymes when separated from their normal partner (Higgins et al. 2001; Heiter et al. 2005; Yunusova et al. 2006; Xu et al. 2007), whereas dimeric enzymes can be mutated to generate a single catalytic site (Stahl et al. 1996; Wende et al. 1996; Morgan et al. 2000; Simoncsits et al. 2001; Heiter et al. 2005). Examples are BbvCI (Heiter et al. 2005), BspD6I (GACTC [4/6]) (Kachalova et al. 2008), BsrDI (GCAATG [2/0]) (Xu et al. 2007), Mva1269I (Armalyte et al. 2005), and BtsCI (GGATG [2/0]) (Too et al. 2010). Such nickases have been used in fusions with zinc fingers, TALE proteins, and methyl CpG binding domains (for further details, see Boch et al. 2009; Moscou and Bogdanove 2009; Hockemeyer et al. 2011; Gabsalilow et al. 2013; Mussolino et al. 2014; Pingoud et al. 2014; Ramalingam et al. 2014; Thanisch et al. 2014; Dreyer et al. 2015; Rogers et al. 2015).

Control of Restriction of Type II Enzymes

Control by C Proteins

Expression of the MTase gene and methylation of the host DNA before synthesis of the REase is essential after entry of a Type II system into the cell. In 1992, the Blumenthal laboratory provided the first evidence for temporal control in a subset of R-M systems, the plasmid-based PvuII system of *Proteus vulgaris* (Tao et al. 1991; Tao and Blumenthal 1992), soon followed by that in the BamHI system (Ives et al. 1992; Sohail et al. 1995).

FIGURE 6. (*Continued.*) (*Left*) wtPvuII, a homodimer in which the DNA-binding module of a TALE protein is fused via a linker of defined length. (*Right*) scPvuII, a monomeric nuclease in which the DNA-binding module of a TALE protein is fused via a linker of defined length. (*b*) Model of a TALE–wtPvuII fusion protein. The fusion protein is a dimer of identical subunits, each composed of a PvuII subunit and a TALE protein. This model was constructed by aligning the structures of the individual proteins PDB 1pvi (Cheng et al. 1994) and PDB 3ugm (Mak et al. 2012) on a DNA composed of the PvuII recognition site and two TALE target sites upstream of and downstream from the PvuII recognition site, separated by 6 bp. The carboxyl termini of the PvuII subunits and the amino termini of the TALE protein are separated by ∼3 nm. This distance must be covered by a peptide linker of suitable length. The image was generated with PyMol. (Reprinted from Yanik et al. 2013.)

A small C gene upstream of, and partially overlapping with, the REase gene is coexpressed from p_{res1}, located within the MTase gene, at low level with the REase after entry of the self-transmissible PvuII plasmid into a new host, whereas the MTase gene is expressed at normal levels from its own two promoters p_{mod1} and p_{mod2} located within the C gene (Fig. 7).

The C protein binds to two palindromic DNA sequences (C boxes) upstream of the C and REase genes: O_L, associated with activation, and O_R, associated with repression. Low basal expression from the pvuIIC promoter leads to accumulation of the activator, which enhances transcription of the C and REase genes (Tao et al. 1991; Tao and Blumenthal 1992; Bart et al. 1999; Knowle et al. 2005; Williams et al. 2013). After this initial low-level expression of C·PvuII protein from the weak promoter p_{res1}, positive feedback by high-affinity binding of a C protein dimer to the distal O_L site later stimulates expression from the second promoter p_{res}, resulting in a leaderless transcript and more C and R protein. The proximal site O_R is a much weaker binding site, but C protein bound at O_L enhances the affinity of O_R for C protein, and at high levels of C protein, the protein–O_R complex down-regulates expression of C and R. In this way, C protein is both an activator and negative regulator of its own transcription.

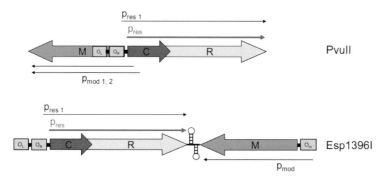

FIGURE 7. Intricate control of restriction in the operons of the Type II R-M systems of PvuII and Esp1396I by controlling C proteins (Loenen et al. 2014b). A small C gene upstream of, and partially overlapping with, R is coexpressed from p_{res1}, located within the M gene, at low level with R after entry of the self-transmissible PvuII plasmid into a new host, whereas M is expressed at normal levels from its own two promoters p_{mod1} and p_{mod2} located within the C gene. A similar C protein operates in Esp1396I, but in this case the genes are convergently transcribed with transcription terminator structures in between, and M is expressed from a promoter under negative control of operator O_R, when engaged by the C protein in a manner similar to that of the PvuII system. C proteins keep both R and M under control and have been tentatively identified in more than 300 R-M systems. See the text for further details. (Reprinted from Loenen et al. 2014b.)

The regulation is similar to gene control in phage lambda: Differential binding affinities for the promoters in turn depend on differential DNA sequence and dual symmetry recognition. C proteins belong to the helix-turn-helix family of transcriptional regulators that include the cI and cro repressor proteins of lambdoid phages. In the wake of PvuII and BamHI, other R-M systems were discovered that were controlled by C proteins, including BglII (A↓GATCT) (Anton et al. 1997), Eco72I (CAC↓GTG) (Rimseliene et al. 1995), EcoRV (Zheleznaya et al. 2003), Esp1396I (CCANNNN↓NTGG) (Cesnaviciene et al. 2003; Bogdanova et al. 2009), SmaI (CCC↓GGG) (Heidmann et al. 1989), and AhdI (McGeehan et al. 2005). In the case of Esp1396I, the genes are convergently transcribed with transcription terminator structures in between, and the MTase gene is expressed from a promoter under negative control of operator O_R, when engaged by C protein in a manner similar to that of the PvuII system (Fig. 7). C·Esp1396I controls O_R, O_L, and O_M in a similar manner as described above. In this way, C proteins keep both R and M under control. This delay of REase expression depends on the rate of C-protein accumulation, and this may help explain the ability of C-regulated R-M systems to spread widely (Williams et al. 2013). By September 2013, REBASE listed 19 characterized C proteins, as well as 432 putatives (http://rebase.neb.com/rebase/rebase.html). The organization of the genes in the system and regulatory details differ from system to system, and some C proteins are fused to their REase genes (http://rebase.neb.com/rebase/rebase.html; Tao et al. 1991; Tao and Blumenthal 1992; Semenova et al. 2005; Bogdanova et al. 2009; Liang and Blumenthal 2013). Whether R-M systems as a whole evolved in concert with C proteins remains to be investigated.

The first structures of C proteins without DNA appeared in 2005: C·AhdI from Geoff Kneale's laboratory (McGeehan et al. 2005) and C·BclI from Ganesaratinam (Bali) K. Balendiran's laboratory in collaboration with NEB (Sawaya et al. 2005). These structures resembled those of helix-turn-helix DNA-binding proteins, as expected. The details of the interactions between C proteins and their C boxes in the DNA came later with the studies on the AhdI operon, and the structures of C·AhdI, C·Esp1396I, and C·BclI (Marks et al. 2003; McGeehan et al. 2004, 2005, 2006; Streeter et al. 2004; Sawaya et al. 2005; Callow et al. 2007; Papapanagiotou et al. 2007; Bogdanova et al. 2008; Ball et al. 2009). With the structure and further experiments, the mechanism behind the genetic switch could be elucidated (McGeehan et al. 2008, 2012; Ball et al. 2009, 2012). C·Esp1396I bound as a tetramer, with two dimers bound adjacently on the 35-bp operator sequence $O_L + O_R$ (McGeehan et al. 2008). This cooperative binding of dimers to the DNA operator controls the switch from activation to repression of the C and R genes. The existence of C proteins explained the difficulty to introduce some R-M genes into *E. coli* (e.g., BamHI [Brooks et al. 1989]; see Loenen et al. 2014b for further details). The C genes belong to different incompatibility groups, which exclude unrelated R-M systems (called "apoptotic mutual exclusion"): For example, the *pvuIIC* and *bamHIC* genes define one exclusion group and prevent entry of *ecoRVC* due to premature activation of the EcoRV REase gene (Nakayama and Kobayashi 1998).

In 2016, the group of Iwona Mruk in Gdansk reported an unexpected regulatory variation on the above theme (Rezulak et al. 2016). The C·Csp231I gene regulates expression of the REase gene like other C-regulated R-M systems, but there is additional novel control. Separate tandem promoters drive most transcription of the Csp231I REase gene, a distinctive property not seen in other tested C-linked R-M systems. Further, the C protein only partially controls REase expression, yet plays a role in viability of the cells within the population by affecting stability and propagation. Deletion of the C gene led to high REase activity and resulted in loss of these cells in mixed cultures with wild-type R-M cells.

Transcriptional Control: The Case of EcoRI

The transcriptional control discussed above via C proteins has been found for many Type II enzymes, but not all Type II enzymes have such multiple (convergent) promoters and controlling C proteins. A prime example is EcoRI, whose enzymatic activity is controlled in a different way until methylation is complete. The Tokyo group of Ichizo Kobayashi investigated the intricate control of the EcoRI gene, *ecoRIR* (Liu and Kobayashi 2007; Liu et al. 2007; Mruk et al. 2011). This gene is upstream of the modification gene, *ecoRIM*. The M gene can be transcribed from two promoters within *ecoRIR*, allowing expression of the MTase gene with and without *ecoRIR*, as there is no transcription terminator between the two genes. In addition, the *ecoRIR* gene has two reverse promoters. These convergent promoters negatively affect each other, as in lambda (Ward and Murray 1979). Transcription from the reverse promoter is terminated by the forward promoters and generates a small antisense RNA. The presence of the antisense RNA gene in *trans* reduced lethality mediated by cleavage of undermethylated chromosomes after loss of the EcoRI plasmid (postsegregational killing) (Heitman et al. 1989; Mruk et al. 2011). This can be viewed as programmed cell death in prokaryotes. Kobayashi compares R-M systems with toxin/antitoxin (TA) systems composed of an intracellular toxin (the REase) and an antitoxin (the MTase) that neutralizes its effect. These systems would limit the genetic flux between lineages with different sequence-specific DNA methylation ("epigenetic identity") but would require intricate control of restriction activity (reviewed in Mruk and Kobayashi 2014).

Control by the Cognate MTase

M·Ecl18kI and M·SsoII are two MTases that act as transcription factors and activate expression of their respective REase genes via binding to the regulatory site in the promoter region of these genes (Karyagina et al. 1997; Shilov et al. 1998; Fedotova et al. 2009). The amino-terminal region of M·Ecl18kI performs the regulatory function, but is also important for methylation activity. Loss of methylation activity per se does not prevent the MTase from performing its regulatory function and even increases its affinity to the regulatory site. However, the presence of the methylation domain is necessary for M·Ecl18kI to perform its regulatory function (Burenina et al. 2013).

PART B: TYPE I ENZYMES

Type I Families and Diversity

As discussed in Chapter 7, a single Type I common ancestor is likely, given the high sequence similarity of confirmed (biochemically analyzed) and putative enzymes and irrespective of the host within subclasses up to 80%–99%, between subclasses ~20%–35%. This section is based on two reviews published in 2014 (Loenen et al. 2014a,b), and the reader is referred to these for more details and references. By 2013, ~50% (1140/2145) of sequenced bacterial and archaeal genomes in REBASE carried one or more *hsdR*, *hsdM*, and *hsdS* genes and 40% appeared to have none, whereas the remainder had some but not all three genes or disrupted or scrambled genes (http://rebase. neb.com/rebase/rebase.html). On average, cells had two systems, although, for example, *Desulfococcus oleovorans* has eight systems. Type I enzymes could undergo specificity changes via TRD exchanges by homologous recombination, unequal crossing-over, or transposition (Chapter 6, Fig. 6). Domain shuffling is not limited to Type IA enzymes but can be observed between members of the same or different families (Loenen et al. 2014a). Within Type I families, HsdS subunits have the same organization, but between families they have different amino and carboxyl termini (circular permutations; see Loenen et al. 2014a for details). Circular permutation of HsdS of EcoAI indicated structural, but not necessarily functional equivalence, as different permutations resulted in an active R-M system, active in methylation only, or inactive, indicating that the HsdS termini interact with the HsdM and HsdR subunits (Janscak and Bickle 1998). Some bacteria have only one or two *hsdR* and *hsdM* genes but many *hsdS* genes (up to 22 in *Mycoplasma* sp.!) allowing multiple specificity changes providing protection against invaders (Sitaraman and Dybvig 1997; Dybvig et al. 1998; Loenen et al. 2014a). Shuffling of those 22 *hsdS* genes could easily result in more than 500 new specificities, "a defensive repertoire reminiscent of the immunoglobulins of higher organisms" (Loenen et al. 2014a). The advent of SMRT sequencing, which allows the localization of methylated bases, has led to a breakthrough in the determination of Type I recognition sites (Eid et al. 2009; Flusberg et al. 2010; Korlach et al. 2010; Clark et al. 2012; Korlach and Turner 2012), which may generate renewed interest in these "sophisticated molecular machines" (Murray 2000). SMRT sequencing not only led to an exponential increase in the number of known Type I recognition specificities (rising from approximately 40 biochemically characterized specificities in 2011 to more than 1100 by 2017) but also the discovery of Type I enzymes that produce m6A on one strand and m4C on the other (Morgan et al. 2016).

Single-Molecule Studies of EcoKI and EcoR124I

AFM and single-molecule studies, together with improved biochemical and biophysical methods, revealed new details about translocation by EcoKI via the motor domains that belong to superfamily 2 (SF2) (Neaves et al. 2009). Mutational analysis of the DEAD-box, RecA-domain-like, motifs of EcoR124I showed long-range effects of various mutations—for example, nuclease mutants could lower translocation and ATP usage rate, there could be a decrease in the off rate, and/or there could be slower restart and turnover (Sisakova et al. 2008a,b). Dimerization appeared to occur preferentially on two-site DNA, whereas DNA looping could occur in the absence of ATP hydrolysis. Would this be a common way to bring distant DNA regions together? Would this mean that SF2-dependent enzyme complexes in higher organisms also use such looping (which are involved in DNA repair, replication, recombination, chromosome remodeling, and RNA metabolism; for discussion, see, e.g., Tuteja and Tuteja 2004; Singleton et al. 2007; Fairman-Williams et al. 2010; Ramanathan and Agarwal 2011; Umate et al. 2011)?

Single-molecule studies using magnetic tweezers were designed to analyze single translocating molecules of EcoR124I in real time (Seidel et al. 2004, 2005, 2008; Stanley et al. 2006; Seidel and Dekker 2007). These experiments provided details on the rate of DNA translocation, as well as the processivity and ATP dependence of the HsdR motors. New facts emerged that may be of consequence for the studies on the eukaryotic SF2-dependent complexes mentioned above: (1) The two motors could work independently and the enzyme tracked along the helical pitch of the DNA on torsionally constrained molecules; (2) translocation could stop and restart by disassembly and reassembly, and (3) the HsdR subunits released the DNA roughly every 500 bp during this process (whereas the MTase remained attached to the recognition site); in other words, about four times over a distance of 2 kb. Concomitantly with this stop and restart process, the enzymes consumed vast amounts of ATP.

A translocation block by collision with another HsdR or the presence of super-coiled DNA resulted in cleavage. The enzyme remained at the site but could be displaced by other proteins (e.g., *E. coli* RecBCD) (Bianco and Hurley 2005).

Type I Enzyme Atomic Structure

In the absence of crystals, the DNA recognition complex of EcoKI, the trimeric M·EcoKI (M_2S_1), had been modeled on 3D structures of other MTases (Chapter 7, Fig. 10). This model suggested a common origin of Type I and Type II MTases. Would this ancestral MTase combine with one or two HsdR molecules allowing translocation to the site of cleavage? Did translocation involve contacts with nonspecific DNA adjacent to the recognition site in a cleft in HsdR, which would close and reopen using ATP, Mg^{2+}, and probably SAM, to fuel and control the conformational changes? Other questions remained to be answered: Would HsdR touch one strand or both strands of the dsDNA via backbone contacts, and what about step size, or the amount of DNA transported per physical step? And why no cleavage during the initial translocation? Was the translocation rate too high, or the catalytic PD region in the wrong conformation to contact the DNA? One thing seemed certain: Two REases were needed for dsDNA breaks—one for each strand. The answer to some of these questions came when finally the first structures of Type I enzymes appeared.

The Structure of the M·EcoKI (M_2S_1) Complex

The first crystal structures of HsdS subunits appeared in 2005 and 2010 (Calisto et al. 2005; Kim et al. 2005). The two TRDs are in inverted orientations, which makes the S subunit functionally symmetric (Fig. 8A, bottom right; see Loenen et al. 2014a for discussion). Each TRD consists of a globular DBD and an α-helical dimerization domain. The long α-helices (D1 and D2) encoded by the two conserved regions of the *hsdS* gene associate to form an antiparallel coiled-coil dimerization helix between the two variable HsdS specificity domains (S1 and S2) (Calisto et al. 2005; Kim et al. 2005). Amino acid side chains down their lengths interlock "like tines of a zipper" and form a hydrophobic core that holds the two helices together and separates the globular specificity domains by a fixed distance. S1 and S2 each recognize one-half of the recognition sequence. Each TRD also associates with one HsdM subunit to form an M_2S trimer. Neither HsdS nor HsdM subunits bind to DNA alone, but the EcoKI trimer methylates both strands of the recognition sequence—the "top" strand of the 5′ half-sequence (Am6AC) and the "bottom" strand of the 3′ half-sequence (CGm6AC) of the bipartite AACNNNNNNGTCG —thus protecting the host DNA during replication (Fig. 8A).

A

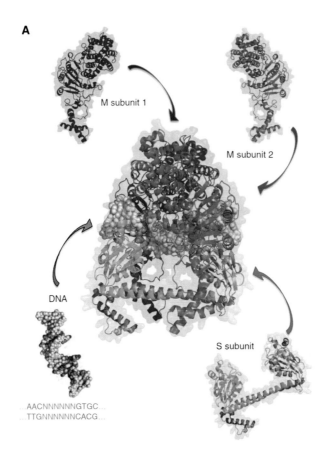

FIGURE 8. (*A*) Model of the M·EcoKI MTase (PDB ID: 2Y7H). The S subunit is composed of two TRDs in inverted orientations. Each TRD comprises a globular DBD and an α-helical dimerization domain. The N-TRD (green) and C-TRD (orange) are specific for the two halves of the recognition sequence (AACNNNNNNGCAG). Zipper-like association of the helices separates the globular domains by a fixed distance and reverses the orientation of the C-TRD. Each TRD also associates with one M subunit (identical, but shown here in different shades of blue for clarity) to form an M_2S trimer, which methylates both strands and protects the resident DNA during DNA replication. (*B*) Structure of the Type IA HsdS protein S-ORF132P (PDB ID: 1YF2). Structure of the Type IA HsdS protein S-MjaXI (PDB ID: 1YF2). The upper diagram shows the domain organization of the protein; arrows represent DBDs, and curly lines represent dimerization α-helices. The amino acid sequence of the protein is shown *below*, with the domains in corresponding colors. *Below* this are three views of the structure, from three perpendicular directions: "sideways," "end-on," and "above." The panels on the *left* depict the protein; those on the *right* depict the protein with modeled DNA positioned approximately as it is bound. The DNA was taken from PDB ID: 2Y7H and transferred by structural alignment of the S subunits. (*A,B*, Reprinted from Loenen et al. 2014a.)

(Figure continued on following page.)

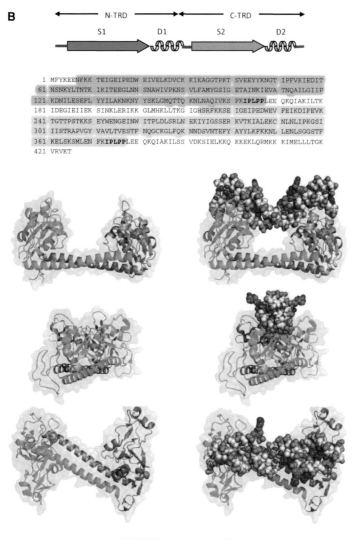

FIGURE 8. *(Continued.)*

The structure and sequence of the S subunit from *Methanocaldococcus jannaschii*, S·MiaXI, is shown in the top part of Figure 8B. The recognition sequence of this protein is unknown. It is closely related to the Type IA family of EcoKI. Below this are three views of the structure from three perpendicular directions. The panels on the left show the protein on its own, and those on the right a model of the protein bound to DNA.

The structure of the trimeric M·EcoKI (M_2S_1) was resolved in 2009, thanks to the product of the *0.3* gene product of phage T7, T7 Ocr (Chapter 6), which proved to be a DNA mimic (see section Antagonists of Type I Action: Antirestriction, starting on page 230). Ocr was used to stabilize the otherwise labile MTase complex (Kennaway et al. 2009). Many different single M·EcoKI-Ocr complexes were imaged in the EM, allowing a reconstruction of the 3D complex to a resolution of 18 Å. A model of M·EcoKI is shown in Figure 8A, which depicts the location of the specificity domains of the S subunit in relation to the two M subunits (see Kennaway et al. 2009 for further details).

The Structure of the EcoKI and EcoR124I ($R_2M_2S_1$) Complexes

The first data on the crystal structure of EcoR124I HsdR were published by Lapkouski et al. (Lapkouski et al. 2007, 2009). This suggested how the pentamer might be assembled and how the motors might translocate dsDNA (Lapkouski et al. 2009). The PD motif was found opposite the translocation domain, which would allow coupling of translocation to restriction. A model was proposed (Lapkouski et al. 2009) based on this HsdR structure, a DNA path across the subunit, and an early, incomplete model of the MTase core (Obarska et al. 2006). This model has much in common with the later model by Kennaway et al. (Kennaway et al. 2012) but differs in the orientation of the HsdR with respect to the MTase core and the path taken by the DNA (see below). Soon afterward, crystal data appeared on the amino-terminal fragment of a putative Type I enzyme from *Vibrio* sp. This contained three globular domains with an endonuclease core and the ATPase site close to the probable DNA-binding site for translocation. The authors suggested the involvement of a linker helix in the transition from DNA motor protein to nuclease (Uyen et al. 2008, 2009).

In 2012, years of efforts by David Dryden and coworkers finally paid off and the structure of the pentameric EcoKI and EcoR124I ($R_2M_2S_1$) was elucidated by computer-assisted EM single-particle reconstructions (Kennaway et al. 2012). Single-particle analysis of negative stain EM images showed large differences between DNA-bound (Fig. 9A) and unbound EcoR124I enzymes (Fig. 9B), with their longest dimensions being ~18 nm versus ~22–26 nm, respectively. (The smaller particles were the $R_1M_2S_1$ form and were analyzed separately as described later.) EcoR124I with DNA was in a closed conformation, whereas the enzyme alone was in an open form without DNA. Apparent twofold symmetry was visible in many image averages. Using these data, a 3D reconstruction was generated (Fig. 9A) of EcoR124I bound to a 30-bp dsDNA fragment with the unmethylated recognition site and the enzyme on its own (Fig. 9B; see Kennaway et al. 2012 for details).

EcoR124I without DNA was highly extended and more flexible, which allowed a low-resolution (~3.5-nm) 3D reconstruction of the enzyme (Fig. 9B). Most particles (~80%) appeared to have their twofold axis roughly

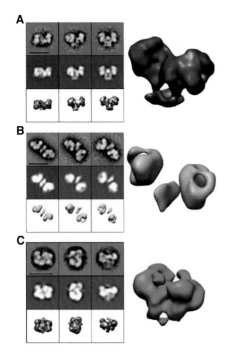

FIGURE 9. Gallery of Type I RM structures and conformations determined by EM and single-particle analysis. (*A*) EcoR124I + DNA (closed state) negative stain EM. (*B*) EcoR124I without DNA (open state) negative stain EM. (*C*) EcoKI + DNA negative stain EM. For each 3 × 3 panel, the *top* rows are image averages, the *middle* rows are their corresponding reprojections, and the *bottom* rows are 3D surface views of the 3D reconstruction (bars, 200 Å); on the *right* is a larger 3D surface perspective view. See Kennaway et al. 2012 for further details. (Reprinted from Kennaway et al. 2012, with permission from Cold Spring Harbor Laboratory Press.)

normal to the plane of the carbon film, but ~5% were seen to be folded up into the closed state, indicating a dynamic equilibrium between states in the absence of cognate DNA. Some very thin connections between the domains were "likely pivot points for flexing to allow the enzyme to close up" (Kennaway et al. 2012). These data showed that the subunits strongly moved in a manner to allow entry of the DNA substrate (Fig. 9A,B). This change from an extended structure to a more compact form in the presence of DNA had been seen previously for the MTase (Kennaway et al. 2009; see Kennaway et al. 2012 for discussion).

Negatively stained particles of EcoKI with DNA bound (Fig. 9C) appeared smaller than EcoR124I with DNA (~16 nm long) and appeared to be more rounded and variable than EcoR124I. The 3D reconstruction of EcoKI with a 75-bp fragment of dsDNA indicated a compact structure with many features similar to EcoR124I with DNA, including recognizable density for the five subunits in a matching arrangement, suggesting a common architecture for Type I enzymes. However, the EcoKI particles were compact and appeared to be identical with and without DNA, and not elongated, as seen for EcoR124I without DNA. The dynamic equilibrium between open and closed forms apparently favored the closed form for EcoKI under the conditions used for EM.

Scattering experiments were used to construct a model showing the location of HsdR and the MTase (Kennaway et al. 2012). In the elongated structure of EcoR124I, the two HsdR subunits were located toward the extreme ends on either side of the MTase core. Fortunately, a fraction of the enzyme existed as a tetramer ($R_1M_2S_1$). In negative stain EM, the particles of EcoR124I with (Fig. 10A) or without (Fig. 10B) DNA were not 100% homogeneous, and further analysis showed a large missing region at the extremity of the smaller particles, which had to be the location of one of the HsdR subunits. The existence of tetrameric complexes with only one

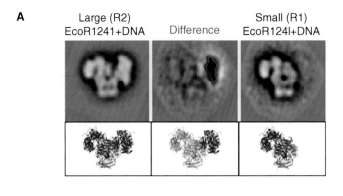

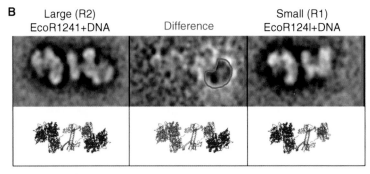

FIGURE 10. 2D difference images from EM data show the position of the HsdR in the EcoR124I complex. (*A*) Difference imaging between image averages of large (*left*) and small (*right*) particles in the EcoR124I + DNA negative stain EM data set reveals a large "negative density" region (red contour at −2.5σ), consistent with a missing HsdR in the small particles. (*B*) Difference imaging of HsdR in the open state of EcoR124I (without DNA). Although the relative flexibility of the open complex gives rise to a less well-defined difference map, a region of negative density consistent with a missing HsdR is visible nevertheless (red contour). The deduced atomic structure of each EM particle is shown *below* the EM image. (Adapted from Kennaway et al. 2012, with permission from Cold Spring Harbor Laboratory Press.)

HsdR was consistent with previous biochemical data on EcoR124I (Janscak et al. 1998).

Although the DNA is not visible in these experiments, the authors were able to use T7 Ocr, which binds very tightly to the DNA-binding site of Type I enzymes (Atanasiu et al. 2002; Walkinshaw et al. 2002). EcoR124I-Ocr complexes adopted a closed conformation and further analysis revealed the position of a banana-shaped object running through the center of the enzyme but tilted at an angle relative to the long axis of the 3D map. This banana-like shape matched well with the structure of the Ocr protein (Walkinshaw et al. 2002). This orientation of Ocr in the EM map plus the structural models of the MTase core of EcoKI (Kennaway et al. 2009) and of EcoR124I (Kennaway et al. 2012) allowed only one possible orientation of the MTase with the dimerization helix of the S subunit exposed to the solvent. This was in agreement with previous observations that this region could accommodate small (Gubler and Bickle 1991) and large (Kannan et al. 1989) amino acid insertions, and even a fusion with green fluorescent protein (Chen et al. 2010), without loss of function. Moreover, limited proteolysis indicated preferential cleavage within the dimerization helix (Webb et al. 1995), and hence surface exposure of this region.

In these studies the carboxyl terminus of the R subunit of EcoR124I (aa residues 893–1038) was not visible, but could be modeled using known crystal structures (Kennaway et al. 2012). Using these crystal structures, the EM data, and scattering analyses, atomic models of complete R subunits for EcoKI and EcoR124I were constructed, backed up by the plethora of published biochemical data on these enzymes.

The authors proposed a model that fit the data and gave the location and directionality of the DNA motor domains by aligning these with those of the dsDNA-bound SWI2/SNF2 chromatin remodeling translocase from *S. solfataricus* (Lapkouski et al. 2009). The direction of DNA translocation of this translocase was known and imposed a similar directionality on each HsdR, and because these had to pull DNA in toward the MTase core of the Type I enzyme, the orientation of each HsdR relative to the MTase core became defined. Based on DNA footprinting experiments (Mernagh et al. 1998; Powell et al. 1998) and the known minimum length of 45 bp of DNA required for ATP hydrolysis (Roberts et al. 2011), the assumption was made that the DNA path between the DNA bound to the HsdR and the DNA bound to the core MTase could not be longer than ~40 bp. This meant that the motor domains of the HsdR had to have their DNA-binding sites close to the DNA-binding site of the MTase core. Placement of the HsdR on either side of the MTase and interacting directly with DNA was further supported by the length of the structure of another DNA mimic protein, ArdA (Nekrasov et al. 2007; McMahon et al. 2009), which occupies the entire

DNA-binding site on Type I enzymes. This then allowed the complete structures for the closed forms of EcoR124I and EcoKI to be constructed as shown in Figure 11.

Placement of the R subunit of EcoR124I forced two large kinks in the DNA to allow the DNA to thread through the MTase core (Fig. 11A). This kinked path shortens the through-space end-to-end distance of a duplex bound to the enzyme by ~10 nm, in line with AFM measurements of complexes of EcoR124I on DNA that showed that binding of the enzyme shortened the length of a long linear DNA molecule by ~11 nm (van Noort et al. 2004). AFM measurements of EcoKI bound to DNA also showed a pronounced kink (Walkinshaw et al. 2002; Neaves et al. 2009), and circular dichroism analysis of EcoR124I also indicated a large structural distortion to the DNA when bound (Taylor et al. 1994).

A fit of subunits into the EcoKI EM density map (Fig. 11B) corresponded closely to that of EcoR124I in the closed state. The thin protrusions at either side of the EM envelope for EcoKI could fit the long coiled-coil amino-terminal extensions of unknown function predicted in the R subunit of EcoKI but absent in EcoR124I (Fig. 11B). Significant sequence differences existed between the two enzymes, and this might account for other structural differences, although the overall architecture remained unchanged.

An optimal fit of subunits into the lower-resolution open EcoR124I map (Fig. 11C) was obtained by moving and rotating each HsdM–HsdR pair as a single rigid body away from HsdS. A relatively simple ~90° rotation and an ~80° twist around a pivot point near the carboxyl terminus of HsdM were sufficient to move between open and closed states. It had previously been shown that HsdR and HsdM can form a complex (Dryden et al. 1997), supporting movement of the two subunits as a rigid body. The carboxy-terminal residues of EcoKI HsdM were disordered in the crystal (PDB ID: 2ar0) (Kennaway et al. 2009) and were sensitive to proteolysis (Cooper and Dryden 1994) and could play the role of the flexible linker proposed. Proteolytic removal of this region inhibited assembly of the pentamer (Powell et al. 2003).

Taken together, the data showed how Type I enzymes assemble, bind, and distort DNA before the initiation of ATP-driven DNA translocation. Although these EM and small-angle scattering structures were of low resolution, the proposed atomic models are in agreement with the extensive data from biochemical, biophysical, and genetic studies (Murray 2000, 2002; Loenen 2003; Tock and Dryden 2005). These provided several further constraints on the subunit orientations and gave confidence in the atomic models shown in Figure 11. These make it clear that there is an equilibrium between open and closed forms of the Type I enzymes, with the equilibrium constant depending on the particular enzyme and the presence or absence of DNA (and

A   EcoR124I+DNA   **B**   EcoKI+DNA

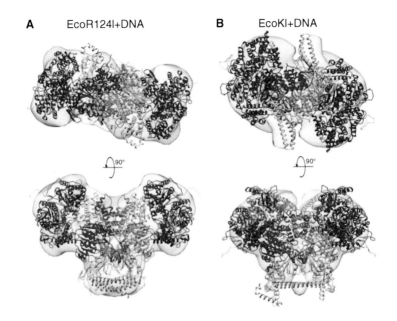

C   EcoR124I (no DNA)

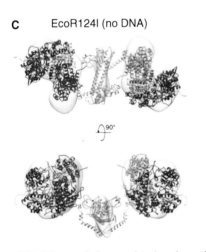

FIGURE 11. Atomic models of EcoR124I + DNA, EcoR124I, and EcoKI + DNA docked into the EM map densities. (*A*) Two views of the EcoR124I + DNA model showing the MTase core closed around DNA (green; DNA bound to each HsdR is not shown for clarity). Adenine bases are flipped out into the active sites of each of the two HsdM (light and dark blue), induced by an ~45° bend in the DNA. (Yellow) HsdS, (red) HsdR, with the β-sheets of the RecA-like motor domains (orange). (Gray) Residues missing from the crystal structures (the 44 and 152 carboxy-terminal residues of HsdM and HsdR, respectively) were modeled de novo. The carboxy-terminal regions of HsdM extend down to bind at the coiled coil of HsdS, and the HsdR carboxy-terminal domains fill some empty density next to the amino terminus of HsdM. (*B*) A model for EcoKI bound to DNA (colors as in *A*). HsdS and HsdM from the MTase structure (PDB ID: 2y2C) were docked in as a single rigid body; HsdR modeled on those from EcoR124I (PDB ID: 2w00) (see Supplemental Material of Kennaway et al. 2012) and placed in a position analogous to the EcoR124I model. (*C*) The model of EcoR124I in the open conformation (i.e., without DNA; colors as in *A*). Although the EM map is at a lower resolution, a full atomic model could be built, aided by the EcoR124I + DNA model, SANS data, and 2D difference imaging. HsdM and HsdR swing out as a unit away from HsdS. The predicted hinge regions in the carboxyl termini of the HsdM (gray) and their connections to HsdS are not well resolved. (Reprinted from Kennaway et al. 2012, with permission from Cold Spring Harbor Laboratory Press.)

presumably the cofactors SAM and ATP). EcoKI prefers to be closed whether DNA is present or not and must therefore transiently open up to allow DNA access to the MTase core. EcoR124I appears to prefer an open form in the absence of DNA but is closed with DNA bound.

It would appear possible for the Type I enzymes to reach the closed "initiation" complex with the S-shaped DNA path via different routes. The model shown in Figure 12A is based on the EcoR124I structure but David Dryden (pers. comm.) states that "it would also apply to other Type I enzymes if we could actually ever see the open conformation—which we did not for EcoKI but this does not mean that it does not exist transiently." The open form can bind DNA nonspecifically using HsdR (left side of Fig. 12A) and diffuse along the DNA until the MTase core recognizes a target sequence or dissociates. The trigger for closing and formation of the initiation

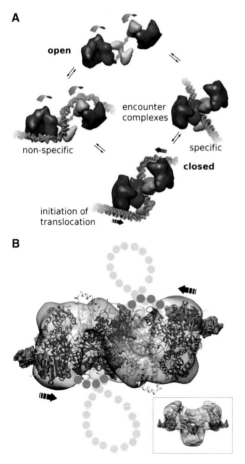

FIGURE 12. Schematic of large-scale conformational change and initiation of DNA looping and translocation by EcoR124I. (*A*) Type I enzymes exist in a dynamic equilibrium between open and closed states (movement is shown by orange arrows, and pivot points in carboxy-terminal regions of HsdM are indicated by pink dots). DNA (green) binding to form encounter complexes can occur nonspecifically to the HsdR (red) or via the target sequence to the MTase core (HsdM is in light and dark blue, and HsdS is in yellow). Complete closure of the enzyme and bending of the DNA around the HsdR produces the initiation complex for DNA translocation. (*B*) The predicted complete path of the DNA (green dots) through the atomic model of EcoR124I with segments of bound DNA. This is the proposed initiation complex (from Fig. 10A). During active translocation, the DNA would then form expanding loops from each side (light-green dots for DNA, and the direction of translocation is shown by black arrows). The *inset* shows the initiation complex turned 90° to the main panel. (Reprinted from Kennaway et al. 2012, with permission from Cold Spring Harbor Laboratory Press.)

complex would be most likely the recognition of the target sequence by the MTase core. Alternatively, the closed form of the enzyme must open up transiently to allow DNA to enter the MTase core, followed by closing of the core around the DNA (right side of Fig. 12A) and diffusion of the enzyme on the DNA until it either recognizes its DNA target sequence or reopens and dissociates. Starting the process of target sequence location and recognition via this pathway means that the motor domains of the HsdR will have to rely on the inherent flexibility of DNA for them to grasp it and force it into the S-like shape shown in the initiation complex.

The introduction of sharp bends in the DNA would require considerable energy to be expended by the enzyme. This may come from the transition between open and closed forms of the RM enzyme, but it may also require the hydrolysis of ATP by the HsdR. The models suggest that once the enzyme has closed around DNA and the motor of an HsdR subunit has a good grip on a segment of DNA, further hydrolysis of ATP (required for translocation and cleavage, although not DNA binding) would push the segment bound to the motor toward the central MTase core, as indicated by large arrows in Figure 12A. Because the MTase core is also tightly bound to the DNA target sequence, DNA at the bend between the segments bound to the motor and to the MTase core would twist and perhaps even buckle, forming the small loop shown in Figure 12B. Formation of this highly strained loop is certain to be energetically unfavorable, in agreement with translocation measurements for Type I enzymes in which it appears that much ATP is used in abortive attempts to initiate translocation (Seidel et al. 2008). Once the loop has formed, further DNA translocation would occur as the motors pump DNA toward the MTase core. Single-molecule experiments make it clear that the motors can work independently (Seidel et al. 2004, 2008), perhaps explaining why early EM studies showed both single- and double-looped structures (Yuan et al. 1980; Endlich and Linn 1985). In light of the large changes occurring upon DNA binding, it is possible that the actively translocating enzyme undergoes further changes in structure (e.g., in the presence of ATP). One may speculate that this great flexibility would allow the enzyme to accommodate the stresses built up during the extensive DNA translocation periods observed for these molecular machines. It is noteworthy, in this respect, that a process of deassembly of the enzymes occurs after DNA cleavage, and some of the subunits—although not all and depending on the particular Type I enzyme—can be reused (Roberts et al. 2011; Simons and Szczelkun 2011).

As mentioned above, the earlier model of Lapkouski et al. (Lapkouski et al. 2009) differs from this new model in two respects: namely, the orientation of the HsdR with respect to the MTase core, and the path taken by the DNA. Previously (Lapkouski et al. 2009), the interface of HsdR with the MTase core was not defined when compared with the new models. More importantly, the DNA was proposed to bend across the motor domains of HsdR, so that it came near to the endonuclease domain in the same HsdR and could be cleaved. If this model was correct (Lapkouski et al. 2009), the partially assembled $R_1M_2S_1$ form of EcoR124I should have been able to cleave DNA, which was, however, not the case (Janscak et al. 1998). The current model suggests that the endonuclease domain of one HsdR is in proximity to DNA translocated by the other HsdR (Fig. 12B). This would explain the absence of DNA cleavage by partially assembled $R_1M_2S_1$ forms of EcoR124I, despite the fact that such an assembly translocates DNA effectively (Janscak et al. 1998; Seidel et al. 2004, 2008). Thus, the current models are a significant improvement on the previously published models (Davies et al. 1999; Lapkouski et al. 2009).

Last, the structural models presented can be compared with the structures of complex Type II REases enzymes in groups IIB and IIG (Roberts et al. 2003), which cleave at defined distances either on both sides (IIB) or only one side (IIG) of the target sequence. As mentioned above, these classes are Type I-like combined R-M systems, with an amino-terminal endonuclease PD domain directly fused to a γ-class MTase domain in a single protein, but without motor domains (Dryden 1999; Nakonieczna et al. 2009; Shen et al. 2011). Whereas Type IIB enzymes have a single HsdS-like subunit with two TRDs, the Type IIG enzymes BpuSI and MmeI have only one TRD. The Type IIB enzyme is effectively a dimeric Type IIG. Thus, a Type IIB enzyme is like a motorless Type I system, and a Type IIG system is like one-half of a motorless Type I RM enzyme. Figure 13 compares the relative locations of one endonuclease domain, one HsdM, and the HsdS from the closed form of EcoR124I with the structures of MmeI and BpuSI (Nakonieczna et al. 2009; Shen et al. 2011). It can be seen how fusion of the endonuclease domain from HsdR to the start of HsdM in EcoR124I would move it to the same location as observed in the Type IIG REases and lead to cleavage downstream from the target sequence. Thus, the proposed role of gene fusions in the evolution of different groups of Type II R-M systems (Mokrishcheva et al. 2011) can be extended to include the evolution of the Type I systems.

Additional Roles for Type I Enzymes

In 1977, Werner Arber wondered whether REases might have additional functions in the cell (Arber 1977). The finding that a Type I enzyme could cleave a replication fork at its branch may indicate that the answer is yes (Ishikawa et al. 2009). This probably happens when the enzyme travels along the DNA and encounters a replication fork, halts, and cuts (Ishikawa et al. 2009). Possibly in line with this, the groups of Hirotada Mori in Nara and Chieko Wada in

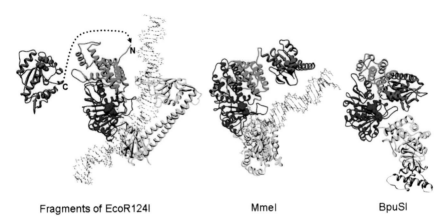

Fragments of EcoR124I   Mmel   BpuSI

FIGURE 13. Structural evolution of Type IIG enzymes from a Type I enzyme undergoing fusion of the carboxyl terminus of an endonuclease domain from HsdR via deletion of the motor domains, to the amino terminus of HsdM. (*Left*) Part of EcoR124I, with one endonuclease domain from HsdR (red), one HsdM (green is the amino-terminal domain, and blue is the MTase catalytic domain), and HsdS (yellow with two TRDs). DNA bound to the MTase core is shown, but DNA bound to HsdR is omitted for clarity. The dashed line shows how the end of the endonuclease domain could join with the amino terminus of HsdM to form a structure similar to the type IIG structures shown on the *right*. The catalytic motifs in the endonuclease domain and HsdM are shown in spacefill. (*Middle*) Model of Mmel with bound DNA with the same colors for the equivalent domains (Nakonieczna et al. 2009; coordinates from ftp://genesilico.pl/iamb/models/RM.Mmel). (*Right*) Crystal structure of BpuSI (PDB ID: 3s1s) with the same coloring of domains as in the other structures and with an inserted extra domain shown in gray (Shen et al. 2011). DNA is absent in this structure, and one can see that the endonuclease domain would be blocking the DNA-binding site on the TRD. Shen et al. (2011) proposed that the endonuclease domain would twist away to allow DNA sequence recognition. (Reprinted from Kennaway et al. 2012, with permission from Cold Spring Harbor Laboratory Press.)

Kyoto identified protein–protein interactions in *E. coli* (Arifuzzaman et al. 2006) between the EcoKI subunits and some other proteins, in a comprehensive pull-down assay using a His-tagged library of ORFs. Some of these interactions may be just a matter of sticky proteins, but others could be of in vivo relevance—for example, ATP-dependent helicase HrpA, the replicative helicase DnaB, DNA polymerase III DnaE, or CTP synthase PyrG, which may point to potential fine-tuning of R-M activity with DNA replication and primary metabolism. Such cross talk between (endo)nucleases and primary metabolism may well be universal, although likely to be much more complex in eukaryotic systems. As such, the *E. coli* system remains a useful model system to study such complex processes.

In an interesting report, Marie Weiserova and colleagues used immunoblotting to show that HsdR is phosphorylated on threonine in vivo only when coproduced with the MTase subunits (HsdM and HsdS) (Cajthamlova et al. 2007). HsdR lacks this phosphorylation when introduced in the cell by itself. Is this as yet unexplained phosphorylation of EcoKI HsdR another way of restriction control or genome maintenance (e.g., involved in recruiting other enzymes to the DNA) (Cajthamlova et al. 2007)? It certainly warrants further investigation.

On a different track, the Type I *ecoprrI* system contains an additional gene, originally identified by the group of Tom Bickle (Tyndall et al. 1994; Kaufmann 2000). The protein encoded by *prrC* proved to be a member of the group of latent anticodon nucleases (ACNs), which are proteins that may be linked to stress responses, and have been called RNA-based innate immunity systems that distinguish self from non-self (Jain et al. 2011).

Most studies on Type I systems concerned enzymes from strains that could be cultured in the laboratory. However, with the advent of whole-genome sequencing, many Type I systems have been identified in benign as well as pathogenic bacteria, in which they may limit genetic exchange between species and contribute to genome and strain stability. Because of the presence in their genomes of sometimes large numbers and different types of REases, these enzymes have proven useful for strain typing of, for example, methicillin-resistant *Staphylococcus aureus* (MRSA) bacteria (Lindsay 2010). REases such as SauI can limit horizontal transfer through conjugation, phage transduction, or transposition of antibiotic-resistance genes or virulence factors, but they cannot completely prevent it (Waldron and Lindsay 2006; Veiga and Pinho 2009). Knowledge of the target sites of different SauI members in a wide range of *S. aureus* lineages will aid studies on these pathogens (Cooper et al. 2017).

In 2011, the Kobayashi group showed that within a gene, stretches of amino acids can move from one position to another (Furuta et al. 2011). The authors suggest that such lateral domain movements within genes may be a novel common route to generate new specificities. Finally, much attention has been paid to pathogens that use phase variation, as discussed in Part E.

Antagonists of Type I Action: Antirestriction

This section is a shorter version of the text by Loenen et al. (Loenen et al. 2014a). Note that this type of antirestriction is rather different from the ClpXP-mediated proteolysis mentioned before (Chapter 7; see also, e.g., Simons et al. 2014), or the lambda Ral protein (or the analogous prophage protein Lar), which alleviates restriction by changing EcoKI from a maintenance into a de novo MTase (Chapter 6; see also, e.g., Loenen 2003). Antirestriction (anti-R) and anti-restriction-modification (anti-RM) systems in phage, plasmids, and transposons enhance their survival in a new host. Much of the early seminal work on anti-R by the T-uneven phages T7 and T3 was carried out by the groups of Bill Studier (Studier 1975, 2013; Studier and Movva 1976; Dunn and Studier 1981; Mark and Studier 1981; Bandyopadhyay et al. 1985; Moffatt and Studier 1988) and Detlev Kruger (Kruger et al. 1977a,b,c, 1983). These phages inject a small part of their DNA carrying the *0.3* gene (which encodes Ocr; see next page). The Ocr protein is produced and inactivates EcoKI and EcoBI, before the remainder of their DNA enters the cell. The best-known "artful dodger" of host restriction is probably phage T4, which encodes multiple functions to escape host defenses, some of them useful to genetic engineers (e.g., polynucleotide kinase) (Kruger and Bickle 1983; Bickle and Kruger 1993; Miller et al. 2003; Rifat et al. 2008; Petrov et al. 2010). Other work on anti-RM was carried out by the groups of Tom Bickle and about 20 papers (mainly in Russian) by Belogurov and Zavil'gel'skii spanning nearly 30 years (reviewed in Kruger and Bickle 1983; Bickle and Kruger 1993; Zavil'gel'skii 2000; Thomas et al. 2003; Tock and Dryden 2005; Dryden 2006), whereas in the United Kingdom the Wilkins laboratory solved the riddle of control of anti-R of the self-transmissible IncI plasmid, thus identifying novel transcriptional regulation from a single-stranded promoter (see the next section; Althorpe et al. 1999; Bates et al. 1999; Wilkins 2002; Nasim et al. 2004). During evolution, new mechanisms and countermechanisms between anti-RM and RM appeared continuously (Kruger and Bickle 1983; Bickle and Kruger 1993; Zavil'gel'skii 2000; Wilkins 2002; Thomas et al. 2003; Putnam and Tainer 2005; Tock and Dryden 2005; Dryden 2006; Zavil'gel'skii and Rastorguev 2009). T7 Ocr proved to be a DNA mimic with a large negatively charged patch on its surface (Atanasiu et al. 2001; Walkinshaw et al. 2002); see the next section. In the wake of Ocr, the structures of several other anti-RM proteins have been elucidated: ArdA from Tn916 from *Enterococcus faecalis* (Serfiotis-Mitsa et al. 2008), ArdB from *E. coli* CFT073 (Oke et al. 2010), and KlcA, an ArdB homolog from plasmid pBP136 from *Bordetella pertussis* (Serfiotis-Mitsa et al. 2010); see the subsection The Structure of ArdA.

Whereas Ocr seems to be confined to phage (particularly T7 and its relatives), ArdA and ArdB proteins are usually encoded by conjugative plasmids and transposons (Gefter et al. 1966; Hausmann 1967; Chilley and Wilkins 1995; Chen et al. 2014). In T7, Ocr is synthesized for the first 2 min of infection, before entry of the remainder of the phage DNA (Gefter et al. 1966; Hausmann 1967, 1988; Hausmann and Messerschmid 1988a,b; Moffatt and Studier 1988; Garcia and Molineux 1996; Molineux 2001). This completely inhibits the resident Type I enzymes, and the rest of the T7 genome can safely enter the cell. Interestingly, Ocr could bind in vitro to *E. coli* RNA polymerase (Ratner 1974), leading to the possibility that Ocr has another role in the cell, like other "moonlighting" proteins (Mani et al. 2015; Jeffery 2016) as, for example, reported for SF2 proteins involved in the skin disease Xeroderma pigmentosum (Le May et al. 2010; Kuper et al. 2014; Compe and Egly 2016). These SF2s are involved in nucleotide excision repair, but also in transcription, leading to active demethylation of CpG islands (Le May et al. 2010).

In the case of ArdA of plasmid ColIb-P9, ssDNA (which is resistant to restriction) enters the cell and forms an unusual promoter via a dsDNA hairpin, which allows transcription of the *ardA* gene (Chilley and Wilkins 1995; Althorpe et al. 1999; Bates et al. 1999; Nasim et al. 2004). Production of ArdA or ArdB rapidly inhibits the resident Type I enzymes. This novel transcription method may well be more common and deserves further research. Importantly, genome sequencing projects indicate that *ard* genes are widespread and often accompanied by antibiotic-resistance genes. The impact of combined transfer of these genes on the rate of spread of resistance in bacterial populations will be obvious.

The Structure of Ocr

Ocr is a striking example of DNA mimicry by a protein (reviewed in Loenen et al. 2014a). It is a dimer of two monomers of 116 aa and shaped like a banana with a length of ~7.5 nm and 2–2.5 nm thick (Dunn et al. 1981; Atanasiu et al. 2001; Blackstock et al. 2001; Walkinshaw et al. 2002; Zavil'gel'skii et al. 2009). In this way, it mimics the shape and surface charge of a section of B-form DNA (Fig. 14). Each monomer contains several α-helices, a long

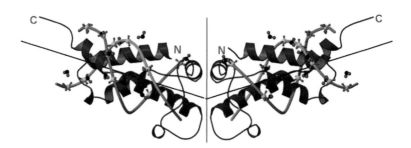

FIGURE 14. Superimposition of two 12-bp B-DNA molecules on the T7 Ocr dimer. The phosphate groups of 12 bases in each DNA dodecamer overlap with 12 carboxyl groups on each Ocr monomer, thus mimicking the shape and surface charge of B-form DNA. Ocr is shown as a blue ribbon with amino (N) and carboxyl (C) termini indicated and the dimer interface shown as a red line. Phosphate groups are colored yellow (phosphorus) and purple (oxygen). The carboxyl groups are colored red (oxygen) and black (carbon). The sugar backbones of the DNA chains are colored in two shades of green with the base pairs omitted for clarity. Vectors for the DNA helical axes are drawn as black lines. (Reprinted from Atanasiu et al. 2002.)

loop, and unstructured flexible amino and carboxyl termini, respectively. The thinness of the structure means that it has a minimal hydrophobic core with many aromatic amino acids, which may resemble the aromatic core of another DNA mimic, Qnr (Hegde et al. 2005). Despite the small core, Ocr is very stable to heat and chemical denaturation (Atanasiu et al. 2001). On the surface of each monomer are 34 negatively charged amino acids and only 6 positive amino acids (~12 of each would be expected for a typical globular protein of this size), although not all of these are required for activity (Stephanou et al. 2009a,b; Kanwar et al. 2016). The negative surface charges are spaced at roughly the same separation as the phosphate groups on 24 bp of B-form DNA containing a bend in its center, which explains its affinity for Type I enzymes (Atanasiu et al. 2002; Su et al. 2005).

The Structure of ArdA

ArdA is also a mimic of B-form DNA, like T7 Ocr (Fig. 15; McMahon et al. 2009). In the crystal structure, ArdA is a dimer but it can exist both as a dimer and as a higher multimer in solution (Serfiotis-Mitsa et al. 2008). ArdA from Tn916 from *E. faecalis* is 166 aa long with a very small dimer interface (like Ocr). The dimer is an elongated bent cylinder of ~15 nm × 2 nm. The thinness of the structure again means that it has a minimal hydrophobic core, but unlike Ocr, ArdA is not very stable to denaturation (Serfiotis-Mitsa et al. 2008). The fold of ArdA is completely different from that of Ocr: Each ArdA monomer has three small, loosely packed domains, suggesting a flexible structure. The domain folds have been found in other protein structures with a mix of α-helices, β-strands, and loops. The surface of each monomer is covered with numerous carboxyl groups such that the dimer mimics ~42 bp of bent B-form DNA. The flexibility of the structure may indicate that the ArdA protein can mold itself to the contorted S-shaped DNA-binding groove on the Type I enzyme, with different domains interacting with the different R, M, and S subunits (Kennaway et al. 2009, 2012).

The Structure of ArdB

The structures of two members of the ArdB family have been solved by both crystallography (ArdB from a pathogenicity island in *E. coli* CFT073) and NMR spectroscopy (KlcA from *B. pertussis* plasmid pBP136 [Oke et al. 2010; Serfiotis-Mitsa et al. 2010]). The ArdB and KlcA amino acid sequences are close homologs with >30% sequence identity. *Klc* genes form part of the *kor* operon involved in a regulatory network of these promiscuous plasmids (Larsen and Figurski 1994). The two ArdB structures are clearly very different from those of Ocr and ArdA and are globular proteins with a novel fold. They are

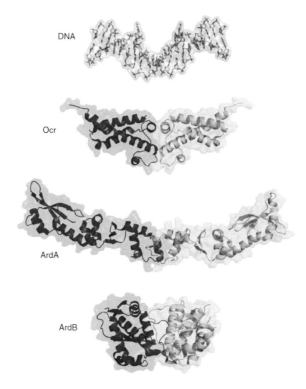

DNA

Ocr

ArdA

ArdB

FIGURE 15. Protein inhibitors of Type I R–M enzymes. (*Top to bottom*) DNA model (hydrogen atoms omitted) from PDB: 2Y7H displayed on the same scale as the proteins for structural comparisons; T7 Ocr (PDB: 1S7Z and 2Y7C), ArdA (PDB: 2W82) from Tn916 of *E. faecalis* (Davies et al. 1999), and ArdB (PDB: 2WJ9) from a pathogenicity island of *E. coli* CFT073, respectively. All three proteins are homodimeric. Their subunits are identical, but are displayed here in different colors. (Reprinted from Loenen et al. 2014a.)

neither elongated nor possess significant charged patches, so they are unlikely to cause anti-R via DNA mimicry (see Loenen et al. 2014a for discussion).

Effect of Protein Inhibitors Ocr and Ard on Restriction and Modification

The effectiveness of Ocr, ArdA, and ArdB in inhibiting Type I R-M systems has been tested in vivo with the classical efficiency of plating test (e.o.p. test; see, e.g., Chapter 1) comparing the titer of phage on an r$^+$ host carrying an anti-RM gene on a plasmid versus the strain lacking the plasmid (Walkinshaw et al. 2002; Serfiotis-Mitsa et al. 2008, 2010; Zavil'gel'skii and Rastorguev 2009; Zavil'gel'skii et al. 2009, 2011). Active anti-R enhances the number of recovered phages, which are then tested for modification by comparing the e.o.p. on the r^+ versus r^- strain: anti-M activity leads to a lower e.o.p. on the former. A novel calibrated in vivo titration assay was designed for EcoKI by the group of Zavil'gel'skii (Zavil'gel'skii et al. 2009), which allows antirestriction proteins to be distinguished based on the quantitative differences seen at different expression levels of Type I enzymes (Zavil'gel'skii and Rastorguev 2009; Zavil'gel'skii et al. 2009, 2011).

The plating assays show that Ocr blocks all Type I R-M systems (Walkinshaw et al. 2002). This is a direct consequence of the extremely strong binding of Ocr to the DNA-binding groove in the MTase core of the enzymes (Atanasiu et al. 2002). ArdA (Serfiotis-Mitsa et al. 2008) and ArdB/KlcA (Serfiotis-Mitsa et al. 2010) block restriction in all Type I families. ArdA discriminates between restriction and modification (Nekrasov et al. 2007). ArdA of plasmid R16 preferentially targets the restriction function of EcoKI (Thomas et al. 2003). This minimal anti-M effect is due to the binding of ArdA to the MTase core being of similar or weaker strength than DNA binding to the core. This weak binding is sufficient to prevent restriction but not modification. ArdA and ArdB differ in their propensity to block modification (Walkinshaw et al. 2002; Zavil'gel'skii and Rastorguev 2009; Zavil'gel'skii et al. 2009, 2011; Serfiotis-Mitsa et al. 2010). ArdB/KlcA wild variants all have strong restriction inhibition but weak effect on modification for four Type I classes (Serfiotis-Mitsa et al. 2010). ArdB also shows little or no anti-M effect in vivo and, in line with this, no interaction has been observed in vitro between ArdB and the MTase core. Furthermore, although ArdB causes anti-R in vivo, no effect could be demonstrated in vitro on restriction. Therefore, the mechanism of anti-R used by ArdB is indirect and requires further investigation. David Dryden states (Loenen et al. 2014a): "Our understanding of anti-RM is still in its infancy. Aside from the three systems described above, few others have been studied beyond their initial discovery. Given their synergistic role with restriction and modification in regulating horizontal gene transfer and the resistome (Wright 2010; Stern and Sorek 2011), this deficiency in our knowledge needs to be addressed."

Type I Single Protein

Type ISP (Type I single protein) are similar to Type IIL REases (like, e.g., MmeI), but have an additional helicase-ATPase domain, which is essential for restriction (Smith et al. 2009a,b,c). The prototype ISP REases are two plasmid-encoded R-M systems in *Lactococcus lactis*, LlaGI and LlaBIII. LlaBIII is of commercial importance as protection against phage infections in milk fermentations. An α-helical coupler domain connects the Mrr-like PD domain–cum–SF2 helicase region to the m6A γ-MTase-TRD domain.

With the exception of the TRD region, LlaBIII is >95% homologous to LlaGI (Sisakova et al. 2013). Structural and single-molecule studies indicated a conformational change after binding to the recognition site; the enzyme leaves the site and translocates DNA without looping, at ~300 bp/sec at 25°C, consuming one to two ATP per base pair (Chand et al. 2015; Kulkarni et al. 2016). As the PD and helicase/ATPase domains are upstream of the direction of translocation, these results indicated that the enzyme could not simply dimerize via its nuclease domain like Type I enzymes (Chand et al. 2015; Kulkarni et al. 2016). Together with data from single cleavage sequence analysis, this led to the proposal that DNA cleavage occurred as a result of multiple nicks by colliding enzymes, roughly halfway between sites, with the nuclease domains distal (Chand et al. 2015). Recent experiments show that translocation activates the nuclease domains via distant interactions of the helicase or MTase-TRD, without requiring direct nuclease dimerization (van Aelst et al. 2015).

Sequence analyses of 552 Type ISP enzymes showed structurally well-conserved elements involved in target recognition of LlaGI and LlaBIII, although the primary sequences of the TRDs were not that well conserved (cf. Type II enzymes) (Kulkarni et al. 2016). This led to a partial consensus code for target recognition by this class of enzymes with specificity changes due to residues that contacted the bases as well as novel contacts (Kulkarni et al. 2016).

PART C: TYPE III ENZYMES

Introduction

As detailed in the previous chapters, research on the Type III REases was basically limited to the enzymes from phage P1 and plasmid P15 by the groups of Bickle, Rao, Kruger and Reuter, and later Szczelkun (McClelland 2004), whereas the structure of EcoP15I would be finally published by Aneel Aggarwal and coworkers (see the next section). The EcoP15I R-M complex acted as MTase in the absence of ATP, and as MTase or REase in the presence of ATP, depending on the methylation state of the recognition site. Important questions remained: Why was the MTase a dimer of two Mod subunits, as methylation occurred on only one strand of the recognition sequence? Did this play a role in the stability of the complex? There appeared to be quite convincing evidence for a translocation and cutting mechanism similar to that of the Type I enzymes, but why the differences with respect to the interaction with ATP and SAM, and the location of the cut site? Did cleavage (always) occur after DNA tracking with ATP as an energy source, and collision by two complexes? Did this involve one Res subunit or two Res subunits, and did this occur in one or both directions? Would collision result in a conformational change and cleavage 25–27 bp downstream from one of the two sites?

This part of the chapter is based on the historical perspective by Rao et al. (Rao et al. 2014), recent papers that addressed the role of ATP hydrolysis in long-distance communication between sites before cleavage could occur, and the different models based on 1D diffusion and/or 3D-DNA looping. Although it had been known that Type III enzymes needed two sites in a specific head-to-head orientation for cleavage, later evidence indicates that the sites can also be in a tail-to-tail configuration (van Aelst et al. 2010). More-over, new data provided evidence for the trimeric nature of EcoP15I, EcoP1I, and PstII: one Res (and not two) and two Mod subunits (Butterer et al. 2014). The long-awaited structure appeared of the first Type III REase, which is also the first structure of a *dimeric* MTase, EcoP15I, which shed unexpected new light on the interactions of this Mod$_2$ dimer with the Res subunit (Gupta et al. 2015); see the next page. Finally, whole-genome sequencing data indicated Type III R-M systems in many sequenced genomes, in which a role for these enzymes in "phase variation" is being unraveled with respect to pathogenicity and virulence of clinically relevant organisms, such as *H. influenzae* and biofilm formation. After its initial discovery in the 1970s by Andrzej Piekarowicz and Stuart Glover in *H. influenzae* (Glover and Piekarowicz 1972; Piekarowicz and Glover 1972; Piekarowicz 1974; Piekarowicz and Kalinowska 1974; Piekarowicz et al. 1974, 1975, 1976, 1981, 1986; Jablonska et al. 1975; Piekarowicz and Baj 1975; Kauc and Piekarowicz 1978; Piekarowicz and Brzezinski 1980; Brzezinski and Piekarowicz 1982; Piekarowicz 1982, 1984), phase variation has become an important phenomenon and has also been studied in, for example, *Neisseria* sp. (Piekarowicz et al. 1988; Kwiatek and Piekarowicz 2007; Adamczyk-Popławska et al. 2009; Kwiatek et al. 2010); see the video from Andrzej's talk at the aforementioned CSHL meeting in 2013 (Piekarowicz 2013), and in Part E.

The Structure of EcoP15I

Different types of experiments addressing the composition of the EcoP15I R-M complex and the mechanism of DNA translocation and looping resulted in different models and much controversy (Peakman and Szczelkun 2004; Raghavendra and Rao 2004; Reich et al. 2004; Crampton et al. 2007a,b; Moncke-Buchner et al. 2009; Ramanathan et al. 2009; van Aelst et al. 2010; Dryden et al. 2011; Szczelkun 2011; Wyszomirski et al. 2012; Schwarz et al. 2013; Rao et al. 2014). Was this related to differences in the composition of the complex (ResMod$_2$ or Res$_2$Mod$_2$ [Rao et al. 2014]) or perhaps even multimeric complexes present in the preparation? There was evidence for translocation and DNA looping by EcoP15I, but was looping essential and/or could ATP drive 1D diffusion of the enzyme on the DNA? In the absence of crystals, Aneel Aggarwal and coworkers used small-angle X-ray scattering (SAXS) and analytical ultracentrifugation to analyze the structure of the EcoP15I R-M complex and the dimeric Mod$_2$MTase (Gupta et al. 2012). Whereas the MTase was relatively compact, the R-M complex was an elongated crescent shape of ~218 Å. Their data were in line with a model in which the MTase dimer was lodged between Res subunits. Did this mean that the Res subunits would come together and form a sliding clamp around the DNA in order to cut the DNA? Three years later (2015), the first crystal structures of the EcoP15I complex with DNA (and AMP) were published. These results came as a surprise and led to novel insights into the way in which the helicase and modification domains of EcoP15I interacted with DNA and each other (Gupta et al. 2015).

DNA Recognition by EcoP15I

Until this DNA-EcoP15I cocrystal structure appeared, by necessity the interpretation of data on dimeric MTase-DNA structures was limited to those obtained with monomeric MTases (all without DNA) (Gupta et al. 2015). The structure was a big surprise: One EcoP15I Mod subunit, ModA, turns out to be involved in specific recognition of the bases in the target site, whereas the other subunit, ModB, has the target adenine (CAGC**A**G) in its catalytic cleft, which is rotated 180° out of the DNA helix (Fig. 16).

In contrast to γ-class MTases (Type I HsdM or Type II M·TaqI) or α-class MTases (Type II M·FokI, EcoDam), in which the TRD is adjacent to the active site cleft, EcoP15I Mod belongs to the β-class MTases, where the TRD lies far off this cleft (Gupta et al. 2015). The authors suggested that, by extension, a similar division of labor (ModA subunit for recognition, ModB subunit for modification) may be used by other, mainly dimeric, β-class MTases (e.g., M·RsrI) (Thomas and Gumport 2006) and even also apply to other DNA or RNA m6A-MTases in other organisms, including mammalian cells (Gupta et al. 2015). Such RNA methylation is very common in both nucleus and cytoplasm and, for example, is implicated in RNA metabolism (transcription/splicing) and stem cell development (possibly involving the β-class MTases METTL3/METTL14) (Gupta et al. 2015). Is this division of labor universal? Is it active in all kingdoms, exemplified, for example, by the plant de novo MTase, DRM2 (domains rearranged MTase 2, which methylates only one DNA strand, like EcoP15I) or SPOUT RNA MTases (in which the RNA binds in a cleft between two monomers, and the target base is in the catalytic pocket of one monomer) (Gupta et al. 2015)?

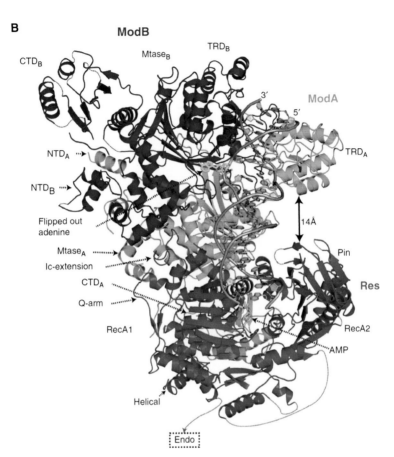

FIGURE 16. Overall structure of EcoP15I/DNA/AMP complex (PDB: 4ZCFI). (*A*) The domain arrangements of Mod and Res subunits. (*B*) An overall view of the Mod and Res subunits (Mod2Res1) bound to DNA and AMP. The two Mod protomers, ModA and ModB, are shown in cyan and blue, respectively, whereas the Res subunit is shown in magenta. The DNA is shown in gray, with the exception of the extrahelical adenine base (yellow). The AMP molecule is shown in yellow. ModA recognizes DNA through base-specific interaction from its TRD (TRDA) and interacts with Res through its MTase domain and CTD. The TRD of ModB (TRDB) does not enter the DNA major groove. CTD of ModA (CTDA) interacts with the Res subunit, whereas the CTD of ModB (CTDB) is exposed to solvent. ModA and ModB dimerize via their NTDs (NTDA/B) and central MTase domains (MTaseA/B). AMP binds in a cleft between the RecA1 and RecA2 motor domains of the Res subunit. The endonuclease domain that ensues the helical spacer is disordered and labeled in a dashed box. The proximity of TRDA of ModA and Pin domain of Res (interdomain distance B14 Å) is highlighted by a double-headed arrow. The intervening loops in the structure that are not modeled because of weak density are represented by colored dashes. (Reprinted from Gupta et al. 2015.)

Restriction by EcoP15I

During the early 1990s, helicases had been considered ATP-dependent DNA and RNA unwinding enzymes, but this view was subsequently challenged by data on Type I and other enzymes indicating translocation without strand separation. It became clear that the specificity of helicases or translocases for different substrates was dictated by additional regions in between the RecA motor domains and/or the amino- or carboxy-terminal flanking regions (Singleton et al. 2007). For example, true helicases contain a wedge-like domain between the RecA domains to disrupt the base pair for unwinding (Singleton et al. 2007). During unwinding or translocation, the motors consume ATP with every step, but why did some enzymes consume very little ATP while traveling long distances on the DNA? The answer came from single-molecule fluorescent microscopy studies. These indicated that ATP hydrolysis of EcoP15I bound to its target site did not result in DNA translocation: The energy generated induced a conformational change that resulted in long-range diffusion of the enzyme on the DNA (Schwarz et al. 2013). Such ATP-triggered change of state, which allows sliding, was named "molecular switching" of the enzyme, which could happen on DNA as well as RNA (Szczelkun et al. 2010; Schwarz et al. 2013; Szczelkun 2013), but could probably also cause other events such as protein–protein interactions, for example, for the clamp loader (Kelch 2016). Subsequent kinetic studies support this notion of two distinct ATPase phases, a rapid consumption of ~10 ATP inducing a conformational change and a slower phase related to the rate of dissociation of the enzyme from the recognition site (Toth et al. 2015).

The novel EcoP15I structure revealed three new substructures, two within the RecA1 segment and one between RecA1 and RecA2: a loop, a β-hairpin-like "Q-arm," and a novel "Pin" domain (Fig. 16; Gupta et al. 2015). These are scattered through the RecA-like domain. Two are in the amino-terminal half and the Pin domain is an insertion that divides RecA1 half (motifs I-III) from RecA2 (motifs IV–VI) (see Supplementary Fig. 5 of Gupta et al. 2015 and Fig. 1a in Mackeldanz et al. 2013). The motor domains bound dsDNA and facilitated DNA sliding via this specialized Pin domain. The Pin domain adopted a tertiary structure that extended toward the ModA TRD subunit and interacted with the translocating strand of the DNA duplex (Gupta et al. 2015). The DNA was severely distorted from B form at two sites along its axis (i.e., where the adenine was ejected from the recognition site and near the ModA-Res interface). The first distortion was due to intrusion of ModA into the DNA major groove, whereas at the ModA-Res junction the DNA was bent ~24° toward the minor groove, in the direction of the ModA TRD and Res Pin domain. The result was a reduction in the distance between these domains to <14 Å that "may facilitate an interaction between the two domains when EcoP15I assumes a diffusive or sliding state on DNA" (Gupta et al. 2015). Therefore, the EcoP15I motor domain interacted predominantly with the translocating strand. The semiclosed configuration of the EcoP15I motor domain differed from the more open structure of another translocase, *S. solfataricus* SF2 translocase, which might indicate that EcoP15I was in an intermediate state, following ATP hydrolysis but before AMP dissociation (see Gupta et al. 2015 for further discussion).

With this structure containing AMP in hand, could one predict what might happen in the presence of ATP? How might the enzyme slide? According to Aneel Aggarwal and coworkers (Gupta et al. 2015), the ModA TRD might move from the DNA major groove to the Pin domain and in this way adopt a "nonspecific" conformation that would result in DNA sliding. Such a movement of ~40° would be possible because of a flexible linker between the TRD and MTase domain, which would prevent the Pin domain reaching the DNA. This structural model is in agreement with single-molecule studies, which suggest that the ResMod$_2$ complex moves along the DNA (like Type ISP, but unlike Type I, in which the M$_2$S complex remains bound to the recognition site). This trimeric complex would slide along the DNA until it collides with another bound complex, which would make it cleavage competent (Schwarz et al. 2013).

PART D: TYPE IV ENZYMES

Introduction

Like the Type IIM REases, Type IV modification-dependent REases (MDEs) recognize a variety of DNA modifications at cytosine or adenine bases (http://rebase.neb.com/rebase/rebase.html; Carlson et al. 1994; Roberts et al. 2003). This section is based on two recent reviews (Loenen and Raleigh 2014; Weigele and Raleigh 2016), and the reader is referred to these for more details and references. Over the years many papers were published reporting phages protecting their DNA against a wide range of Type I, II, and III REases by base modification such as methylation. This did not protect them against Type IV enzymes that preferentially or exclusively attack modified Cs and As. In the laboratory strain *E. coli* K12, McrA and McrBC recognize hm5C DNA and m5C, but the pathogenic *E. coli* CT596 also carries the *gmrS* and *gmrD* (glucose-modified hm5C restriction) genes allowing restriction of ghm5C DNA in, for example, wild-type T4 DNA (Bair and Black 2007). Interestingly, this activity in turn could be inhibited by T4 IPI*, a small protein encoded by this champion dodger of bacterial defense systems (Bair et al. 2007; Rifat et al. 2008).

Another Type IV REase, Mrr (modified DNA rejection and restriction), was identified in *E. coli* K12, because it caused cloning trouble by recognizing both m5C and m6A (Heitman and Model 1987; Waite-Rees et al. 1991). In addition to m5C, m6A, and m4C, many other modifications exist in both prokaryotes and eukaryotes with probable roles in defense and stress situations, in part via regulation of replication and transcription (Freitag and Selker 2005; Lobner-Olesen et al. 2005; Zhou et al. 2005; Borst and Sabatini 2008; Kaminska and Bujnicki 2008; Low and Casadesus 2008; Iyer et al. 2009; Kriaucionis and Heintz 2009; Ou et al. 2009; Tahiliani et al. 2009; Xu et al. 2009; Prohaska et al. 2010; Wang et al. 2011; Loenen and Raleigh 2014). The Type IV REases are hard to identify as they lack a cognate MTase, and therefore one cannot use Pacific Biosciences (PacBio) SMRT sequencing to find the recognition sites. Their discovery relies on a genetic system and suitable phages or plasmids as challengers. Cell extracts to digest DNA in vitro and analyze on gel are of little use because of severe degradation of the DNA (Eid et al. 2009; Korlach and Turner 2012; Roberts et al. 2015).

It is not known whether all Type IV enzymes flip the methylated base out of the helix, but that may be common. Figure 17 shows the structure of McrBC compared to other base-flipping proteins.

Fusions of DNA Binding and Cleavage Domains

All characterized Type IV proteins are fusions of various cleavage and DNA recognition domains. By 2014, six groups of enzymes had been identified that recognize modified DNA with low sequence selectivity, with the prototypes McrA, McrBC, Mrr, SauUSI, MspJI, PvuRts1I, and GmrSD (Loenen and Raleigh 2014). Five more groups have various fusions of the DBD and the cleavage domain, whereas they may be ATP- or GTP-dependent for recognition or cleavage (Weigele and Raleigh 2016). As PvuRts1I is qualitatively similar to MspJI family enzymes, and in turn to DpnI (i.e., typical IIM REases), it is debatable whether some enzymes should be classified as Type IIM or Type IV: The key feature discriminating IIM and IV enzymes is cleavage position, although this position is known (and fixed) for IIM enzymes (DpnI, PvuRts1I family, MspJI family), the cleavage position for the Type IV enzymes is either variable (McrBC) or unknown (McrA, Mrr).

McrA

McrA (recognition sequence YCGR) of *E. coli* K12 has been extensively analyzed by bioinformatics and mutagenesis. McrA has an amino-terminal

A

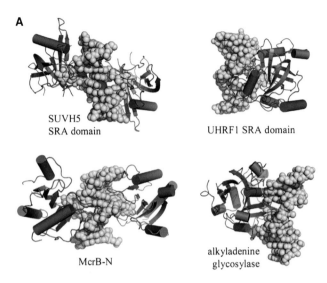

B

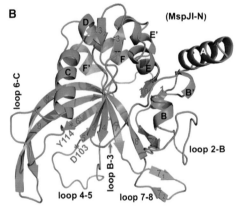

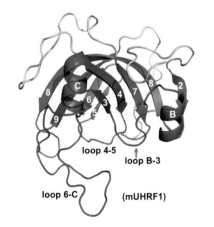

FIGURE 17. McrB-N in comparison to other base-flipping proteins. (*A*) SRA domains SUVH5 (3Q0C) and UHRF1 (2ZKD) use loops extending from a crescent formed from two beta sheets to flip C or m5C from undeformed B-form DNA into a pocket (*top row*), whereas McrB-N (3SSC; *bottom row*) uses loops from one beta sheet to distort the DNA and flip the base. It resembles the human alkyladenine glycosylase (1BNK; *bottom row*) in bending the DNA toward the major groove, while flipping the base via the minor groove (Sukackaite et al. 2012). (*B*) The SRA-like hemimethylated m5C recognition domains. A ribbon model of the amino-terminal domain of the MspJI structure (4F0Q and 4F0P; *top*) compared with the SRA domain of URHF1 (PDB: 3FDE; *bottom*). The crescent shape formed by interacting beta sheets and helices αB and αC are the conserved features of the SRA domain highlighted here. Loops on the concave side of UHRF1 participate in flipping the base, and similar loops presumably do so for MspJI. Two of these vary in length among family members and may play roles in sequence context specificity (Horton et al. 2012). (Reprinted from Loenen and Raleigh 2014; *A*, originally adapted from Sukackaite et al. 2012; *B*, originally adapted from Horton et al. 2012.)

DBD, and a carboxy-terminal HNH domain required for restriction in vivo (Bujnicki et al. 2000; Anton and Raleigh 2004). It recognizes m5C and hm5C (but not ghm5C), with a preference for C or T at the 5' position, (Y>R)m5CGR; in vitro it binds m5CpG DNA, but does not restrict (Mulligan and Dunn 2008; Mulligan et al. 2010; Loenen and Raleigh 2014; Loenen et al. 2014b). Interestingly, a mutation in the DBD enabled in vivo discrimination between m5C (still recognized) and hm5C (not recognized) (Anton and Raleigh 2004). A protein with a similar HNH nuclease domain, but different DBD, in *Streptomyces coelicolor* A3, ScoA3McrA, cuts DNA modified by the *E. coli* Dcm protein (Cm5CWGG) or phosphorothioate (PT)-modified sites (or both) at a variable distance, and cleavage depends on Mn^{2+} or Co^{2+} (Gonzalez-Ceron et al. 2009; Xu et al. 2009; Liu et al. 2010).

McrBC

McrBC of *E. coli* K12 has been well characterized (Raleigh 1992; Sutherland et al. 1992; Gast et al. 1997; Pieper et al. 1997, 2002; Stewart and Raleigh 1998; Stewart et al. 2000; Panne et al. 2001; Pieper and Pingoud 2002; Sukackaite et al. 2012; for details, see Loenen and Raleigh 2014; Weigele and Raleigh 2016). The *mcrB* gene also encodes an additional protein, McrB (S), which starts at an internal translation initiation codon (thus lacking the first 161 aa) and appears to have a regulatory function (Beary et al. 1997). Figure 18 shows a model for the assembly of the McrBC complex (Loenen and Raleigh 2014). McrBC is a GTP-dependent heteroheptamer in which McrC (with the PD nuclease motif) binds a complex of McrB with GTP and DNA. McrBC makes a double-strand cut near one Rm5C site but requires the cooperation of two sites or a translocation block. The sites may be on different daughters across a fork. These are separated by 30-3000 bp, may be on either strand, and minor cleavage clusters are found ~40, ~50, and ~60 nt from the m5C (Pieper et al. 2002).

Mrr

Mrr of *E. coli* K12 recognizes m6A and m5C (with uncertain specificity), which prevented cloning of some R-M systems (e.g., PstI [CTGm6AG], HhaII [Gm6ANTC], and StyLTI [CAGm6AG]) and other genes (Heitman and Model 1987; Kelleher and Raleigh 1991; Waite-Rees et al. 1991; Loenen and Raleigh 2014; Weigele and Raleigh 2016). Mrr contains a predicted variant of the PD motif in the carboxy-terminal domain, with a presumed amino-terminal winged-helix DBD, like the MspJI family (Loenen and Raleigh 2014; Weigele and Raleigh 2016).

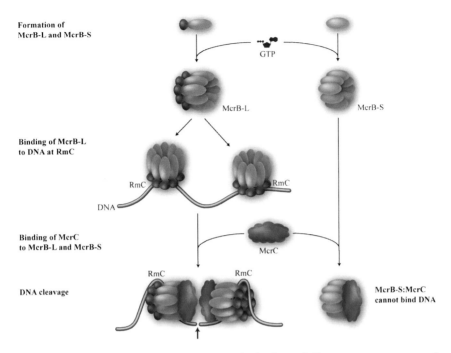

FIGURE 18. McrBC assembly model (Loenen and Raleigh 2014). Two proteins are expressed from mcrB in vivo. Both the complete protein (McrB-L) and a small one missing the amino terminus (McrB-S; *top row*) bind GTP, forming high-order multimers detected by gel filtration (second row). When visualized by scanning transmission EM, these appear as heptameric rings with a central channel. Rings of McrB-L in top views show projections that may correspond to the amino-terminal DBD (red segment). Both forms can then associate with McrC, judged again by gel filtration. McrB-L: GTP can bind to its specific substrate (RmC) in the absence of McrC (third row); in its presence, the substrate is cleaved (fourth row). GTP hydrolysis is required for cleavage (arrow): A supershifted binding complex forms in the presence of GTP-γ-S, but no cleavage occurs. Translocation accompanies GTP hydrolysis; dsDNA cleavage requires collaboration between two complexes, or a translocation block. The path of the DNA in the figure is arbitrary, as is the conformation of McrC. (Modified from Bourniquel and Bickle 2002, with permission from Elsevier Masson SAS.)

SauUSI

The (d)ATP-dependent SauUSI (SCNGS) of *S. aureus* recognizes Sm5CNGS (S = C or G) (Loenen and Raleigh 2014). It has a PLD nuclease domain (like BfiI and other REases), a carboxy-terminal DBD, a helicase/translocase domain in between, and a cleavage domain (Xu et al. 2011) with a reaction mechanism resembling that of topoisomerases and transposases (Interthal et al. 2001; Sasnauskas et al. 2003, 2007), but not quite the same: PLD nucleases employ a covalent protein–DNA intermediate, but, unlike topoisomerases and some (e.g., serine) transposases, the covalent linkage is made by a histidine rather than serine or tyrosine.

PvuRts1I

PvuRts1I from *P. vulgaris Rts1* restricted glucosylated T-even phages in vivo (Janosi et al. 1994; Loenen and Raleigh 2014). It recognizes (g)mC(N11-13/N9-10)G—that is, it cuts 11–13 nt downstream from a modified C, which is 20–23 nt before a G (hence, not very specific). It has approximately 20 active homologs, including AbaSI form *Acinetobacter baumanii*, which recognize m5C slightly and hm5C and ghm5C better, with differing preferences and weak and variable selectivity for the sequence surrounding the modified base (Szwagierczak et al. 2011; Borgaro and Zhu 2013). These enzymes require two modified sites for dsDNA cleavage ~22 nt apart, with incisions ~11–13 nt 3′ to the modified base on the one strand and 9–10 nt 3′ on the other. Bioinformatics, mutational evidence, and crystal structures indicate a carboxy-terminal SRA-like (SET and ring finger–associated) DBD, and an amino-terminal cleavage domain that is a divergent member of the PD family (Kazrani et al. 2014; Shao et al. 2014; Weigele and Raleigh 2016). The structure of AbaSI confirmed the results with PvuRts1I, with an amino-terminal nuclease domain resembling that of VSR, whereas the carboxy-terminal domain resembles SRA family members (Horton et al. 2014a; Weigele and Raleigh 2016).

GmrSD

The aforementioned GmrSD enzyme, encoded by the *gmrS* and *gmrD* genes from a pathogenic *E. coli* strain, also exists as a single fusion protein with similar characteristics (Bair et al. 2007; He et al. 2015; Machnicka et al. 2015). In vitro, the enzyme prefers ghm5C DNA to unmodified m5C. GmrD is the nuclease, whereas the GmrS domain has a ParB/Srx fold, present in conjugative plasmids (He et al. 2015; Machnicka et al. 2015; see Weigele and Raleigh 2016 for details). The enzyme has a strong preference for UTP over GTP and CTP (Loenen and Raleigh 2014; Weigele and Raleigh 2016).

MspJI

MspJI from *Mycobacterium* sp. JLS, was the first member of a family of Type IV enzymes, which cleave with a four-base 5′ extension 12 nt from the m5C, and 16–17 nt on the opposite strand (Bujnicki and Rychlewski 2001; Zheng et al. 2010; Cohen-Karni et al. 2011; Horton et al. 2012, 2014b,c). Members recognize m5C and hm5C, but not ghm5C. MspJI has a SRA-like amino-terminal domain, which flips m5C out of the helix, like other SRA-like proteins. Swapping experiments with short protein loops contacting nearby bases, which varied among three members, significantly reduced selectivity for sequences flanking the modified base (Sasnauskas et al. 2015; Weigele and Raleigh 2016).

The endonuclease domain cuts a different DNA strand than the DNA bound by that polypeptide. This is reminiscent of M·EcoP15I in which the DBD of one subunit binds the recognition sequence, whereas the catalytic domains of, in this case, *two* other subunits, effect dsDNA cleavage. MspJI (and the isoschizomer AspBHI [Horton et al. 2014b,c]) can be thought of as Type IIS enzymes (albeit enzymes that bind only when the recognition site is modified) and their tetrameric structure and domain cooperation could be typical of many "normal" Type IIS enzymes, as well as the Type IIB, IIC, and IIG enzymes. The convention is to think of these acting as transient homo-dimers, but they might well act as tetramers (and even as multimers of tet-ramers, e.g., BcgI) instead.

PART E: PHASE VARIATION

Introduction

Bacterial pathogens not only infect the host, but also try to maintain themselves ("colonize") by hiding from the immune system and/or confusing it. One way of doing this is by hypermutation at simple sequence repeats or homopolymeric tracks located within the reading frame or in the promoter of a subset of genes (Moxon et al. 2006). This is due to polymerase slippage (called slipped-strand mispairing [SSM]), which could be an important evolutionary strategy (Levinson and Gutman 1987a,b). Genetic variation in the population of patho-genic bacteria via SSM would result in two or (many) more different pheno-types that allow the strain to evade the host immune system (Robertson and Meyer 1992). SSM changes the number of repeats or bases, which switches promoters "on" or "off" (e.g., by changing the distance between the -35 and -10 regions), causes frameshifts in coding regions, and/or dictates alternative usage of multiple translation initiation codons in different reading frames, thus altering or abolishing DNA recognition (see, e.g., van Ham et al. 1993; van Belkum et al. 1998; De Bolle et al. 2000; van der Woude and Baumler 2004; Srikhanta et al. 2005, 2010; Moxon et al. 2006; van der Woude 2006; Dixon et al. 2007).

The groups of Richard Moxon, Andrzej Piekarowicz, and Michael Jen-nings have studied repeat variation that created this so-called "phase variation" in three pathogens, *H. influenzae*, *H. pylori*, and *Neisseria* sp., revealing altered expression of up to 80 genes, including genes important for iron uptake, DNA repair, electron transport, amino acid transport, and growth (reviewed in Srikhanta et al. 2010) but also for MTases. In the latter cases, the MTases can function as an on–off switch for multiple genes that allow the pathogen to combat host immunity (De Bolle et al. 2000; Seib et al. 2002; Srikhanta et al. 2005, 2010; Casadesus and Low 2006; Moxon et al. 2006; Wion and Casadesus 2006; Marinus and Casadesus 2009). Various other clinical isolates also contain MTases with repeats of variable length, which are either lone MTases or associated with Type II systems (Kong et al. 2000; Xu et al. 2000a,b; Lin et al. 2001; Vitkute et al. 2001; Skoglund et al. 2007). These genetic systems have been called "phasevarions" or "phase-variable regulons," and appear to be a common strategy to randomly switch between distinct cell types and create phenotypic heterogeneity in the bacterial pathogenic population (Weiser et al. 1990; van Ham et al. 1993; Hallet 2001; Srikhanta et al. 2010). Such methylation-driven alternative gene expression can be very complex because of the presence of multiple phase-variable genes (see Srikhanta et al. 2010 for further details).

Taken together, different types of variation may alter gene expression: (1) reversible changes (i.e., alternative states of the same genes result in expression or not, or expression of different genes [e.g., flagella or tail fibers], via inversion of promoters or amino termini, or slipped mispairing); (2) diversity selected changes, in which a rare genotype is favored because of changing circumstances (involving outside forces); and (3) replacement or addition variation, in which a gene is replaced by another gene or is added extra.

Variable Type II systems

Around the turn of the century, two dozen potential R-M systems were identified in two completely sequenced *H. pylori* strains, 26695 and J99, amounting for >4% of the total genome (i.e., much more than in other sequenced bacterial genomes) (Kong et al. 2000; Lin et al. 2001). Although nearly 90% of the R-M genes were present in both strains, <30% of the Type II R-M systems were functional in both strains with different sets active in each strain. An interesting observation was that all strain-specific R-M genes were active, whereas most shared genes were inactive. Did this indicate that these active strain-specific genes had been acquired recently via horizontal transfer from other bacteria? And did these pathogenic strains constantly acquire new R-M systems and inactivate and delete the old ones (Lin et al. 2001)? Were these multiple R-M systems a "primitive bacterial immune system, by alternatively turning on/off a subset of numerous R-M systems" (Kong et al. 2000)? Support for this idea came from other *H. pylori* strains, in which the R-M systems also proved to be highly diverse (Xu et al. 2000a). In addition, several

REases had novel specificities: Hpy178III (TCNNGA), Hpy99I (GGWCG), and Hpy188I (TCNGA) (Xu et al. 2000a). The latter system was absent in the 26695 and J99 strains, whereas the GC content implied that the Hpy188I system had been recently introduced into the *H. pylori* genome (Xu et al. 2000b).

In 2016, phase variation of a Type IIG enzyme was reported in *Campylobacter jejuni* NCTC11168 (Anjum et al. 2016). This IIG protein methylates adenine in CCCGA and CCTGA sequences. Using both inhibition of restriction and PacBio-derived methylome analyses of mutants and phase variants, the cj0031c allele in this strain was demonstrated to alter site-specific methylation patterns and gene expression, which "may indirectly change adaptive traits" (Anjum et al. 2016).

Phase-Variable Type III systems

Phase-variable Type III *mod* genes have been identified in *H. influenzae*, *H. pylori*, and *Neisseria* sp., which contain tandem repeats that may be homopolymeric, or repeat tracks of 2, 3, 4, or 5 nt (Srikhanta et al. 2010). The *mod*-like gene of *H. influenzae* Rd has 40 AGTC repeats within its ORF. This *mod* gene was found in 21 out of 23 other *H. influenzae* strains, and in 13 of those the locus contained repeats of variable length (De Bolle et al. 2000). These repeats comprised a hypervariable region in the central region of the *mod* gene of 22 nontypeable *H. influenzae* strains, whereas the *res* gene was conserved (Bayliss et al. 2006). Moreover, similar *mod* genes with similar hypervariable regions were identified in pathogenic *Neisseria* sp., suggesting horizontal transfer of these genes between different species. This high phase variability of these MTases would not only protect against phage infections but might "also have implications for other fitness attributes of these bacterial species" (Bayliss et al. 2006). An example of phase variation that allows typing of these so-called untypeable *H. influenzae* ("NTHi") strains is shown in Figure 19, redesigned from Figure 3 in Fox et al. 2007. Variation of the number of repeats generates ON/OFF switches. Switching within a clonal population results in subpopulations with and without RM activity. In *Haemophilus* and other taxa, this results in variable expression of distant genes as well, presumably regulated by the presence of modification in regulatory regions (e.g., Tan et al. 2016; see, for a review, Sanchez-Romero et al. 2015). The variable regulation at distant sites does not require activity of the restriction function (Fox et al. 2007). Also observed in this figure is the presence of variable segments (TRDs) that give distinct recognition specificities. These variations are generated by horizontal transfer between strains, species, and even genera (Bayliss et al. 2006). Further studies indicated that in *H. influenzae* Rd, phase variation of *modA* was

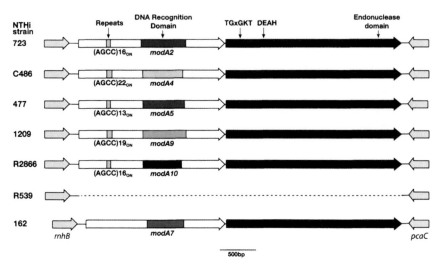

FIGURE 19. Phase variation. A representation of phase-variable methyltransferase genes present in different strains of nontypeable *H. influenzae* (NTHi). The prototypical strain of each NTHi strain in which each allele/arrangement is present is shown on the *left* side. Each modA gene is represented as a white arrow, with the DNA recognition domain (DRD) represented by a colored box. Downstream from each modA is the cognate restriction endonuclease gene, res, with the locations of the ATP-binding motif (TGxGKT), the ATP-hydrolysis motif (DEAH), and the endonuclease motif, PD···(D/E)XK, indicated above. Phase-variable modA genes are represented by the five most common modA alleles found in NTHi isolates from patients with middle ear infection (otitis media [OM]). These alleles—modA2, 4, 5, 9, and 10—all contain a simple-sequence repeat (SSR) tract; in this case the sequence AGCC repeats n times. The SSR tract is represented by a gray box in these alleles, with the number of AGCC repeats and the expression status of this number of repeats shown underneath each allele. For example, modA2 in NTHi strain 723 contains 16 AGCC repeats, which leads to expression (ON) of the gene. NTHi strain R539 contains a deletion of the entire mod-res region. NTHi strain 162 contains the modA7 allele that does not contain SSRs and therefore is not phase variable. ModA alleles that are not phase-variable contain the 12-nt sequence 5'-TCAGATAGTCAG-3' in place of a SSR. In all cases, the genes flanking the mod-res locus are conserved: upstream is the gene *rnhB* (represented by the blue arrow to the *left* of each modA gene); downstream is the gene *pcaC* (represented by the yellow arrow to the *right* of each res gene). (Courtesy of John Atack.)

calculated to occur at a frequency of 4×10^{-6} mutations/division/repeat unit for off-to-on switches and 7×10^{-6} mutations/division/repeat unit for on-to-off switches (De Bolle et al. 2000). The restriction phenotype of a Type III system in *N. gonorrhoeae*, NgoAXP, switched randomly because of a change in the number of pentanucleotides (CCAAC/G) present at the 5' end of the coding region of the *ngoAXPmod* gene (Adamczyk-Popławska et al. 2009). The *mod* gene in another *N. gonorrhoeae* strain, FA1090, was linked to biofilm formation, adhesion to human cells and epithelial cell invasion, and hence to pathogenicity and systemic infection (Kwiatek et al. 2015). Various serotypes of *Pasteurella haemolytica* have CACAG repeats within the 5′ end of a Type III R-M system, repeats which could change in length upon serial subculture and most likely also occurred as a result of DNA SSM (Ryan and Lo 1999).

Phase-Variable Type I systems

Phase variation also occurs in Type I systems. HindI of *H. influenzae* has a $(GACGA)_4$ repeat, and changes in the number of pentanucleotides (which is influenced by Dam methylation) within the coding sequence of *hsdM* were linked to protection against phage (Zaleski et al. 2005). In *Mycoplasma pulmonis*, two *hsd* loci each contain two *hsdS* genes with complex, site-specific DNA inversion systems (Dybvig et al. 1998). This generates a complete family of related *hsdS* genes with extensive sequence variations that recognize different DNA sequences, suggestive of additional roles in genome maintenance (Dybvig et al. 1998).

In the Type IC NgoAV system from *N. gonorrhoeae*, the length of tandem repeats of four amino acids were involved in the generation of a truncated or full-length HsdS protein, and only the long protein could complement other Type IC systems (Adamczyk-Popławska et al. 2011). Similar tetra-amino acid repeats (either TAEL, LEAT, SEAL, or TSEL) were identified in other Type IC systems in distantly related bacteria (Adamczyk-Popławska et al. 2003). Was this a common special characteristic of Type IC systems (Adamczyk-Popławska et al. 2003), and is there a tale to tell? Do these data shed light on the origin of the switch between the EcoR124I and EcoR124II systems mentioned in Chapter 6 (Fig. 6)? These proteins recognize GAA (N6) RTGC and GAA (N7) RTGC, respectively, and the presence of either six or seven bases between the two specific DNA regions was shown to relate to the presence of either two or three TAEL repeats in the respective HsdS subunits (Price et al. 1989). Price et al. (1989) suggested that the switch between the two specificities could be due to unequal crossing-over in these repeats (Price et al. 1989). Is this a coincidence or are these repeats perhaps the end result of prolonged SSM? Were these TAEL repeats once much longer, and were they slowly eliminated during continued growth under laboratory conditions, because they were no longer needed?

Such an assumption could fit in with LacZ fusion experiments by Bolle et al. (2000) with respect to the above-mentioned *mod*-like gene of *H. influenzae* Rd (with 40 AGTC repeats within its ORF). These authors fused a *lacZ* reporter to a chromosomal copy of *mod* downstream from the repeats, which resulted in high-level phase variation. Changing the number of repeats changed mutation rates. Phase variation occurred at a high frequency in strains with the wild-type number of repeats. Rates increased linearly with tract length over the range 17–38 repeat units. The majority of tract alterations were insertions or deletions of one repeat unit with a 2:1 bias toward contractions of the tract (De Bolle et al. 2000). This could be interpreted to mean that the shorter the track, the higher the chance that the track would become shorter faster. As under laboratory conditions the pressure is absent to protect against either phage or host immunity, the bacteria with shorter tracts would have a favorable advantage over bacteria with longer tracts. That advantage would eventually be lost once the number of repeats became very small, and the length of the spacer between the two recognition domains could no longer be decreased without loss of the DNA recognition function. The end result would be two active enzymes, EcoR124I and EcoR124II, with only two or three TAEL repeats left.

Monika Adamczyk-Popławska et al. (Adamczyk-Popławska et al. 2011) analyzed a poly(G) tract in the *hsdS(NgoAV1)* gene in *N. gonorrhoeae*. Deletion of 1 nt in this tract with seven guanines led to a frameshift at the 3′ end of the *hsdS(NgoAV1)* gene and fusion to a second downstream *hsdS* gene, *hsdS (NgoAV2)* (Adamczyk-Popławska et al. 2011). This resulted in a longer HsdS protein with two TRDs, rather than the original truncated HsdS protein with a single TRD derived from *hsdS(NgoAV1)* (Adamczyk-Popławska et al. 2011). Such a contraction of the poly(G) tract that caused this frameshift might well occur in vivo, as the authors found a minor subpopulation of cells that appeared to have only six guanines. Thus, it could be argued that the strain could switch this Type I system "on" (two TRDs = protection against phage and/or other foreign DNA) or "off" (one TRD and the possibility of DNA exchange via horizontal transfer).

FINAL THOUGHTS

It has proven very difficult to generate mutant or hybrid restriction enzymes with long DNA specificity sites that would serve as good tools for gene therapy. The new RNA-based method of genome editing using the CRISPR–Cas9 system may present a better alternative, but also has its drawbacks. To answer the question of how foolproof the CRISPR system is, Bull and Malik (2017) state in their discussion of a recent paper by Champer et al. (2017) that "the easy targeting of CRISPR, the very property that has led to its current popularity, may also be its downfall as a practical means to control populations or suppress disease transmission." This would fit in with the research described in this book, which proves that organisms go to incredible lengths to keep genome alterations under tight control to avoid chromosomal instability and allow only a low level of heterogeneity.

The year of the 5th NEB meeting in Bristol in 2004 (Fig. 20) was also the year of the publication of the first and only book totally dedicated to restriction enzymes (edited by Alfred Pingoud [Pingoud 2004]). It reflects the growing realization of the importance of restriction-modification systems in "cells, not eppendorfs" (King and Murray 1994), and their role outside the laboratory in communities with benign and pathogenic bacteria and archaea.

The large current number of restriction enzymes and the 50-odd structures indicate overlap between types and subtypes in different ways. The enzymes may share a common ancestor with three separate domains for DNA recognition, restriction, and methylation, whereas Type I and III enzymes also contain the ATP-dependent SF2-related molecular motor domains. The cooperation of restriction enzymes between sites with or without looping and with or without collision/stalling may be more the rule than the exception. Different catalytic nuclease domains, mainly with the PD$\cdots$(D/E)XK motif, but also HNH, GIY-YIG, or PLD domains, seem to have been "mixed and matched" during the course of evolution. The structures indicate careful control of positioning of the nuclease domain and large conformational changes before cleavage can occur, plus a variety of other control systems to avoid rampant nuclease activity, which would result in genome instability. Such other control may inhibit synthesis of the restriction enzymes in the cell at the level of transcription at the operon promoter or via C proteins, at the level of translation and/or posttranslation, and finally via DNA mimics employed by various plasmids and phages that inhibit Type I enzymes. The structure of the EcoP15I Type III enzyme indicates a division of labor of the two modification subunits—one for DNA recognition and the other for methylation, which may be a more general mechanism and may also apply to EcoKI (see below). EcoP15I hydrolyses approximately 30 ATP molecules in two steps (a fast consumption of approximately 10 ATP molecules followed by a slower consumption of approximately 20 ATP, which switches the enzyme into another rather distinct structural state that can diffuse on DNA over long distances (Schwarz et al. 2013). This resembles the situation with EcoKI, in which ATP also acts as allosteric effector (Burckhardt et al. 1981). The big difference with EcoP15I is that the EcoKI methyltransferase complex remains bound to the recognition site, whereas the HsdR subunits translocate DNA and hydrolyze ATP, and the whole EcoP15I complex slides away from the site (thus consuming less ATP). Both the ATP-triggered thermal diffusion and ATP-dependent initiation of translocation are important observations for studies into the much more complex eukaryotic methyltransferases and helicases.

5th New England Biolabs Meeting on Restriction/Modification, Wills Hall, Bristol - 4th to 8th Sept 2004

General Information

Inaugural New England Biolabs Workshop on Biological DNA Modification
Gloucester, MA (USA)
20-23 May 1988

2nd New England Biolabs Workshop on Biological DNA Modification
Berlin, Germany
2-7 Sept 1990

3rd New England Biolabs Workshop on Biological DNA Modification
Vilnius, Lithuania
22-28 May 1994

4th New England Biolabs Workshop on Biological DNA Modification
Innsbruck, Austria
2-7 September 1997

5th New England Biolabs Meeting on Restriction-Modification
Bristol, UK
4-8 September 2004

Previous New England Biolabs workshops have been characterised by science of the highest quality and a friendly, collegiate atmosphere. We hope to continue this tradition with the Bristol meeting, RM2004. The change in name this year reflects a change in the aims of meeting, to focus on prokaryotic restriction-modification systems as a whole. The meeting will cover a broad range of topics from the mechanistic details of individual enzymes through to whole genome and proteome analysis.

5th New England Biolabs Meeting on Restriction / Modification

Saturday 4th - Wednesday 8th September 2004
Wills Hall, University of Bristol
Bristol, United Kingdom

6th New England Biolabs Meeting on DNA Restriction and Modification

1st to 6th August 2010
Campus of the Jacobs University
Bremen, Germany

Organized by Albert Jeltsch, Alfred Pingoud & Wolfgang Wende

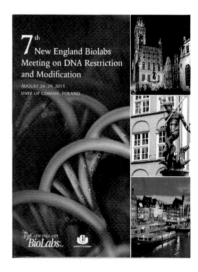

FIGURE 20. The NEB meetings between 1988 and 2015. (*Top left*) The first four NEB meetings on restriction and modification enzymes. (*Top right*) The 5th NEB meeting on Restriction/Modification in Bristol (2004) was organized by Mark Sczcelkun, Bernard Connolly, and David Dryden. (*Bottom left*) The 6th NEB meeting on DNA Restriction and Modification in Bremen (2010) was organized by Albert Jeltsch, Alfred Pingoud, and Wolfgang Wende. (*Bottom right*) The 7th NEB meeting on DNA Restriction and Modification in Gdansk (2015) was organized by Iwona Mruk, Geoffrey Wilson, and Richard Morgan. (Courtesy of Rich Roberts.)

One of the most important applications of restriction enzymes with concomitant impact on science and society has been the development of the DNA fingerprinting technique by Alec Jeffreys. Alec started his "Zoo blots" in the 1970s next door to the author of this book in Leicester, where he continued his work on restriction fragment length polymorphisms (RFLPs), which research could fill a book by itself (Fig. 21; Jeffreys 2006).

A final note about the intriguing story of EcoKI, hemimethylation, and cofactor SAM, a subject close to the author's heart (Loenen and Murray 1986; Loenen 2003, 2006, 2010, 2017, 2018): As already suggested in 1981 (Burckhardt et al. 1981), EcoKI has two types of binding sites for SAM—one for DNA recognition (the effector site) and the other for methylation (the methyl donor site). How does this relate to the ability of EcoKI to switch from a maintenance methyltransferase (with a strong preference for methylation of hemimethylated DNA) to a de novo methyltransferase? This ability is a property of Type IA enzymes (and not Type IC enzymes such as EcoR124I) that has been observed in the presence of the lambda Ral protein (Zabeau et al. 1980; Loenen and Murray 1986). How can a small protein like Ral cause such an important switch? Are there similar proteins to be discovered in eukaryotic systems? And what about the de novo methylation EcoKI mutants (e.g., the HsdM L113Q mutant made in Noreen Murray's laboratory) (Kelleher et al. 1991)? In contrast to wild-type EcoKI, which has two indistinguishable high-affinity SAM-binding sites, mutants like L113Q have only one high-affinity site, whereas the second site has a low affinity for SAM (Winter 1997). Is it the asymmetry of the complex with DNA that controls the switch between methylation and restriction? How the ability to bind only one SAM

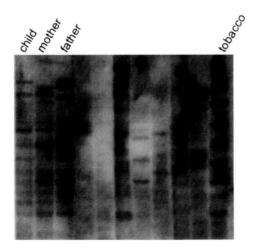

FIGURE 21. The first DNA fingerprint developed by Alec Jeffreys in the Genetics Department in Leicester, September 10, 1984. (Courtesy of Alec Jeffreys.)

molecule properly turns L113Q into a de novo enzyme, without apparently affecting the methylation reaction as such, requires further investigation, as the ability to distinguish between unmethylated and hemimethylated DNA is of fundamental importance to cellular activities.

REFERENCES

Abrosimova LA, Monakhova MV, Migur AY, Wolfgang W, Pingoud A, Kubareva EA, Oretskaya TS. 2013. Thermo-switchable activity of the restriction endonuclease SsoII achieved by site-directed enzyme modification. *IUBMB Life* **65:** 1012–1016.

Adamczyk-Popławska M, Kondrzycka A, Urbanek K, Piekarowicz A. 2003. Tetra-amino-acid tandem repeats are involved in HsdS complementation in type IC restriction-modification systems. *Microbiology* **149:** 3311–3319.

Adamczyk-Popławska M, Lower M, Piekarowicz A. 2009. Characterization of the NgoAXP: phase-variable type III restriction-modification system in *Neisseria gonorrhoeae*. *FEMS Microbiol Lett* **300:** 25–35.

Adamczyk-Popławska M, Lower M, Piekarowicz A. 2011. Deletion of one nucleotide within the homonucleotide tract present in the hsdS gene alters the DNA sequence specificity of type I restriction-modification system NgoAV. *J Bacteriol* **193:** 6750–6759.

Aiken C, Milarski-Brown K., Gumport R.I. 1986. RsrI and EcoRI are two different restriction endonucleases that recognise the same DNA sequence. *Fed Proc* **45:** 1914.

Althorpe NJ, Chilley PM, Thomas AT, Brammar WJ, Wilkins BM. 1999. Transient transcriptional activation of the IncI1 plasmid anti-restriction gene (*ardA*) and SOS inhibition gene (*psiB*) early in conjugating recipient bacteria. *Mol Microbiol* **31:** 133–142.

Anjum A, Brathwaite KJ, Aidley J, Connerton PL, Cummings NJ, Parkhill J, Connerton I, Bayliss CD. 2016. Phase variation of a Type IIG restriction-modification enzyme alters site-specific methylation patterns and gene expression in *Campylobacter jejuni* strain NCTC11168. *Nucleic Acids Res* **44:** 4581–4594.

Anton BP, Raleigh EA. 2004. Transposon-mediated linker insertion scanning mutagenesis of the *Escherichia coli* McrA endonuclease. *J Bacteriol* **186:** 5699–5707.

Anton BP, Heiter DF, Benner JS, Hess EJ, Greenough L, Moran LS, Slatko BE, Brooks JE. 1997. Cloning and characterization of the BglII restriction-modification system reveals a possible evolutionary footprint. *Gene* **187:** 19–27.

Arber W. 1977. What is the function of restriction enzymes? *Trends Biochem Sci* **2:** N176–N178.

Arifuzzaman M, Maeda M, Itoh A, Nishikata K, Takita C, Saito R, Ara T, Nakahigashi K, Huang HC, Hirai A, et al. 2006. Large-scale identification of protein–protein interaction of *Escherichia coli* K-12. *Genome Res* **16:** 686–691.

Armalyte E, Bujnicki JM, Giedriene J, Gasiunas G, Kosinski J, Lubys A. 2005. Mva1269I: a monomeric type IIS restriction endonuclease from *Micrococcus varians* with two EcoRI- and FokI-like catalytic domains. *J Biol Chem* **280:** 41584–41594.

Atanasiu C, Byron O, McMiken H, Sturrock SS, Dryden DT. 2001. Characterisation of the structure of ocr, the gene 0.3 protein of bacteriophage T7. *Nucleic Acids Res* **29:** 3059–3068.

Atanasiu C, Su TJ, Sturrock SS, Dryden DT. 2002. Interaction of the *ocr* gene 0.3 protein of bacteriophage T7 with EcoKI restriction/modification enzyme. *Nucleic Acids Res* **30:** 3936–3944.

Bair CL, Black LW. 2007. A type IV modification dependent restriction nuclease that targets glucosylated hydroxymethyl cytosine modified DNAs. *J Mol Biol* **366:** 768–778.

Bair CL, Rifat D, Black LW. 2007. Exclusion of glucosyl-hydroxymethylcytosine DNA containing bacteriophages is overcome by the injected protein inhibitor IPI*. *J Mol Biol* **366:** 779–789.

Baker O, Tsurkan S, Fu J, Klink B, Rump A, Obst M, Kranz A, Schrock E, Anastassiadis K, Stewart AF. 2017. The contribution of homology arms to nuclease-assisted genome engineering. *Nucleic Acids Res* **45:** 8105–8115.

Ball N, Streeter SD, Kneale GG, McGeehan JE. 2009. Structure of the restriction-modification controller protein C.Esp1396I. *Acta Crystallogr, Sect D: Biol Crystallogr* **65:** 900–905.

Ball NJ, McGeehan JE, Streeter SD, Thresh SJ, Kneale GG. 2012. The structural basis of differential DNA sequence recognition by restriction-modification controller proteins. *Nucleic Acids Res* **40:** 10532–10542.

Bandyopadhyay PK, Studier FW, Hamilton DL, Yuan R. 1985. Inhibition of the type I restriction-modification enzymes EcoB and EcoK by the gene 0.3 protein of bacteriophage T7. *J Mol Biol* **182:** 567–578.

Bart A, Dankert J, van der Ende A. 1999. Operator sequences for the regulatory proteins of restriction modification systems. *Mol Microbiol* **31:** 1277–1278.

Bates S, Roscoe RA, Althorpe NJ, Brammar WJ, Wilkins BM. 1999. Expression of leading region genes on IncI1 plasmid ColIb-P9: genetic evidence for single-stranded DNA transcription. *Microbiology* **145** (Pt 10): 2655–2662.

Baumann K. 2014. Epigenetics: Enhancers under TET control. *Nat Rev Mol Cell Biol* **15:** 699.

Bayliss CD, Callaghan MJ, Moxon ER. 2006. High allelic diversity in the methyltransferase gene of a phase variable type III restriction-modification system has implications for the fitness of *Haemophilus influenzae*. *Nucleic Acids Res* **34:** 4046–4059.

Beary TP, Braymer HD, Achberger EC. 1997. Evidence of participation of McrB(S) in McrBC restriction in *Escherichia coli* K-12. *J Bacteriol* **179:** 7768–7775.

Bellamy SR, Milsom SE, Scott DJ, Daniels LE, Wilson GG, Halford SE. 2005. Cleavage of individual DNA strands by the different subunits of the heterodimeric restriction endonuclease BbvCI. *J Mol Biol* **348:** 641–653.

Bellamy SR, Milsom SE, Kovacheva YS, Sessions RB, Halford SE. 2007. A switch in the mechanism of communication between the two DNA-binding sites in the SfiI restriction endonuclease. *J Mol Biol* **373:** 1169–1183.

Bellamy SR, Mina P, Retter SE, Halford SE. 2008. Fidelity of DNA sequence recognition by the SfiI restriction endonuclease is determined by communications between its two DNA-binding sites. *J Mol Biol* **384:** 557–563.

Bellamy SR, Kovacheva YS, Zulkipli IH, Halford SE. 2009. Differences between Ca^{2+} and Mg^{2+} in DNA binding and release by the SfiI restriction endonuclease: implications for DNA looping. *Nucleic Acids Res* **37:** 5443–5453.

Bhakta MS, Henry IM, Ousterout DG, Das KT, Lockwood SH, Meckler JF, Wallen MC, Zykovich A, Yu Y, Leo H, et al. 2013. Highly active zinc-finger nucleases by extended modular assembly. *Genome Res* **23:** 530–538.

Bianco PR, Hurley EM. 2005. The type I restriction endonuclease EcoR124I, couples ATP hydrolysis to bidirectional DNA translocation. *J Mol Biol* **352:** 837–859.

Bibikova M, Golic M, Golic KG, Carroll D. 2002. Targeted chromosomal cleavage and mutagenesis in *Drosophila* using zinc-finger nucleases. *Genetics* **161:** 1169–1175.

Bickle TA, Kruger DH. 1993. Biology of DNA restriction. *Microbiol Rev* **57:** 434–450.

Biebricher A, Wende W, Escude C, Pingoud A, Desbiolles P. 2009. Tracking of single quantum dot labeled EcoRV sliding along DNA manipulated by double optical tweezers. *Biophys J* **96:** L50–L52.

Bitinaite J, Schildkraut I. 2002. Self-generated DNA termini relax the specificity of SgrAI restriction endonuclease. *Proc Natl Acad Sci* **99:** 1164–1169.

Bitinaite J, Wah DA, Aggarwal AK, Schildkraut I. 1998. FokI dimerization is required for DNA cleavage. *Proc Natl Acad Sci* **95:** 10570–10575.

Blackstock JJ, Egelhaaf SU, Atanasiu C, Dryden DT, Poon WC. 2001. Shape of Ocr, the gene 0.3 protein of bacteriophage T7: modeling based on light scattering experiments. *Biochemistry* **40:** 9944–9949.

Boch J, Scholze H, Schornack S, Landgraf A, Hahn S, Kay S, Lahaye T, Nickstadt A, Bonas U. 2009. Breaking the code of DNA binding specificity of TAL-type III effectors. *Science* **326:** 1509–1512.

Bochtler M, Szczepanowski RH, Tamulaitis G, Gražulis S, Czapinska H, Manakova E, Šikšnys V. 2006. Nucleotide flips determine the specificity of the Ecl18kI restriction endonuclease. *EMBO J* **25:** 2219–2229.

Bogdanova E, Djordjevic M, Papapanagiotou I, Heyduk T, Kneale G, Severinov K. 2008. Transcription regulation of the type II restriction-modification system AhdI. *Nucleic Acids Res* **36:** 1429–1442.

Bogdanova E, Zakharova M, Streeter S, Taylor J, Heyduk T, Kneale G, Severinov K. 2009. Transcription regulation of restriction-modification system Esp1396I. *Nucleic Acids Res* **37:** 3354–3366.

Bogdanove AJ, Bohm A, Mille JC, Morgan RD, Stoddard BL. 2018. Engineering altered protein–DNA recognition specificity. *Nucleic Acids Res* **46:** 4845–4871.

Bonnet I, Biebricher A, Porte PL, Loverdo C, Benichou O, Voituriez R, Escude C, Wende W, Pingoud A, Desbiolles P. 2008. Sliding and jumping of single EcoRV restriction enzymes on non-cognate DNA. *Nucleic Acids Res* **36:** 4118–4127.

Borgaro JG, Zhu Z. 2013. Characterization of the 5-hydroxymethylcytosine-specific DNA restriction endonucleases. *Nucleic Acids Res* **41:** 4198–4206.

Borst P, Sabatini R. 2008. Base J: discovery, biosynthesis, and possible functions. *Ann Rev Microbiol* **62:** 235–251.

Bourniquel AA, Bickle TA. 2002. Complex restriction enzymes: NTP-driven molecular motors. *Biochimie* **84:** 1047–1059.

Boyd AC, Charles IG, Keyte JW, Brammar WJ. 1986. Isolation and computer-aided characterization of MmeI, a type II restriction endonuclease from *Methylophilus methylotrophus*. *Nucleic Acids Res* **14:** 5255–5274.

Bozic D, Gražulis S, Šikšnys V, Huber R. 1996. Crystal structure of *Citrobacter freundii* restriction endonuclease Cfr10I at 2.15 A resolution. *J Mol Biol* **255:** 176–186.

Brooks JE, Benner JS, Heiter DF, Silber KR, Sznyter LA, Jager-Quinton T, Moran LS, Slatko BE, Wilson GG, Nwankwo DO. 1989. Cloning the BamHI restriction modification system. *Nucleic Acids Res* **17:** 979–997.

Brzezinski R, Piekarowicz A. 1982. Steps in the reaction mechanism of the *Haemophilus influenzae* Rf restriction endonuclease. *J Mol Biol* **154:** 615–627.

Bujnicki JM. 2004. Molecular phylogenetics of restriction endonucleases. In *Restriction endonucleases* (ed. Pingoud A), pp. 63–93. Springer, Berlin.

Bujnicki JM, Rychlewski L. 2001. Grouping together highly diverged PD-(D/E)XK nucleases and identification of novel superfamily members using structure-guided alignment of sequence profiles. *J Mol Microbiol Biotechnol* **3:** 69–72.

Bujnicki JM, Radlinska M, Rychlewski L. 2000. Atomic model of the 5-methylcytosine-specific restriction enzyme McrA reveals an atypical zinc finger and structural similarity to ββαMe endonucleases. *Mol Microbiol* **37**: 1280–1281.

Bujnicki JM, Radlinska M, Rychlewski L. 2001. Polyphyletic evolution of type II restriction enzymes revisited: two independent sources of second-hand folds revealed. *Trends Biochem Sci* **26**: 9–11.

Bull JJ, Malik HS. 2017. The gene drive bubble: new realities. *PLoS Genet* **13**: e1006850.

Burckhardt J, Weisemann J, Yuan R. 1981. Characterization of the DNA methylase activity of the restriction enzyme from *Escherichia coli* K. *J Biol Chem* **256**: 4024–4032.

Burenina OY, Fedotova EA, Ryazanova AY, Protsenko AS, Zakharova MV, Karyagina AS, Solonin AS, Oretskaya TS, Kubareva EA. 2013. Peculiarities of the regulation of gene expression in the Ecl18kI restriction-modification system. *Acta Nat* **5**: 70–80.

Butterer A, Pernstich C, Smith RM, Sobott F, Szczelkun MD, Toth J. 2014. Type III restriction endonucleases are heterotrimeric: comprising one helicase-nuclease subunit and a dimeric methyltransferase that binds only one specific DNA. *Nucleic Acids Res* **42**: 5139–5150.

Cajthamlova K, Sisakova E, Weiser J, Weiserova M. 2007. Phosphorylation of Type IA restriction-modification complex enzyme EcoKI on the HsdR subunit. *FEMS Microbiol Lett* **270**: 171–177.

Calisto BM, Pich OQ, Pinol J, Fita I, Querol E, Carpena X. 2005. Crystal structure of a putative type I restriction-modification S subunit from *Mycoplasma genitalium*. *J Mol Biol* **351**: 749–762.

Callahan SJ, Morgan RD, Jain R, Townson SA, Wilson GG, Roberts RJ, Aggarwal AK. 2011. Crystallization and preliminary crystallographic analysis of the type IIL restriction enzyme MmeI in complex with DNA. *Acta Crystallogr, Sect F: Struct Biol Cryst Commun* **67**: 1262–1265.

Callahan SJ, Luyten YA, Gupta YK, Wilson GG, Roberts RJ, Morgan RD, Aggarwal AK. 2016. Structure of Type IIL restriction-modification enzyme MmeI in complex with DNA has implications for engineering new specificities. *PLoS Biol* **14**: e1002442.

Callow P, Sukhodub A, Taylor JE, Kneale GG. 2007. Shape and subunit organisation of the DNA methyltransferase M.AhdI by small-angle neutron scattering. *J Mol Biol* **369**: 177–185.

Capoluongo E, Giglio A, Leonetti F, Belardi M, Giannetti A, Caprilli F, Ameglio F. 2000. DNA heterogeneity of *Staphylococcus aureus* strains evaluated by SmaI and SgrAI pulsed-field gel electrophoresis in patients with impetigo. *Res Microbiol* **151**: 53–61.

Carlson K, Raleigh EA, Hattman S. 1994. In *Molecular biology of bacteriophage T4* (ed. Karam JD, et al.). American Society for Microbiology, Washington, DC.

Carroll D. 2011a. Genome engineering with zinc-finger nucleases. *Genetics* **188**: 773–782.

Carroll D. 2011b. Zinc-finger nucleases: a panoramic view. *Curr Gene Ther* **11**: 2–10.

Carroll D. 2014. Genome engineering with targetable nucleases. *Ann Rev Biochem* **83**: 409–439.

Carroll D, Beumer KJ. 2014. Genome engineering with TALENs and ZFNs: repair pathways and donor design. *Methods* **69**: 137–141.

Casadesus J, Low D. 2006. Epigenetic gene regulation in the bacterial world. *Microbiol Mol Biol Rev* **70**: 830–856.

Catto LE, Ganguly S, Milsom SE, Welsh AJ, Halford SE. 2006. Protein assembly and DNA looping by the FokI restriction endonuclease. *Nucleic Acids Res* **34**: 1711–1720.

Catto LE, Bellamy SR, Retter SE, Halford SE. 2008. Dynamics and consequences of DNA looping by the FokI restriction endonuclease. *Nucleic Acids Res* **36**: 2073–2081.

Cesnaviciene E, Mitkaite G, Stankevicius K, Janulaitis A, Lubys A. 2003. Esp1396I restriction-modification system: structural organization and mode of regulation. *Nucleic Acids Res* **31:** 743–749.

Champer J, Reeves R, Oh SY, Liu C, Liu J, Clark AG, Messer PW. 2017. Novel CRISPR/Cas9 gene drive constructs reveal insights into mechanisms of resistance allele formation and drive efficiency in genetically diverse populations. *PLoS Genet* **13:** e1006796.

Chan SH, Stoddard BL, Xu SY. 2011. Natural and engineered nicking endonucleases—from cleavage mechanism to engineering of strand-specificity. *Nucleic Acids Res* **39:** 1–18.

Chand MK, Nirwan N, Diffin FM, van Aelst K, Kulkarni M, Pernstich C, Szczelkun MD, Saikrishnan K. 2015. Translocation-coupled DNA cleavage by the Type ISP restriction-modification enzymes. *Nat Chem Biol* **11:** 870–877.

Chandrasegaran S, Smith J. 1999. Chimeric restriction enzymes: what is next? *Biol Chem* **380:** 841–848.

Chandrashekaran S, Saravanan M, Radha DR, Nagaraja V. 2004. Ca^{2+}-mediated site-specific DNA cleavage and suppression of promiscuous activity of KpnI restriction endonuclease. *J Biol Chem* **279:** 49736–49740.

Chen K, Roberts GA, Stephanou AS, Cooper LP, White JH, Dryden DT. 2010. Fusion of GFP to the M·EcoKI DNA methyltransferase produces a new probe of Type I DNA restriction and modification enzymes. *Biochem Biophys Res Commun* **398:** 254–259.

Chen K, Reuter M, Sanghvi B, Roberts GA, Cooper LP, Tilling M, Blakely GW, Dryden DT. 2014. ArdA proteins from different mobile genetic elements can bind to the EcoKI Type I DNA methyltransferase of *E. coli* K12. *Biochim Biophys Acta* **1844:** 505–511.

Cheng X, Roberts RJ. 2001. AdoMet-dependent methylation, DNA methyltransferases and base flipping. *Nucleic Acids Res* **29:** 3784–3795.

Cheng X, Balendiran K, Schildkraut I, Anderson JE. 1994. Structure of PvuII endonuclease with cognate DNA. *EMBO J* **13:** 3927–3935.

Chilley PM, Wilkins BM. 1995. Distribution of the ardA family of antirestriction genes on conjugative plasmids. *Microbiology* **141** (Pt 9): 2157–2164.

Chuluunbaatar T, Ivanenko-Johnston T, Fuxreiter M, Meleshko R, Rasko T, Simon I, Heitman J, Kiss A. 2007. An EcoRI-RsrI chimeric restriction endonuclease retains parental sequence specificity. *Biochim Biophys Acta* **1774:** 583–594.

Clark TA, Murray IA, Morgan RD, Kislyuk AO, Spittle KE, Boitano M, Fomenkov A, Roberts RJ, Korlach J. 2012. Characterization of DNA methyltransferase specificities using single-molecule, real-time DNA sequencing. *Nucleic Acids Res* **40:** e29.

Cohen-Karni D, Xu D, Apone L, Fomenkov A, Sun Z, Davis PJ, Kinney SR, Yamada-Mabuchi M, Xu SY, Davis T, et al. 2011. The MspJI family of modification-dependent restriction endonucleases for epigenetic studies. *Proc Natl Acad Sci* **108:** 11040–11045.

Compe E, Egly JM. 2016. Nucleotide excision repair and transcriptional regulation: TFIIH and beyond. *Ann Rev Biochem* **85:** 265–290.

Cooper LP, Dryden DT. 1994. The domains of a type I DNA methyltransferase. Interactions and role in recognition of DNA methylation. *J Mol Biol* **236:** 1011–1021.

Cooper LP, Roberts GA, White JH, Luyten YA, Bower EKM, Morgan RD, Roberts RJ, Lindsay JA, Dryden DTF. 2017. DNA target recognition domains in the Type I restriction and modification systems of *Staphylococcus aureus*. *Nucleic Acids Res* **45:** 3395–3406.

Crampton N, Roes S, Dryden DT, Rao DN, Edwardson JM, Henderson RM. 2007a. DNA looping and translocation provide an optimal cleavage mechanism for the type III restriction enzymes. *EMBO J* **26:** 3815–3825.

Crampton N, Yokokawa M, Dryden DT, Edwardson JM, Rao DN, Takeyasu K, Yoshimura SH, Henderson RM. 2007b. Fast-scan atomic force microscopy reveals that the type III restriction enzyme EcoP15I is capable of DNA translocation and looping. *Proc Natl Acad Sci* **104:** 12755–12760.

Csefalvay E, Lapkouski M, Guzanova A, Csefalvay L, Baikova T, Shevelev I, Bialevich V, Shamayeva K, Janscak P, Kuta Smatanova I, et al. 2015. Functional coupling of duplex translocation to DNA cleavage in a type I restriction enzyme. *PloS One* **10:** e0128700. doi:10.1371/journal.pone.0128700.

Cymerman IA, Obarska A, Skowronek KJ, Lubys A, Bujnicki JM. 2006. Identification of a new subfamily of HNH nucleases and experimental characterization of a representative member, HphI restriction endonuclease. *Proteins* **65:** 867–876.

Daniels LE, Wood KM, Scott DJ, Halford SE. 2003. Subunit assembly for DNA cleavage by restriction endonuclease SgrAI. *J Mol Biol* **327:** 579–591.

Davies GP, Martin I, Sturrock SS, Cronshaw A, Murray NE, Dryden DT. 1999. On the structure and operation of type I DNA restriction enzymes. *J Mol Biol* **290:** 565–579.

De Bolle X, Bayliss CD, Field D, van de Ven T, Saunders NJ, Hood DW, Moxon ER. 2000. The length of a tetranucleotide repeat tract in *Haemophilus influenzae* determines the phase variation rate of a gene with homology to type III DNA methyltransferases. *Mol Microbiol* **35:** 211–222.

Dedkov VS, Degtyarev SK. 1998. *Actinobacillus* and *Streptococcus*: producers of isoschizomers of the restriction endonucleases R.HphI, R.SauI, R.NheI, R.MboI and R.SwaI. *Biol Chem* **379:** 573–574.

Deibert M, Gražulis S, Sasnauskas G, Šikšnys V, Huber R. 2000. Structure of the tetrameric restriction endonuclease NgoMIV in complex with cleaved DNA. *Nat Struct Biol* **7:** 792–799.

Deng D, Yan C, Pan X, Mahfouz M, Wang J, Zhu JK, Shi Y, Yan N. 2012. Structural basis for sequence-specific recognition of DNA by TAL effectors. *Science* **335:** 720–723.

Denjmukhametov MM, Brevnov MG, Zakharova MV, Repyk AV, Solonin AS, Petrauskene OV, Gromova ES. 1998. The Ecl18kI restriction-modification system: cloning, expression, properties of the purified enzymes. *FEBS Lett* **433:** 233–236.

Dixon K, Bayliss CD, Makepeace K, Moxon ER, Hood DW. 2007. Identification of the functional initiation codons of a phase-variable gene of *Haemophilus influenzae*, lic2A, with the potential for differential expression. *J Bacteriol* **189:** 511–521.

Dreyer AK, Hoffmann D, Lachmann N, Ackermann M, Steinemann D, Timm B, Siler U, Reichenbach J, Grez M, Moritz T, et al. 2015. TALEN-mediated functional correction of X-linked chronic granulomatous disease in patient-derived induced pluripotent stem cells. *Biomaterials* **69:** 191–200.

Dryden DTF. 1999. Bacterial methyltransferases. In *S-adenosylmethionine-dependent methyltransferases: structures and functions* (ed. Cheng X, Blumenthal RM), pp. 283–340. World Scientific, Singapore.

Dryden DT. 2006. DNA mimicry by proteins and the control of enzymatic activity on DNA. *Trends Biotechnol* **24:** 378–382.

Dryden DT, Cooper LP, Thorpe PH, Byron O. 1997. The in vitro assembly of the EcoKI type I DNA restriction/modification enzyme and its in vivo implications. *Biochemistry* **36:** 1065–1076.

Dryden DT, Edwardson JM, Henderson RM. 2011. DNA translocation by type III restriction enzymes: a comparison of current models of their operation derived from ensemble and single-molecule measurements. *Nucleic Acids Res* **39:** 4525–4531.

Dunn DB, Smith JD. 1955a. The occurrence of 6-methylaminopurine in microbial deoxyribonucleic acids. *Biochem J* **60:** pxvii.

Dunn DB, Smith JD. 1955b. Occurrence of a new base in the deoxyribonucleic acid of a strain of *Bacterium coli. Nature* **175:** 336–337.

Dunn JJ, Studier FW. 1981. Nucleotide sequence from the genetic left end of bacteriophage T7 DNA to the beginning of gene 4. *J Mol Biol* **148:** 303–330.

Dunn JJ, Elzinga M, Mark KK, Studier FW. 1981. Amino acid sequence of the gene 0.3 protein of bacteriophage T7 and nucleotide sequence of its mRNA. *J Biol Chem* **256:** 2579–2585.

Dunten PW, Little EJ, Gregory MT, Manohar VM, Dalton M, Hough D, Bitinaite J, Horton NC. 2008. The structure of SgrAI bound to DNA; recognition of an 8 base pair target. *Nucleic Acids Res* **36:** 5405–5416.

Dunten PW, Little EJ, Horton NC. 2009. The restriction enzyme SgrAI: structure solution via combination of poor MIRAS and MR phases. *Acta Crystallogr, Sect D: Biol Crystallogr* **65:** 393–398.

Durai S, Mani M, Kandavelou K, Wu J, Porteus MH, Chandrasegaran S. 2005. Zinc finger nucleases: custom-designed molecular scissors for genome engineering of plant and mammalian cells. *Nucleic Acids Res* **33:** 5978–5990.

Dybvig K, Sitaraman R, French CT. 1998. A family of phase-variable restriction enzymes with differing specificities generated by high-frequency gene rearrangements. *Proc Natl Acad Sci* **95:** 13923–13928.

Ehrlich M, Gama-Sosa MA, Carreira LH, Ljungdahl LG, Kuo KC, Gehrke CW. 1985. DNA methylation in thermophilic bacteria: N4-methylcytosine, 5-methylcytosine, and N6-methyladenine. *Nucleic Acids Res* **13:** 1399–1412.

Ehrlich M, Wilson GG, Kuo KC, Gehrke CW. 1987. N4-methylcytosine as a minor base in bacterial DNA. *J Bacteriol* **169:** 939–943.

Eid J, Fehr A, Gray J, Luong K, Lyle J, Otto G, Peluso P, Rank D, Baybayan P, Bettman B, et al. 2009. Real-time DNA sequencing from single polymerase molecules. *Science* **323:** 133–138.

Embleton ML, Šikšnys V, Halford SE. 2001. DNA cleavage reactions by type II restriction enzymes that require two copies of their recognition sites. *J Mol Biol* **311:** 503–514.

Embleton ML, Vologodskii AV, Halford SE. 2004. Dynamics of DNA loop capture by the SfiI restriction endonuclease on supercoiled and relaxed DNA. *J Mol Biol* **339:** 53–66.

Endlich B, Linn S. 1985. The DNA restriction endonuclease of *Escherichia coli* B. I. Studies of the DNA translocation and the ATPase activities. *J Biol Chem* **260:** 5720–5728.

Fairman-Williams ME, Guenther UP, Jankowsky E. 2010. SF1 and SF2 helicases: family matters. *Curr Opin Struct Biol* **20:** 313–324.

Fedotova EA, Protsenko AS, Zakharova MV, Lavrova NV, Alekseevsky AV, Oretskaya TS, Karyagina AS, Solonin AS, Kubareva EA. 2009. SsoII-like DNA-methyltransferase Ecl18kI: interaction between regulatory and methylating functions. *Biochem Biokhim* **74:** 85–91.

Flusberg BA, Webster DR, Lee JH, Travers KJ, Olivares EC, Clark TA, Korlach J, Turner SW. 2010. Direct detection of DNA methylation during single-molecule, real-time sequencing. *Nat Meth* **7:** 461–465.

Fox KL, Dowideit SJ, Erwin AL, Srikhanta YN, Smith AL, Jennings MP. 2007. *Haemophilus influenzae* phasevarions have evolved from type III DNA restriction systems into epigenetic regulators of gene expression. *Nucleic Acids Res* **35:** 5242–5252.

Freitag M, Selker EU. 2005. Controlling DNA methylation: many roads to one modification. *Curr Opin Genet Dev* **15:** 191–199.

Friedhoff P, Franke I, Meiss G, Wende W, Krause KL, Pingoud A. 1999. A similar active site for non-specific and specific endonucleases. *Nat Struct Biol* **6:** 112–113.

Friedrich T, Fatemi M, Gowhar H, Leismann O, Jeltsch A. 2000. Specificity of DNA binding and methylation by the M.*FokI* DNA methyltransferase. *Biochim Biophys Acta* **1480:** 145–159.

Furuta Y, Kawai M, Uchiyama I, Kobayashi I. 2011. Domain movement within a gene: a novel evolutionary mechanism for protein diversification. *PloS One* **6:** e18819.

Gabriel R, Lombardo A, Arens A, Miller JC, Genovese P, Kaeppel C, Nowrouzi A, Bartholomae CC, Wang J, Friedman G, et al. 2011. An unbiased genome-wide analysis of zinc-finger nuclease specificity. *Nat Biotechnol* **29:** 816–823.

Gabsalilow L, Schierling B, Friedhoff P, Pingoud A, Wende W. 2013. Site- and strand-specific nicking of DNA by fusion proteins derived from MutH and I-SceI or TALE repeats. *Nucleic Acids Res* **41:** e83. doi:10.1093/nar/gkt1080.

Galburt EA, Chevalier B, Tang W, Jurica MS, Flick KE, Monnat RJ Jr, Stoddard BL. 1999. A novel endonuclease mechanism directly visualized for I-PpoI. *Nat Struct Biol* **6:** 1096–1099.

Galetto R, Duchateau P, Paques F. 2009. Targeted approaches for gene therapy and the emergence of engineered meganucleases. *Expert Opin Biol Ther* **9:** 1289–1303.

Garcia LR, Molineux IJ. 1996. Transcription-independent DNA translocation of bacteriophage T7 DNA into *Escherichia coli*. *J Bacteriol* **178:** 6921–6929.

Gasiunas G, Sasnauskas G, Tamulaitis G, Urbanke C, Razaniene D, Šikšnys V. 2008. Tetrameric restriction enzymes: expansion to the GIY-YIG nuclease family. *Nucleic Acids Res* **36:** 938–949.

Gast FU, Brinkmann T, Pieper U, Kruger T, Noyer-Weidner M, Pingoud A. 1997. The recognition of methylated DNA by the GTP-dependent restriction endonuclease McrBC resides in the N-terminal domain of McrB. *Biol Chem* **378:** 975–982.

Gefter M, Hausmann R, Gold M, Hurwitz J. 1966. The enzymatic methylation of ribonucleic acid and deoxyribonucleic acid. X. Bacteriophage T3-induced *S*-adenosylmethionine cleavage. *J Biol Chem* **241:** 1995–2006.

Gemmen GJ, Millin R, Smith DE. 2006. Tension-dependent DNA cleavage by restriction endonucleases: two-site enzymes are "switched off" at low force. *Proc Natl Acad Sci* **103:** 11555–11560.

Gilmore JL, Suzuki Y, Tamulaitis G, Šikšnys V, Takeyasu K, Lyubchenko YL. 2009. Single-molecule dynamics of the DNA-EcoRII protein complexes revealed with high-speed atomic force microscopy. *Biochemistry* **48:** 10492–10498.

Glover SW, Piekarowicz A. 1972. Host specificity of DNA in *Haemophilus influenzae*: restriction and modification in strain Rd. *Biochem Biophys Res Commun* **46:** 1610–1617.

Golovenko D, Manakova E, Tamulaitienė G, Gražulis S, Šikšnys V. 2009. Structural mechanisms for the 5′-CCWGG sequence recognition by the N- and C-terminal domains of EcoRII. *Nucleic Acids Res* **37:** 6613–6624.

Golovenko D, Manakova E, Zakrys L, Zaremba M, Sasnauskas G, Gražulis S, Šikšnys V. 2014. Structural insight into the specificity of the B3 DNA-binding domains provided by the co-crystal structure of the C-terminal fragment of BfiI restriction enzyme. *Nucleic Acids Res* **42:** 4113–4122.

González-Cerón G, Miranda-Olivares OJ, Servin-González L. 2009. Characterization of the methyl-specific restriction system of *Streptomyces coelicolor* A3(2) and of the role played by laterally acquired nucleases. *FEMS Microbiol Lett* **301:** 35–43.

Goszczynski B, McGhee JD. 1991. Resolution of sequencing ambiguities: a universal FokI adapter permits Maxam-Gilbert re-sequencing of single-stranded phagemid DNA. *Gene* **104:** 71–74.

Gowers DM, Halford SE. 2003. Protein motion from non-specific to specific DNA by three-dimensional routes aided by supercoiling. *EMBO J* **22:** 1410–1418.

Gowers DM, Bellamy SR, Halford SE. 2004. One recognition sequence, seven restriction enzymes, five reaction mechanisms. *Nucleic Acids Res* **32:** 3469–3479.

Gowers DM, Wilson GG, Halford SE. 2005. Measurement of the contributions of 1D and 3D pathways to the translocation of a protein along DNA. *Proc Natl Acad Sci* **102:** 15883–15888.

Gražulis S, Deibert M, Rimseliene R, Skirgaila R, Sasnauskas G, Lagunavicius A, Repin V, Urbanke C, Huber R, Šikšnys V. 2002. Crystal structure of the Bse634I restriction endonuclease: comparison of two enzymes recognizing the same DNA sequence. *Nucleic Acids Res* **30:** 876–885.

Gražulis S, Manakova E, Roessle M, Bochtler M, Tamulaitienė G, Huber R, Šikšnys V. 2005. Structure of the metal-independent restriction enzyme BfiI reveals fusion of a specific DNA-binding domain with a nonspecific nuclease. *Proc Natl Acad Sci* **102:** 15797–15802.

Gubler M, Bickle TA. 1991. Increased protein flexibility leads to promiscuous protein–DNA interactions in type IC restriction-modification systems. *EMBO J* **10:** 951–957.

Guilinger JP, Pattanayak V, Reyon D, Tsai SQ, Sander JD, Joung JK, Liu DR. 2014a. Broad specificity profiling of TALENs results in engineered nucleases with improved DNA-cleavage specificity. *Nat Methods* **11:** 429–435.

Guilinger JP, Thompson DB, Liu DR. 2014b. Fusion of catalytically inactive Cas9 to FokI nuclease improves the specificity of genome modification. *Nat Biotechnol* **32:** 577–582.

Guo J, Gaj T, Barbas CF 3rd. 2010. Directed evolution of an enhanced and highly efficient FokI cleavage domain for zinc finger nucleases. *J Mol Biol* **400:** 96–107.

Gupta R, Nagarajan A, Wajapeyee N. 2010. Advances in genome-wide DNA methylation analysis. *BioTechniques* **49:** ii–ixi.

Gupta YK, Yang L, Chan SH, Samuelson JC, Xu SY, Aggarwal AK. 2012. Structural insights into the assembly and shape of Type III restriction-modification (R-M) EcoP15I complex by small-angle X-ray scattering. *J Mol Biol* **420:** 261–268.

Gupta YK, Chan SH, Xu SY, Aggarwal AK. 2015. Structural basis of asymmetric DNA methylation and ATP-triggered long-range diffusion by EcoP15I. *Nat Commun* **6:** 7363. doi: 10.1038/ncomms8363.

Halford SE. 2001. Hopping, jumping and looping by restriction enzymes. *Biochem Soc Trans* **29:** 363–374.

Halford SE. 2013. http://library.cshl.edu/Meetings/restriction-enzymes/Halford.php.

Halford SE, Marko JF. 2004. How do site-specific DNA-binding proteins find their targets? *Nucleic Acids Res* **32:** 3040–3052.

Halford SE, Bilcock DT, Stanford NP, Williams SA, Milsom SE, Gormley NA, Watson MA, Bath AJ, Embleton ML, Gowers DM, et al. 1999. Restriction endonuclease reactions requiring two recognition sites. *Biochem Soc Trans* **27:** 696–699.

Halford SE, Welsh AJ, Szczelkun MD. 2004. Enzyme-mediated DNA looping. *Ann Rev Biophys Biomol Struct* **33:** 1–24.

Halford SE, Catto LE, Pernstich C, Rusling DA, Sanders KL. 2011. The reaction mechanism of FokI excludes the possibility of targeting zinc finger nucleases to unique DNA sites. *Biochem Soc Trans* **39:** 584–588.

Hallet B. 2001. Playing Dr. Jekyll and Mr. Hyde: combined mechanisms of phase variation in bacteria. *Curr Opin Microbiol* **4:** 570–581.

Handel EM, Cathomen T. 2011. Zinc-finger nuclease based genome surgery: it's all about specificity. *Curr Gene Ther* **11:** 28–37.

Handel EM, Gellhaus K, Khan K, Bednarski C, Cornu TI, Muller-Lerch F, Kotin RM, Heilbronn R, Cathomen T. 2012. Versatile and efficient genome editing in human cells by combining zinc-finger nucleases with adeno-associated viral vectors. *Hum Gene Ther* **23:** 321–329.

Hasan N, Kim SC, Podhajska AJ, Szybalski W. 1986. A novel multistep method for generating precise unidirectional deletions using BspMI, a class-IIS restriction enzyme. *Gene* **50:** 55–62.

Hausmann R. 1967. Synthesis of an *S*-adenosylmethionine-cleaving enzyme in T3-infected *Escherichia coli* and its disturbance by co-infection with enzymatically incompetent bacteriophage. *J Virol* **1:** 57–63.

Hausmann R, Messerschmid M. 1988. Inhibition of gene expression of T7-related phages by prophage P1. *Mol Gen Genet* **212:** 543–547.

Hausmann R, Messerschmid M. 1988. The T7 group. In *The bacteriophages* (ed. Calendar R), Vol.1, 259–290. Plenum, New York.

He X, Hull V, Thomas JA, Fu X, Gidwani S, Gupta YK, Black LW, Xu SY. 2015. Expression and purification of a single-chain Type IV restriction enzyme Eco94GmrSD and determination of its substrate preference. *Sci Rep* **5:** 9747. doi:9710.1038/srep09747.

Hegde SS, Vetting MW, Roderick SL, Mitchenall LA, Maxwell A, Takiff HE, Blanchard JS. 2005. A fluoroquinolone resistance protein from *Mycobacterium tuberculosis* that mimics DNA. *Science* **308:** 1480–1483.

Heidmann S, Seifert W, Kessler C, Domdey H. 1989. Cloning, characterization and heterologous expression of the SmaI restriction-modification system. *Nucleic Acids Res* **17:** 9783–9796.

Heiter DF, Lunnen KD, Wilson GG. 2005. Site-specific DNA-nicking mutants of the heterodimeric restriction endonuclease R·BbvCI. *J Mol Biol* **348:** 631–640.

Heiter DF, Lunnen KD, Morgan RD, Wilson GG. 2015. When less is more: AspCNI and other REases that cleave poorly at high concentrations. In *7th NEB meeting on restriction and modification, Gdansk: Poster P20.*

Heitman J, Model P. 1987. Site-specific methylases induce the SOS DNA repair response in *Escherichia coli. J Bacteriol* **169:** 3243–3250.

Heitman J, Zinder ND, Model P. 1989. Repair of the *Escherichia coli* chromosome after in vivo scission by the EcoRI endonuclease. *Proc Natl Acad Sci* **86:** 2281–2285.

Hendel A, Fine EJ, Bao G, Porteus MH. 2015. Quantifying on- and off-target genome editing. *Trends Biotechnol* **33:** 132–140.

Hendrickson PG, Cairns BR. 2016. Tet proteins enhance the developmental hourglass. *Nat Genet* **48:** 345–347.

Hershey AD, Dixon J, Chase M. 1953. Nucleic acid economy in bacteria infected with bacteriophage T2. I. Purine and pyrimidine composition. *J Gen Physiol* **36:** 777–789.

Hien le T, Zatsepin TS, Schierling B, Volkov EM, Wende W, Pingoud A, Kubareva EA, Oretskaya TS. 2011. Restriction endonuclease SsoII with photoregulated activity—a "molecular gate" approach. *Bioconjug Chem* **22:** 1366–1373.

Higgins LS, Besnier C, Kong H. 2001. The nicking endonuclease N·BstNBI is closely related to type IIs restriction endonucleases MlyI and PleI. *Nucleic Acids Res* **29:** 2492–2501.

Hingorani-Varma K, Bitinaite J. 2003. Kinetic analysis of the coordinated interaction of SgrAI restriction endonuclease with different DNA targets. *J Biol Chem* **278:** 40392–40399.

Hirsch JA, Wah DA, Dorner LF, Schildkraut I, Aggarwal AK. 1997. Crystallization and preliminary X-ray analysis of restriction endonuclease FokI bound to DNA. *FEBS Lett* **403:** 136–138.

Hockemeyer D, Wang H, Kiani S, Lai CS, Gao Q, Cassady JP, Cost GJ, Zhang L, Santiago Y, Miller JC, et al. 2011. Genetic engineering of human pluripotent cells using TALE nucleases. *Nat Biotechnol* **29:** 731–734.

Horton NC. 2015. Allosteric control of DNA cleavage rate and DNA sequence specificity via run-on oligomerization. In *7th NEB meeting on restriction and modification (Gdansk),* Talk T2.

Horton JR, Blumenthal RM, Cheng X. 2004. Restriction endonucleases: structure of the conserved catalytic core and the role of metal ions in DNA cleavage. In *Restriction endonucleases* (ed. Pingoud A), pp. 361–392. Springer, Berlin.

Horton JR, Zhang X, Maunus R, Yang Z, Wilson GG, Roberts RJ, Cheng X. 2006. DNA nicking by HinP1I endonuclease: bending, base flipping and minor groove expansion. *Nucleic Acids Res* **34:** 939–948.

Horton JR, Mabuchi MY, Cohen-Karni D, Zhang X, Griggs RM, Samaranayake M, Roberts RJ, Zheng Y, Cheng X. 2012. Structure and cleavage activity of the tetrameric MspJI DNA modification-dependent restriction endonuclease. *Nucleic Acids Res* **40:** 9763–9773.

Horton JR, Borgaro JG, Griggs RM, Quimby A, Guan S, Zhang X, Wilson GG, Zheng Y, Zhu Z, Cheng X. 2014a. Structure of 5-hydroxymethylcytosine-specific restriction enzyme, AbaSI, in complex with DNA. *Nucleic Acids Res* **42:** 7947–7959.

Horton JR, Nugent RL, Li A, Mabuchi MY, Fomenkov A, Cohen-Karni D, Griggs RM, Zhang X, Wilson GG, Zheng Y, et al. 2014b. Structure and mutagenesis of the DNA modification-dependent restriction endonuclease AspBHI. *Sci Rep* **4:** 4246. doi:10.1038/srep04246.

Horton JR, Wang H, Mabuchi MY, Zhang X, Roberts RJ, Zheng Y, Wilson GG, Cheng X. 2014c. Modification-dependent restriction endonuclease, MspJI, flips 5-methylcytosine out of the DNA helix. *Nucleic Acids Res* **42:** 12092–12101.

Huai Q, Colandene JD, Chen Y, Luo F, Zhao Y, Topal MD, Ke H. 2000. Crystal structure of NaeI-an evolutionary bridge between DNA endonuclease and topoisomerase. *EMBO J* **19:** 3110–3118.

Ibryashkina EM, Zakharova MV, Baskunov VB, Bogdanova ES, Nagornykh MO, Den'mukha-medov MM, Melnik BS, Kolinski A, Gront D, Feder M, et al. 2007. Type II restriction endonuclease R.Eco29kI is a member of the GIY-YIG nuclease superfamily. *BMC Struct Biol* **7:** 48.

Imanishi M, Negi S, Sugiura Y. 2010. Non-FokI-based zinc finger nucleases. *Meth Mol Biol* **649:** 337–349.

Interthal H, Pouliot JJ, Champoux JJ. 2001. The tyrosyl-DNA phosphodiesterase Tdp1 is a member of the phospholipase D superfamily. *Proc Natl Acad Sci* **98:** 12009–12014.

Ishikawa K, Watanabe M, Kuroita T, Uchiyama I, Bujnicki JM, Kawakami B, Tanokura M, Kobayashi I. 2005. Discovery of a novel restriction endonuclease by genome comparison and application of a wheat-germ-based cell-free translation assay: PabI (5′-GTA/C) from the hyperthermophilic archaeon *Pyrococcus abyssi. Nucleic Acids Res* **33:** e112. doi:110 .1093/nar/gni1113.

Ishikawa K, Handa N, Kobayashi I. 2009. Cleavage of a model DNA replication fork by a Type I restriction endonuclease. *Nucleic Acids Res* **37:** 3531–3544.

Ives CL, Nathan PD, Brooks JE. 1992. Regulation of the BamHI restriction-modification system by a small intergenic open reading frame, bamHIC, in both *Escherichia coli* and *Bacillus subtilis. J Bacteriol* **174:** 7194–7201.

Iyer LM, Tahiliani M, Rao A, Aravind L. 2009. Prediction of novel families of enzymes involved in oxidative and other complex modifications of bases in nucleic acids. *Cell Cycle* **8:** 1698–1710.

Jablonska E, Kauc L, Piekarowicz A. 1975. An *Haemophilus influenzae* mutant which inhibits the growth of HP1c1 phage. *Mol Gen Genet* **139:** 157–166.

Jain R, Poulos MG, Gros J, Chakravarty AK, Shuman S. 2011. Substrate specificity and mutational analysis of *Kluyveromyces lactis* γ-toxin, a eukaryal tRNA anticodon nuclease. *RNA* **17:** 1336–1343.

Janosi L, Yonemitsu H, Hong H, Kaji A. 1994. Molecular cloning and expression of a novel hydroxymethylcytosine-specific restriction enzyme (PvuRts1I) modulated by glucosylation of DNA. *J Mol Biol* **242:** 45–61.

Janscak P, Bickle TA. 1998. The DNA recognition subunit of the type IB restriction-modification enzyme EcoAI tolerates circular permutions of its polypeptide chain. *J Mol Biol* **284:** 937–948.

Janscak P, Dryden DT, Firman K. 1998. Analysis of the subunit assembly of the type IC restriction-modification enzyme EcoR124I. *Nucleic Acids Res* **26:** 4439–4445.

Janulaitis A, Klimasauskas S, Petrusyte M, Butkus V. 1983. Cytosine modification in DNA by BcnI methylase yields N4-methylcytosine. *FEBS Lett* **161:** 131–134.

Janulaitis A, Petrusyte M, Maneliene Z, Klimasauskas S, Butkus V. 1992a. Purification and properties of the Eco57I restriction endonuclease and methylase—prototypes of a new class (type IV). *Nucleic Acids Res* **20:** 6043–6049.

Janulaitis A, Vaisvila R, Timinskas A, Klimasauskas S, Butkus V. 1992b. Cloning and sequence analysis of the genes coding for Eco57I type IV restriction-modification enzymes. *Nucleic Acids Res* **20:** 6051–6056.

Jeffery CJ. 2016. Protein species and moonlighting proteins: Very small changes in a protein's covalent structure can change its biochemical function. *J Proteomics* **134:** 19–24.

Jeffreys AJ. 2006. See, e.g., https://www.knaw.nl/en/awards/laureates/dr-h-p-heinekenprijs-voor-biochemie-en-biofysica/alec-j-jeffreys-1950-groot-brittannia; https://www.ncbi.nlm.nih.gov/pubmed/20011117 (interview with Jane Gitschier 2009); https://www.ncbi.nlm.nih.gov/pmc/articles/PMC3831583/pdf/2041-2223-4-21.pdf (2013).

Jeltsch A, Wenz C, Wende W, Selent U, Pingoud A. 1996. Engineering novel restriction endonucleases: principles and applications. *Trends Biotechnol* **14:** 235–238.

Jeschke J, Collignon E, Fuks F. 2016. Portraits of TET-mediated DNA hydroxymethylation in cancer. *Curr Opin Genet Dev* **36:** 16–26.

Ji M, Tan L, Jen-Jacobson L, Saxena S. 2014. Insights into copper coordination in the EcoRI-DNA complex by ESR spectroscopy. *Mol Phys* **112:** 3173–3182.

Johnson TB, Coghill RD. 1925. Researches on pyrimidines. C111. The discovery of *S*-methylcytosine in tuberculinic acid, the nucleic acid of the tubercle bacillus. *J Am Chem Soc* **47**(11): 2838–2844.

Joung JK, Sander JD. 2013. TALENs: a widely applicable technology for targeted genome editing. *Nat Rev Mol Cell Biol* **14:** 49–55.

Jurenaite-Urbanaviciene S, Serksnaite J, Kriukiene E, Giedriene J, Venclovas C, Lubys A. 2007. Generation of DNA cleavage specificities of type II restriction endonucleases by reassortment of target recognition domains. *Proc Natl Acad Sci* **104:** 10358–10363.

Jurica MS, Stoddard BL. 1999. Homing endonucleases: structure, function and evolution. *Cell Mol Life Sci* **55:** 1304–1326.

Kachalova GS, Rogulin EA, Yunusova AK, Artyukh RI, Perevyazova TA, Matvienko NI, Zheleznaya LA, Bartunik HD. 2008. Structural analysis of the heterodimeric type IIS restriction endonuclease R.BspD6I acting as a complex between a monomeric site-specific nickase and a catalytic subunit. *J Mol Biol* **384:** 489–502.

Kaczorowski T, Skowron P, Podhajska AJ. 1989. Purification and characterization of the *Fok*I restriction endonuclease. *Gene* **80:** 209–216.

Kaminska KH, Bujnicki JM. 2008. Bacteriophage Mu Mom protein responsible for DNA modification is a new member of the acyltransferase superfamily. *Cell Cycle* **7:** 120–121.

Kaminska KH, Kawai M, Boniecki M, Kobayashi I, Bujnicki JM. 2008. Type II restriction endonuclease R.Hpy188I belongs to the GIY-YIG nuclease superfamily, but exhibits an unusual active site. *BMC Struct Biol* **8:** 48. doi:10.1186/1472-6807-1188-1148.

Kannan P, Cowan GM, Daniel AS, Gann AA, Murray NE. 1989. Conservation of organization in the specificity polypeptides of two families of type I restriction enzymes. *J Mol Biol* **209:** 335–344.

Kanwar N, Roberts GA, Cooper LP, Stephanou AS, Dryden DT. 2016. The evolutionary pathway from a biologically inactive polypeptide sequence to a folded, active structural mimic of DNA. *Nucleic Acids Res* **44:** 4289–4303.

Karyagina A, Shilov I, Tashlitskii V, Khodoun M, Vasil'ev S, Lau PC, Nikolskaya I. 1997. Specific binding of sso II DNA methyltransferase to its promoter region provides the regulation of Sso II restriction-modification gene expression. *Nucleic Acids Res* **25:** 2114–2120.

Kauc L, Piekarowicz A. 1978. Purification and properties of a new restriction endonuclease from *Haemophilus influenzae* Rf. *Eur J Biochem/FEBS* **92:** 417–426.

Kaufmann G. 2000. Anticodon nucleases. *Trends Biochem Sci* **25:** 70–74.

Kaus-Drobek M, Czapinska H, Sokolowska M, Tamulaitis G, Szczepanowski RH, Urbanke C, Šikšnys V, Bochtler M. 2007. Restriction endonuclease MvaI is a monomer that recognizes its target sequence asymmetrically. *Nucleic Acids Res* **35:** 2035–2046.

Kazrani AA, Kowalska M, Czapinska H, Bochtler M. 2014. Crystal structure of the 5hmC specific endonuclease PvuRts1I. *Nucleic Acids Res* **42:** 5929–5936.

Kelch BA. 2016. Review: The lord of the rings: Structure and mechanism of the sliding clamp loader. *Biopolymers* **105:** 532–546.

Kelleher JE, Raleigh EA. 1991. A novel activity in *Escherichia coli* K-12 that directs restriction of DNA modified at CG dinucleotides. *J Bacteriol* **173:** 5220–5223.

Kelleher JE, Daniel AS, Murray NE. 1991. Mutations that confer de novo activity upon a maintenance methyltransferase. *J Mol Biol* **221:** 431–440.

Kennaway CK, Obarska-Kosinska A, White JH, Tuszynska I, Cooper LP, Bujnicki JM, Trinick J, Dryden DT. 2009. The structure of M.EcoKI Type I DNA methyltransferase with a DNA mimic antirestriction protein. *Nucleic Acids Res* **37:** 762–770.

Kennaway CK, Taylor JE, Song CF, Potrzebowski W, Nicholson W, White JH, Swiderska A, Obarska-Kosinska A, Callow P, Cooper LP, et al. 2012. Structure and operation of the DNA-translocating type I DNA restriction enzymes. *Genes Dev* **26:** 92–104.

Khan F, Furuta Y, Kawai M, Kaminska KH, Ishikawa K, Bujnicki JM, Kobayashi I. 2010. A putative mobile genetic element carrying a novel type IIF restriction-modification system (PluTI). *Nucleic Acids Res* **38:** 3019–3030.

Kim SC, Podhajska AJ, Szybalski W. 1988. Cleaving DNA at any predetermined site with adapter-primers and class-IIS restriction enzymes. *Science* **240:** 504–506.

Kim YC, Grable JC, Love R, Greene PJ, Rosenberg JM. 1990. Refinement of Eco RI endonuclease crystal structure: a revised protein chain tracing. *Science* **249:** 1307–1309.

Kim YG, Li L, Chandrasegaran S. 1994. Insertion and deletion mutants of *Fok*I restriction endonuclease. *J Biol Chem* **269:** 31978–31982.

Kim SC, Skowron PM, Szybalski W. 1996a. Structural requirements for *Fok*I-DNA interaction and oligodeoxyribonucleotide-instructed cleavage. *J Mol Biol* **258:** 638–649.

Kim YG, Cha J, Chandrasegaran S. 1996b. Hybrid restriction enzymes: zinc finger fusions to *Fok*I cleavage domain. *Proc Natl Acad Sci* **93:** 1156–1160.

Kim YG, Shi Y, Berg JM, Chandrasegaran S. 1997. Site-specific cleavage of DNA–RNA hybrids by zinc finger/*Fok*I cleavage domain fusions. *Gene* **203:** 43–49.

Kim YG, Smith J, Durgesha M, Chandrasegaran S. 1998. Chimeric restriction enzyme: Gal4 fusion to FokI cleavage domain. *Biol Chem* **379:** 489–495.

Kim JS, DeGiovanni A, Jancarik J, Adams PD, Yokota H, Kim R, Kim SH. 2005. Crystal structure of DNA sequence specificity subunit of a type I restriction-modification enzyme and its functional implications. *Proc Natl Acad Sci* **102:** 3248–3253.

King G, Murray NE. 1994. Restriction enzymes in cells, not eppendorfs. *Trends Microbiol* **2:** 465–469.

Kingston IJ, Gormley NA, Halford SE. 2003. DNA supercoiling enables the type IIS restriction enzyme BspMI to recognise the relative orientation of two DNA sequences. *Nucleic Acids Res* **31:** 5221–5228.

Kita K, Kotani H, Hiraoka N, Nakamura T, Yonaha K. 1989a. Overproduction and crystallization of FokI restriction endonuclease. *Nucleic Acids Res* **17:** 8741–8753.

Kita K, Kotani H, Sugisaki H, Takanami M. 1989b. The *Fok*I restriction-modification system. I. Organization and nucleotide sequences of the restriction and modification genes. *J Biol Chem* **264:** 5751–5756.

Kita K, Kotani H, Ohta H, Yanase H, Kato N. 1992a. StsI, a new *Fok*I isoschizomer from *Streptococcus sanguis* 54, cleaves 5′ GGATG(N)10/14 3′. *Nucleic Acids Res* **20:** 618. PMCID: 310441.

Kita K, Suisha M, Kotani H, Yanase H, Kato N. 1992b. Cloning and sequence analysis of the StsI restriction-modification gene: presence of homology to *Fok*I restriction-modification enzymes. *Nucleic Acids Res* **20:** 4167–4172.

Kleinstiver BP, Berube-Janzen W, Fernandes AD, Edgell DR. 2011. Divalent metal ion differentially regulates the sequential nicking reactions of the GIY-YIG homing endonuclease I-BmoI. *PloS One* **6:** e23804. doi:10.1371/journal.pone.0023804.

Kleinstiver BP, Wolfs JM, Edgell DR. 2013. The monomeric GIY-YIG homing endonuclease I-BmoI uses a molecular anchor and a flexible tether to sequentially nick DNA. *Nucleic Acids Res* **41:** 5413–5427.

Klug A. 2010a. The discovery of zinc fingers and their applications in gene regulation and genome manipulation. *Annu Rev Biochem* **79:** 213–231.

Klug A. 2010b. The discovery of zinc fingers and their development for practical applications in gene regulation and genome manipulation. *Q Rev Biophys* **43:** 1–21.

Knowle D, Lintner RE, Touma YM, Blumenthal RM. 2005. Nature of the promoter activated by C·PvuII, an unusual regulatory protein conserved among restriction-modification systems. *J Bacteriol* **187:** 488–497.

Kojima KK, Kobayashi I. 2015. Transmission of the PabI family of restriction DNA glycosylase genes: mobility and long-term inheritance. *BMC Genomics* **16:** 817. doi:810.1186/s12864-12015-12021-12863.

Kong H. 1998. "Characterization of a new R-M system, the BcgI system of *Bacillus coagulans*" PhD thesis, Boston University.

Kong H, Smith CL. 1997. Substrate DNA and cofactor regulate the activities of a multi-functional restriction-modification enzyme, BcgI. *Nucleic Acids Res* **25:** 3687–3692.

Kong H, Smith CL. 1998. Does BcgI, a unique restriction endonuclease, require two recognition sites for cleavage? *Biol Chem* **379:** 605–609.

Kong H, Morgan RD, Maunus RE, Schildkraut I. 1993. A unique restriction endonuclease, BcgI, from *Bacillus coagulans*. *Nucleic Acids Res* **21:** 987–991.

Kong H, Roemer SE, Waite-Rees PA, Benner JS, Wilson GG, Nwankwo DO. 1994. Characterization of BcgI, a new kind of restriction-modification system. *J Biol Chem* **269:** 683–690.

Kong H, Lin LF, Porter N, Stickel S, Byrd D, Posfai J, Roberts RJ. 2000. Functional analysis of putative restriction-modification system genes in the *Helicobacter pylori* J99 genome. *Nucleic Acids Res* **28:** 3216–3223.

Korlach J, Turner SW. 2012. Going beyond five bases in DNA sequencing. *Curr Opin Struct Biol* **22:** 251–261.

Korlach J, Bjornson KP, Chaudhuri BP, Cicero RL, Flusberg BA, Gray JJ, Holden D, Saxena R, Wegener J, Turner SW. 2010. Real-time DNA sequencing from single polymerase molecules. *Methods Enzymol* **472:** 431–455.

Kostiuk G, Dikic J, Sasnauskas G, Seidel R, Šikšnys V. 2015. Single-molecule analysis of the monomeric REase BcnI. In *7th NEB meeting on restriction and modification (Gdansk)*: Poster P30.

Kostiuk G, Sasnauskas G, Tamulaitienė G, Šikšnys V. 2011. Degenerate sequence recognition by the monomeric restriction enzyme: single mutation converts BcnI into a strand-specific nicking endonuclease. *Nucleic Acids Res* **39:** 3744–3753.

Kostiuk G, Dikic J, Schwarz FW, Sasnauskas G, Seidel R, Šikšnys V. 2017. The dynamics of the monomeric restriction endonuclease BcnI during its interaction with DNA. *Nucleic Acids Res* **45:** 5968–5979.

Kostrewa D, Winkler FK. 1995. Mg^{2+} binding to the active site of EcoRV endonuclease: a crystallographic study of complexes with substrate and product DNA at 2 A resolution. *Biochemistry* **34:** 683–696.

Kovall RA, Matthews BW. 1999. Type II restriction endonucleases: structural, functional and evolutionary relationships. *Curr Opin Chem Biol* **3:** 578–583.

Kriaucionis S, Heintz N. 2009. The nuclear DNA base 5-hydroxymethylcytosine is present in Purkinje neurons and the brain. *Science* **324:** 929–930.

Kriukiene E. 2006. Domain organization and metal ion requirement of the Type IIS restriction endonuclease MnlI. *FEBS Lett* **580:** 6115–6122.

Kriukiene E, Lubiene J, Lagunavicius A, Lubys A. 2005. MnlI—The member of H-N-H subtype of Type IIS restriction endonucleases. *Biochim Biophys Acta* **1751:** 194–204.

Kruger DH, Bickle TA. 1983. Bacteriophage survival: multiple mechanisms for avoiding the deoxyribonucleic acid restriction systems of their hosts. *Microbiol Rev* **47:** 345–360.

Kruger DH, Reuter M. 2005. Reliable detection of DNA cytosine methylation at CpNpG sites using the engineered restriction enzyme EcoRII-C. *BioTechniques* **38:** 855–856.

Kruger DH, Hansen S, Schroeder C, Presber W. 1977a. Host-dependent modification of bacteriophage T7 and SAMase-negative T3 derivatives affecting their adsorption ability. *Mol Gen Genet* **153:** 107–110.

Kruger DH, Presber W, Hansen S, Rosenthal HA. 1977b. Different restriction of bacteriophages T3 and T7 by P1-lysogenic cells and the role of the T3-coded SAMase. *Z Allg Mikrobiol* **17:** 581–591.

Kruger DH, Schroeder C, Hansen S, Rosenthal HA. 1977c. Active protection by bacteriophages T3 and T7 against *E. coli* B- and K-specific restriction of their DNA. *Mol General Genet* **153:** 99–106.

Kruger DH, Reuter M, Schroeder C, Glatman LI, Chernin LS. 1983. Restriction of bacteriophage T3 and T7 ocr+ strains by the type II restriction endonuclease EcoRV. *Mol Gen Genet* **190:** 349–351.

Kruger DH, Barcak GJ, Reuter M, Smith HO. 1988. EcoRII can be activated to cleave refractory DNA recognition sites. *Nucleic Acids Res* **16:** 3997–4008.

Kubareva EA, Petrauskene OV, Karyagina AS, Tashlitsky VN, Nikolskaya II, Gromova ES. 1992. Cleavage of synthetic substrates containing non-nucleotide inserts by restriction endonucleases. Change in the cleavage specificity of endonuclease SsoII. *Nucleic Acids Res* **20:** 4533–4538.

Kubareva EA, Thole H, Karyagina AS, Oretskaya TS, Pingoud A, Pingoud V. 2000. Identification of a base-specific contact between the restriction endonuclease SsoII and its recognition sequence by photocross-linking. *Nucleic Acids Res* **28:** 1085–1091.

Kulkarni M, Nirwan N, van Aelst K, Szczelkun MD, Saikrishnan K. 2016. Structural insights into DNA sequence recognition by Type ISP restriction-modification enzymes. *Nucleic Acids Res* **44:** 4396–4408.

Kuper J, Braun C, Elias A, Michels G, Sauer F, Schmitt DR, Poterszman A, Egly JM, Kisker C. 2014. In TFIIH, XPD helicase is exclusively devoted to DNA repair. *PLoS Biol* **12:** e1001954. doi:10.1371/journal.pbio.1001954.

Kurpiewski MR, Engler LE, Wozniak LA, Kobylanska A, Koziolkiewicz M, Stec WJ, Jen-Jacobson L. 2004. Mechanisms of coupling between DNA recognition specificity and catalysis in EcoRI endonuclease. *Structure* **12:** 1775–1788.

Kwiatek A, Luczkiewicz M, Bandyra K, Stein DC, Piekarowicz A. 2010. *Neisseria gonorrhoeae* FA1090 carries genes encoding two classes of Vsr endonucleases. *J Bacteriol* **192:** 3951–3960.

Kwiatek A, Mrozek A, Bacal P, Piekarowicz A, Adamczyk-Popławska M. 2015. Type III methyltransferase M.NgoAX from *Neisseria gonorrhoeae* FA1090 regulates biofilm formation and interactions with human cells. *Frontiers Microbiol* **6:** 1426. doi:1410.3389/fmicb.2015.01426.

Lacks S, Greenberg B. 1975. A deoxyribonuclease of *Diplococcus pneumoniae* specific for methylated DNA. *J Biol Chem* **250:** 4060–4066.

Lacks S, Greenberg B. 1977. Complementary specificity of restriction endonucleases of *Diplococcus pneumoniae* with respect to DNA methylation. *J Mol Biol* **114:** 153–168.

Lagunavicius A, Sasnauskas G, Halford SE, Šikšnys V. 2003. The metal-independent type IIs restriction enzyme BfiI is a dimer that binds two DNA sites but has only one catalytic centre. *J Mol Biol* **326:** 1051–1064.

Lambert AR, Sussman D, Shen B, Maunus R, Nix J, Samuelson J, Xu SY, Stoddard BL. 2008. Structures of the rare-cutting restriction endonuclease NotI reveal a unique metal binding fold involved in DNA binding. *Structure* **16:** 558–569.

Landry D, Looney MC, Feehery GR, Slatko BE, Jack WE, Schildkraut I, Wilson GG. 1989. M·FokI methylates adenine in both strands of its asymmetric recognition sequence. *Gene* **77:** 1–10.

Lanio T, Jeltsch A, Pingoud A. 2000. On the possibilities and limitations of rational protein design to expand the specificity of restriction enzymes: a case study employing EcoRV as the target. *Protein Eng* **13:** 275–281.

Lapkouski M, Panjikar S, Kuta Smatanova I, Csefalvay E. 2007. Purification, crystallization and preliminary X-ray analysis of the HsdR subunit of the EcoR124I endonuclease from *Escherichia coli*. *Acta Crystallogr, Sect F: Struct Biol Cryst Commun* **63:** 582–585.

Lapkouski M, Panjikar S, Janscak P, Smatanova IK, Carey J, Ettrich R, Csefalvay E. 2009. Structure of the motor subunit of type I restriction-modification complex EcoR124I. *Nat Struct Mol Biol* **16:** 94–95.

Larsen MH, Figurski DH. 1994. Structure, expression, and regulation of the *kilC* operon of promiscuous IncPα plasmids. *J Bacteriol* **176:** 5022–5032.

Laue F, Ankenbauer W, Schmitz GG, Kessler C. 1990. The selective inhibitory effect of netrop-sin on relaxation of sequence specificity of restriction endonuclease *SgrAI* recognizing the octanucleotide sequence 5′-CR decreases CCGGYG-3′. *Nucleic Acids Res* **18:** 3421.

Laurens N, Bellamy SR, Harms AF, Kovacheva YS, Halford SE, Wuite GJ. 2009. Dissecting pro-tein-induced DNA looping dynamics in real time. *Nucleic Acids Res* **37:** 5454–5464.

Laurens N, Rusling DA, Pernstich C, Brouwer I, Halford SE, Wuite GJ. 2012. DNA looping by FokI: the impact of twisting and bending rigidity on protein-induced looping dynamics. *Nucleic Acids Res* **40:** 4988–4997.

Le May N, Egly JM, Coin F. 2010. True lies: the double life of the nucleotide excision repair factors in transcription and DNA repair. *J Nucleic Acids* **2010.** doi:10.4061/2010/616342.

Leismann O, Roth M, Friedrich T, Wende W, Jeltsch A. 1998. The *Flavobacterium okeanokoites* adenine-N^6-specific DNA-methyltransferase M·FokI is a tandem enzyme of two independ-ent domains with very different kinetic properties. *Eur J Biochem/FEBS* **251:** 899–906.

Lepikhov K, Tchernov A, Zheleznaja L, Matvienko N, Walter J, Trautner TA. 2001. Character-ization of the type IV restriction modification system BspLU11III from *Bacillus* sp. LU11. *Nucleic Acids Res* **29:** 4691–4698.

Levinson G, Gutman GA. 1987a. High frequencies of short frameshifts in poly-CA/TG tandem repeats borne by bacteriophage M13 in *Escherichia coli* K-12. *Nucleic Acids Res* **15:** 5323–5338.

Levinson G, Gutman GA. 1987b. Slipped-strand mispairing: a major mechanism for DNA sequence evolution. *Mol Biol Evol* **4:** 203–221.

Li L, Chandrasegaran S. 1993. Alteration of the cleavage distance of *Fok* I restriction endonu-clease by insertion mutagenesis. *Proc Natl Acad Sci* **90:** 2764–2768.

Li L, Wu LP, Chandrasegaran S. 1992. Functional domains in *Fok* I restriction endonuclease. *Proc Natl Acad Sci* **89:** 4275–4279.

Li L, Wu LP, Clarke R, Chandrasegaran S. 1993. C-terminal deletion mutants of the FokI restric-tion endonuclease. *Gene* **133:** 79–84.

Li T, Huang S, Jiang WZ, Wright D, Spalding MH, Weeks DP, Yang B. 2011. TAL nucleases (TALNs): hybrid proteins composed of TAL effectors and FokI DNA-cleavage domain. *Nucleic Acids Res* **39:** 359–372.

Liang J, Blumenthal RM. 2013. Naturally-occurring, dually-functional fusions between restric-tion endonucleases and regulatory proteins. *BMC Evol Biol* **13:** 218.

Lin LF, Posfai J, Roberts RJ, Kong H. 2001. Comparative genomics of the restriction-modifica-tion systems in *Helicobacter pylori*. *Proc Natl Acad Sci* **98:** 2740–2745.

Lindsay JA. 2010. Genomic variation and evolution of *Staphylococcus aureus*. *Int J Med Microbiol* **300:** 98–103.

Little EJ, Dunten PW, Bitinaite J, Horton NC. 2011. New clues in the allosteric activation of DNA cleavage by SgrAI: structures of SgrAI bound to cleaved primary-site DNA and uncleaved secondary-site DNA. *Acta Crystallogr, Sect D: Biol Crystallogr* **67:** 67–74.

Liu Y, Kobayashi I. 2007. Negative regulation of the EcoRI restriction enzyme gene is associated with intragenic reverse promoters. *J Bacteriol* **189:** 6928–6935.

Liu Y, Ichige A, Kobayashi I. 2007. Regulation of the EcoRI restriction-modification system: Identification of ecoRIM gene promoters and their upstream negative regulators in the ecoRIR gene. *Gene* **400:** 140–149.

Liu G, Ou HY, Wang T, Li L, Tan H, Zhou X, Rajakumar K, Deng Z, He X. 2010. Cleavage of phosphorothioated DNA and methylated DNA by the type IV restriction endonuclease ScoMcrA. *PLoS Genet* **6:** e1001253. doi:10.1371/journal.pgen.1001253.

Lobner-Olesen A, Skovgaard O, Marinus MG. 2005. Dam methylation: coordinating cellular processes. *Curr Opin Microbiol* **8:** 154–160.

Loenen WAM. 2003. Tracking EcoKI and DNA fifty years on: a golden story full of surprises. *Nucleic Acids Res* **31:** 7059–7069.

Loenen WAM. 2006. *S*-adenosylmethionine: jack of all trades and master of everything? *Biochem Soc Trans* **34:** 330–333.

Loenen WAM. 2010. *S*-adenosylmethionine: simple agent of methylation and secret to aging and metabolism? In *Epigenetics of aging* (ed. Tollefsbol TO), pp 107–131. Springer, Berlin.

Loenen WAM. 2017. *S*-adenosylmethionine: a promising avenue in anti-aging medicine? In *Anti-aging drugs: from basic research to clinical practice* (ed. Vaiserman, Alexander M), RSC Drug Discovery Series, Vol. 57, pp. 435–473. Royal Society of Chemistry, Cambridge.

Loenen WAM. 2018. *S*-adenosylmethionine metabolism and aging. In *Epigenetics of aging and longevity* (ed. Moskalev A, Vaiserman A), Ch. 3, pp. 59–93. Elsevier, New York.

Loenen WA, Murray NE. 1986. Modification enhancement by the restriction alleviation protein (Ral) of bacteriophage λ. *J Mol Biol* **190:** 11–22.

Loenen WA, Raleigh EA. 2014. The other face of restriction: modification-dependent enzymes. *Nucleic Acids Res* **42:** 56–69.

Loenen WA, Dryden DT, Raleigh EA, Wilson GG. 2014a. Type I restriction enzymes and their relatives. *Nucleic Acids Res* **42:** 20–44.

Loenen WA, Dryden DT, Raleigh EA, Wilson GG, Murray NE. 2014b. Highlights of the DNA cutters: a short history of the restriction enzymes. *Nucleic Acids Res* **42:** 3–19.

Looney MC, Moran LS, Jack WE, Feehery GR, Benner JS, Slatko BE, Wilson GG. 1989. Nucleotide sequence of the *Fok*I restriction-modification system: separate strand-specificity domains in the methyltransferase. *Gene* **80:** 193–208.

Low DA, Casadesus J. 2008. Clocks and switches: bacterial gene regulation by DNA adenine methylation. *Curr Opin Microbiol* **11:** 106–112.

Lu X, Zhao BS, He C. 2015. TET family proteins: oxidation activity, interacting molecules, and functions in diseases. *Chem Rev* **115:** 2225–2239.

Lukacs CM, Kucera R, Schildkraut I, Aggarwal AK. 2000. Understanding the immutability of restriction enzymes: crystal structure of *Bgl*II and its DNA substrate at 1.5 Å resolution. *Nat Struct Biol* **7:** 134–140.

Luria SE, Human ML. 1952. A nonhereditary, host-induced variation of bacterial viruses. *J Bacteriol* **64:** 557–569.

Lyumkis D, Talley H, Stewart A, Shah S, Park CK, Tama F, Potter CS, Carragher B, Horton NC. 2013. Allosteric regulation of DNA cleavage and sequence-specificity through run-on oligomerization. *Structure* **21:** 1848–1858.

Ma L, Chen K, Clarke DJ, Nortcliffe CP, Wilson GG, Edwardson JM, Morton AJ, Jones AC, Dryden DT. 2013a. Restriction endonuclease TseI cleaves A:A and T:T mismatches in CAG and CTG repeats. *Nucleic Acids Res* **41:** 4999–5009.

Ma X, Shah S, Zhou M, Park CK, Wysocki VH, Horton NC. 2013b. Structural analysis of activated SgrAI-DNA oligomers using ion mobility mass spectrometry. *Biochemistry* **52:** 4373–4381.

Machnicka MA, Kaminska KH, Dunin-Horkawicz S, Bujnicki JM. 2015. Phylogenomics and sequence-structure-function relationships in the GmrSD family of Type IV restriction enzymes. *BMC Bioinf* **16:** 336. doi:310.1186/s12859-12015-10773-z.

Mackeldanz P, Alves J, Moncke-Buchner E, Wyszomirski KH, Kruger DH, Reuter M. 2013. Functional consequences of mutating conserved SF2 helicase motifs in the Type III restriction endonuclease EcoP15I translocase domain. *Biochimie* **95:** 817–823.

Mak AN, Lambert AR, Stoddard BL. 2010. Folding, DNA recognition, and function of GIY-YIG endonucleases: crystal structures of R.Eco29kI. *Structure* **18:** 1321–1331.

Mak AN, Bradley P, Cernadas RA, Bogdanove AJ, Stoddard BL. 2012. The crystal structure of TAL effector PthXo1 bound to its DNA target. *Science* **335:** 716–719.

Mak AN, Bradley P, Bogdanove AJ, Stoddard BL. 2013. TAL effectors: function, structure, engineering and applications. *Curr Opin Struct Biol* **23:** 93–99.

Manakova E, Gražulis S, Zaremba M, Tamulaitienė G, Golovenko D, Šikšnys V. 2012. Structural mechanisms of the degenerate sequence recognition by Bse634I restriction endonuclease. *Nucleic Acids Res* **40:** 6741–6751.

Manakova E, Tamulaitienė G, Tamulaitis G, Mikutenaite M, Šikšnys V. 2015. Crystallographic and functional studies of restriction enzymes recognizing 5′-CCGG tetranucleotide core: Kpn2I and PfoI. In *7th NEB meeting on restriction and modification (Gdansk): Poster P1.*

Mandecki W, Bolling TJ. 1988. *Fok*I method of gene synthesis. *Gene* **68:** 101–107.

Mani M, Chen C, Amblee V, Liu H, Mathur T, Zwicke G, Zabad S, Patel B, Thakkar J, Jeffery CJ. 2015. MoonProt: a database for proteins that are known to moonlight. *Nucleic Acids Res* **43:** D277–D282.

Marinus MG, Casadesus J. 2009. Roles of DNA adenine methylation in host-pathogen interactions: mismatch repair, transcriptional regulation, and more. *FEMS Microbiol Rev* **33:** 488–503.

Mark KK, Studier FW. 1981. Purification of the gene 0.3 protein of bacteriophage T7, an inhibitor of the DNA restriction system of *Escherichia coli. J Biol Chem* **256:** 2573–2578.

Marks P, McGeehan J, Wilson G, Errington N, Kneale G. 2003. Purification and characterisation of a novel DNA methyltransferase, M.AhdI. *Nucleic Acids Res* **31:** 2803–2810.

Marshall JJ, Halford SE. 2010. The type IIB restriction endonucleases. *Biochem Soc Trans* **38:** 410–416.

Marshall JJ, Gowers DM, Halford SE. 2007. Restriction endonucleases that bridge and excise two recognition sites from DNA. *J Mol Biol* **367:** 419–431.

Marshall JJ, Smith RM, Ganguly S, Halford SE. 2011. Concerted action at eight phosphodiester bonds by the BcgI restriction endonuclease. *Nucleic Acids Res* **39:** 7630–7640.

McClarin JA, Frederick CA, Wang BC, Greene P, Boyer HW, Grable J, Rosenberg JM. 1986. Structure of the DNA-Eco RI endonuclease recognition complex at 3 A resolution. *Science* **234:** 1526–1541.

McClelland SE, Szczelkun MD. 2004. The Type I and III restriction endonucleases: structural elements in molecular motors that process DNA. In *Restriction endonucleases* (ed. Pingoud A), pp. 111–135. Springer, Berlin.

McGeehan JE, Streeter S, Cooper JB, Mohammed F, Fox GC, Kneale GG. 2004. Crystallization and preliminary X-ray analysis of the controller protein C.AhdI from *Aeromonas hydrophilia. Acta Crystallogr, Sect D: Biol Crystallogr* **60:** 323–325.

McGeehan JE, Streeter SD, Papapanagiotou I, Fox GC, Kneale GG. 2005. High-resolution crystal structure of the restriction-modification controller protein C.AhdI from *Aeromonas hydrophila. J Mol Biol* **346:** 689–701.

McGeehan JE, Papapanagiotou I, Streeter SD, Kneale GG. 2006. Cooperative binding of the C.AhdI controller protein to the C/R promoter and its role in endonuclease gene expression. *J Mol Biol* **358:** 523–531.

McGeehan JE, Streeter SD, Thresh SJ, Ball N, Ravelli RB, Kneale GG. 2008. Structural analysis of the genetic switch that regulates the expression of restriction-modification genes. *Nucleic Acids Res* **36:** 4778–4787.

McGeehan JE, Ball NJ, Streeter SD, Thresh SJ, Kneale GG. 2012. Recognition of dual symmetry by the controller protein C.Esp1396I based on the structure of the transcriptional activation complex. *Nucleic Acids Res* **40:** 4158–4167.

McMahon SA, Roberts GA, Johnson KA, Cooper LP, Liu H, White JH, Carter LG, Sanghvi B, Oke M, Walkinshaw MD, et al. 2009. Extensive DNA mimicry by the ArdA antirestriction protein and its role in the spread of antibiotic resistance. *Nucleic Acids Res* **37:** 4887–4897.

Mernagh DR, Janscak P, Firman K, Kneale GG. 1998. Protein-protein and protein-DNA interactions in the type I restriction endonuclease R.EcoR124I. *Biol Chem* **379:** 497–503.

Mierzejewska K, Siwek W, Czapinska H, Kaus-Drobek M, Radlinska M, Skowronek K, Bujnicki JM, Dadlez M, Bochtler M. 2014. Structural basis of the methylation specificity of R.DpnI. *Nucleic Acids Res* **42:** 8745–8754.

Miller J, McLachlan AD, Klug A. 1985. Repetitive zinc-binding domains in the protein transcription factor IIIA from *Xenopus* oocytes. *EMBO J* **4:** 1609–1614.

Miller ES, Kutter E, Mosig G, Arisaka F, Kunisawa T, Ruger W. 2003. Bacteriophage T4 genome. *Microbiol Mol Biol Rev* **67:** 86–156.

Miller JC, Holmes MC, Wang J, Guschin DY, Lee YL, Rupniewski I, Beausejour CM, Waite AJ, Wang NS, Kim KA, et al. 2007. An improved zinc-finger nuclease architecture for highly specific genome editing. *Nat Biotechnol* **25:** 778–785.

Miller JC, Tan S, Qiao G, Barlow KA, Wang J, Xia DF, Meng X, Paschon DE, Leung E, Hinkley SJ, et al. 2011. A TALE nuclease architecture for efficient genome editing. *Nat Biotechnol* **29:** 143–148.

Mino T, Aoyama Y, Sera T. 2009. Efficient double-stranded DNA cleavage by artificial zinc-finger nucleases composed of one zinc-finger protein and a single-chain FokI dimer. *J Biotechnol* **140:** 156–161.

Mino T, Mori T, Aoyama Y, Sera T. 2014. Inhibition of DNA replication of human papillomavirus by using zinc finger-single-chain FokI dimer hybrid. *Mol Biotechnol* **56:** 731–737.

Miyazono K, Watanabe M, Kosinski J, Ishikawa K, Kamo M, Sawasaki T, Nagata K, Bujnicki JM, Endo Y, Tanokura M, et al. 2007. Novel protein fold discovered in the PabI family of restriction enzymes. *Nucleic Acids Res* **35:** 1908–1918.

Miyazono K, Furuta Y, Watanabe-Matsui M, Miyakawa T, Ito T, Kobayashi I, Tanokura M. 2014. A sequence-specific DNA glycosylase mediates restriction-modification in *Pyrococcus abyssi*. *Nat Commun* **5:** 3178.

Moffatt BA, Studier FW. 1988. Entry of bacteriophage T7 DNA into the cell and escape from host restriction. *J Bacteriol* **170:** 2095–2105.

Mokrishcheva ML, Solonin AS, Nikitin DV. 2011. Fused eco29kIR- and M genes coding for a fully functional hybrid polypeptide as a model of molecular evolution of restriction-modification systems. *BMC Evol Biol* **11:** 35. doi:10.1186/1471-2148-1111-1135.

Molineux IJ. 2001. No syringes please, ejection of phage T7 DNA from the virion is enzyme driven. *Mol Microbiol* **40:** 1–8.

Moncke-Buchner E, Reich S, Mucke M, Reuter M, Messer W, Wanker EE, Kruger DH. 2002. Counting CAG repeats in the Huntington's disease gene by restriction endonuclease EcoP15I cleavage. *Nucleic Acids Res* **30:** e83. PMID: 12177311.

Moncke-Buchner E, Rothenberg M, Reich S, Wagenfuhr K, Matsumura H, Terauchi R, Kruger DH, Reuter M. 2009. Functional characterization and modulation of the DNA cleavage efficiency of type III restriction endonuclease EcoP15I in its interaction with two sites in the DNA target. *J M Biol* **387:** 1309–1319.

Morgan RD, Luyten YA. 2009. Rational engineering of type II restriction endonuclease DNA binding and cleavage specificity. *Nucleic Acids Res* **37**: 5222–5233.

Morgan RD, Calvet C, Demeter M, Agra R, Kong H. 2000. Characterization of the specific DNA nicking activity of restriction endonuclease N.BstNBI. *Biol Chem* **381**: 1123–1125.

Morgan RD, Bhatia TK, Lovasco L, Davis TB. 2008. MmeI: a minimal Type II restriction-modification system that only modifies one DNA strand for host protection. *Nucleic Acids Res* **36**: 6558–6570.

Morgan RD, Dwinell EA, Bhatia TK, Lang EM, Luyten YA. 2009. The MmeI family: type II restriction-modification enzymes that employ single-strand modification for host protection. *Nucleic Acids Res* **37**: 5208–5221.

Morgan RD, Luyten YA, Johnson SA, Clough EM, Clark TA, Roberts RJ. 2016. Novel m4C modification in type I restriction-modification systems. *Nucleic Acids Res* **44**: 9413–9425.

Mori T, Kagatsume I, Shinomiya K, Aoyama Y, Sera T. 2009. Sandwiched zinc-finger nucleases harboring a single-chain FokI dimer as a DNA-cleavage domain. *Biochem Biophys Res Commun* **390**: 694–697.

Moscou MJ, Bogdanove AJ. 2009. A simple cipher governs DNA recognition by TAL effectors. *Science* **326**: 1501.

Moxon R, Bayliss C, Hood D. 2006. Bacterial contingency loci: the role of simple sequence DNA repeats in bacterial adaptation. *Ann Rev Genet* **40**: 307–333.

Mruk I, Kobayashi I. 2014. To be or not to be: regulation of restriction-modification systems and other toxin-antitoxin systems. *Nucleic Acids Res* **42**: 70–86.

Mruk I, Liu Y, Ge L, Kobayashi I. 2011. Antisense RNA associated with biological regulation of a restriction-modification system. *Nucleic Acids Res* **39**: 5622–5632.

Mulligan EA, Dunn JJ. 2008. Cloning, purification and initial characterization of *E. coli* McrA, a putative 5-methylcytosine-specific nuclease. *Protein Expression Purif* **62**: 98–103.

Mulligan EA, Hatchwell E, McCorkle SR, Dunn JJ. 2010. Differential binding of *Escherichia coli* McrA protein to DNA sequences that contain the dinucleotide m5CpG. *Nucleic Acids Res* **38**: 1997–2005.

Murray NE. 2000. Type I restriction systems: sophisticated molecular machines (a legacy of Bertani and Weigle). *Microbiol Mol Biol Rev* **64**: 412–434.

Murray NE. 2002. 2001 Fred Griffith review lecture. Immigration control of DNA in bacteria: self versus non-self. *Microbiology* **148**: 3–20.

Murray IA, Stickel SK, Roberts RJ. 2010. Sequence-specific cleavage of RNA by Type II restriction enzymes. *Nucleic Acids Res* **38**: 8257–8268.

Mussolino C, Alzubi J, Fine EJ, Morbitzer R, Cradick TJ, Lahaye T, Bao G, Cathomen T. 2014. TALENs facilitate targeted genome editing in human cells with high specificity and low cytotoxicity. *Nucleic Acids Res* **42**: 6762–6773.

Nakayama Y, Kobayashi I. 1998. Restriction-modification gene complexes as selfish gene entities: roles of a regulatory system in their establishment, maintenance, and apoptotic mutual exclusion. *Proc Natl Acad Sci* **95**: 6442–6447.

Nakonieczna J, Zmijewski JW, Banecki B, Podhajska AJ. 2007. Binding of MmeI restriction–modification enzyme to its specific recognition sequence is stimulated by S-adenosyl-L-methionine. *Mol Biotechnol* **37**: 127–135.

Nakonieczna J, Kaczorowski T, Obarska-Kosinska A, Bujnicki JM. 2009. Functional analysis of MmeI from methanol utilizer *Methylophilus methylotrophus*, a subtype IIC restriction-modification enzyme related to type I enzymes. *Appl Environ Microbiol* **75**: 212–223.

Nasim MT, Eperon IC, Wilkins BM, Brammar WJ. 2004. The activity of a single-stranded promoter of plasmid ColIb-P9 depends on its secondary structure. *Mol Microbiol* **53**: 405–417.

Neaves KJ, Cooper LP, White JH, Carnally SM, Dryden DT, Edwardson JM, Henderson RM. 2009. Atomic force microscopy of the EcoKI Type I DNA restriction enzyme bound to DNA shows enzyme dimerization and DNA looping. *Nucleic Acids Res* **37:** 2053–2063.

Nekrasov SV, Agafonova OV, Belogurova NG, Delver EP, Belogurov AA. 2007. Plasmid-encoded antirestriction protein ArdA can discriminate between type I methyltransferase and complete restriction-modification system. *J Mol Biol* **365:** 284–297.

Newman M, Strzelecka T, Dorner LF, Schildkraut I, Aggarwal AK. 1994. Structure of restriction endonuclease BamHI and its relationship to EcoRI. *Nature* **368:** 660–664.

Niv MY, Ripoll DR, Vila JA, Liwo A, Vanamee ES, Aggarwal AK, Weinstein H, Scheraga HA. 2007. Topology of Type II REases revisited; structural classes and the common conserved core. *Nucleic Acids Res* **35:** 2227–2237.

Noyer-Weidner M, Diaz R, Reiners L. 1986. Cytosine-specific DNA modification interferes with plasmid establishment in *Escherichia coli* K12: involvement of rglB. *Mol Gen Genet* **205:** 469–475.

Nwankwo D, Wilson G. 1987. Cloning of two type II methylase genes that recognise asymmetric nucleotide sequences: FokI and HgaI. *Mol Gen Genet* **209:** 570–574.

Obarska A, Blundell A, Feder M, Vejsadova S, Sisakova E, Weiserova M, Bujnicki JM, Firman K. 2006. Structural model for the multisubunit Type IC restriction-modification DNA methyltransferase M.EcoR124I in complex with DNA. *Nucleic Acids Res* **34:** 1992–2005.

Oke M, Carter LG, Johnson KA, Liu H, McMahon SA, Yan X, Kerou M, Weikart ND, Kadi N, Sheikh MA, et al. 2010. The Scottish Structural Proteomics Facility: targets, methods and outputs. *J Struct Funct Genomics* **11:** 167–180.

Orlowski J, Bujnicki JM. 2008. Structural and evolutionary classification of Type II restriction enzymes based on theoretical and experimental analyses. *Nucleic Acids Res* **36:** 3552–3569.

Ou HY, He X, Shao Y, Tai C, Rajakumar K, Deng Z. 2009. *dnd*DB: a database focused on phosphorothioation of the DNA backbone. *PloS One* **4:** e5132. doi:10.1371/journal.pone.0005132.

Panne D, Muller SA, Wirtz S, Engel A, Bickle TA. 2001. The McrBC restriction endonuclease assembles into a ring structure in the presence of G nucleotides. *EMBO J* **20:** 3210–3217.

Papapanagiotou I, Streeter SD, Cary PD, Kneale GG. 2007. DNA structural deformations in the interaction of the controller protein C.AhdI with its operator sequence. *Nucleic Acids Res* **35:** 2643–2650.

Park CK, Stiteler AP, Shah S, Ghare MI, Bitinaite J, Horton NC. 2010. Activation of DNA cleavage by oligomerization of DNA-bound SgrAI. *Biochemistry* **49:** 8818–8830.

Pastor WA, Aravind L, Rao A. 2013. TETonic shift: biological roles of TET proteins in DNA demethylation and transcription. *Nat Rev Mol Cell Biol* **14:** 341–356.

Pattanayak V, Ramirez CL, Joung JK, Liu DR. 2011. Revealing off-target cleavage specificities of zinc-finger nucleases by in vitro selection. *Nat Methods* **8:** 765–770.

Peakman LJ, Szczelkun MD. 2004. DNA communications by Type III restriction endonucleases —confirmation of 1D translocation over 3D looping. *Nucleic Acids Res* **32:** 4166–4174.

Penn NW, Suwalski R, O'Riley C, Bojanowski K, Yura R. 1972. The presence of 5-hydroxymethylcytosine in animal deoxyribonucleic acid. *Biochem J* **126:** 781–790.

Perez-Pinera P, Ousterout DG, Brown MT, Gersbach CA. 2012a. Gene targeting to the ROSA26 locus directed by engineered zinc finger nucleases. *Nucleic Acids Res* **40:** 3741–3752.

Perez-Pinera P, Ousterout DG, Gersbach CA. 2012b. Advances in targeted genome editing. *Curr Opin Chem Biol* **16:** 268–277.

Pernstich C, Halford SE. 2012. Illuminating the reaction pathway of the FokI restriction endonuclease by fluorescence resonance energy transfer. *Nucleic Acids Res* **40:** 1203–1213.

Pertzev AV, Kravetz AN, Mayorov SG, Zakharova MV, Solonin AS. 1997. Isolation of a strain overproducing endonuclease Eco29kI: purification and characterization of the homogeneous enzyme. *Biochem Biokhim* **62:** 732–741.

Petrov VM, Ratnayaka S, Nolan JM, Miller ES, Karam JD. 2010. Genomes of the T4-related bacteriophages as windows on microbial genome evolution. *Virol J* **7:** 292. doi:10.1186/1743-422X-7-292.

Piekarowicz A. 1974. The influence of methionine deprivation on restriction properties of *Haemophilus influenzae* Rd and Ra strains. *Acta Microbiol Pol, Ser A* **6:** 71–74.

Piekarowicz A. 1982. HineI is an isoschizomer of *Hin*fIII restriction endonuclease. *J Mol Biol* **157:** 373–381.

Piekarowicz A. 1984. Preferential cleavage by restriction endonuclease *Hin*fIII. *Acta Biochim Pol* **31:** 453–464.

Piekarowicz A. 2013. REases and DNA MTases in *H. influenzae* and *N. gonorrhoeae*. *Talk at CSHL meeting on the history of Restriction and Modification*.

Piekarowicz A, Baj J. 1975. Host specificity of DNA in *haemophilus influenzae*: The physiological and genetical bases of instability of restriction and modification of DNA in strain Rd. *Acta Microbiol Pol, Ser A* **8:** 119–130.

Piekarowicz A, Brzezinski R. 1980. Cleavage and methylation of DNA by the restriction endonuclease HinfIII isolated from *Haemophilus influenzae* Rf. *J Mol Biol* **144:** 415–429.

Piekarowicz A, Glover SW. 1972. Host specificity of DNA in *Haemophilus influenzae*: the two restriction and modification systems in strain Ra. *Mol Gen Genet* **116:** 11–25.

Piekarowicz A, Kalinowska J. 1974. Host specificity of DNA in *Haemophilus influenzae*: similarity between host-specificity types of *Haemophilus influenzae* Re and Rf. *J Gen Microbiol* **81:** 405–411.

Piekarowicz A, Kauc L, Glover SW. 1974. Host specificity of DNA in *Haemophilus influenzae*: the restriction and modification systems in strains Rb and Rf. *J Gen Microbiol* **81:** 391–403.

Piekarowicz A, Brzezinski R, Kauc L. 1975. Host specificity of DNA in *Haemophilus influenzae*: the in vivo action of the restriction endonucleases on phage and bacterial DNA. *Acta Microbiol Pol, Ser A* **7:** 51–65.

Piekarowicz A, Brzezinski R, Kauc L. 1976. Host specificity of DNA in *Haemophilus influenzae*: DNA restriction enzyme from *H. influenzae* Rf232. *Acta Microbiol Pol* **25:** 307–312.

Piekarowicz A, Bickle TA, Shepherd JC, Ineichen K. 1981. The DNA sequence recognised by the *Hin*fIII restriction endonuclease. *J Mol Biol* **146:** 167–172.

Piekarowicz A, Brzezinski R, Smorawinska M, Kauc L, Skowronek K, Lenarczyk M, Golembiowska M, Siwinska M. 1986. Major spontaneous genomic rearrangements in *Haemophilus influenzae* S2 and HP1c1 bacteriophages. *Gene* **49:** 111–118.

Piekarowicz A, Yuan R, Stein DC. 1988. Identification of a new restriction endonuclease, R. NgoBI, from *Neisseria gonorrhoeae*. *Nucleic Acids Res* **16:** 9868. PMID: 3141904.

Pieper U, Pingoud A. 2002. A mutational analysis of the PD...D/EXK motif suggests that McrC harbors the catalytic center for DNA cleavage by the GTP-dependent restriction enzyme McrBC from *Escherichia coli*. *Biochemistry* **41:** 5236–5244.

Pieper U, Brinkmann T, Kruger T, Noyer-Weidner M, Pingoud A. 1997. Characterization of the interaction between the restriction endonuclease McrBC from *E. coli* and its cofactor GTP. *J Mol Biol* **272:** 190–199.

Pieper U, Groll DH, Wunsch S, Gast FU, Speck C, Mucke N, Pingoud A. 2002. The GTP-dependent restriction enzyme McrBC from *Escherichia coli* forms high-molecular mass complexes with DNA and produces a cleavage pattern with a characteristic 10-base pair repeat. *Biochemistry* **41:** 5245–5254.

Pingoud A. 2004. *Restriction endonucleases.* Springer, Berlin.

Pingoud A, Silva GH. 2007. Precision genome surgery. *Nat Biotechnol* **25:** 743–744.

Pingoud V, Kubareva E, Stengel G, Friedhoff P, Bujnicki JM, Urbanke C, Sudina A, Pingoud A. 2002. Evolutionary relationship between different subgroups of restriction endonucleases. *J Biol Chem* **277:** 14306–14314.

Pingoud V, Conzelmann C, Kinzebach S, Sudina A, Metelev V, Kubareva E, Bujnicki JM, Lurz R, Luder G, Xu SY, et al. 2003. PspGI, a type II restriction endonuclease from the extreme thermophile *Pyrococcus* sp.: structural and functional studies to investigate an evolutionary relationship with several mesophilic restriction enzymes. *J Mol Biol* **329:** 913–929.

Pingoud A, Fuxreiter M, Pingoud V, Wende W. 2005a. Type II restriction endonucleases: structure and mechanism. *Cell Mol Life Sci* **62:** 685–707.

Pingoud V, Geyer H, Geyer R, Kubareva E, Bujnicki JM, Pingoud A. 2005b. Identification of base-specific contacts in protein-DNA complexes by photocrosslinking and mass spectrometry: a case study using the restriction endonuclease SsoII. *Mol BioSyst* **1:** 135–141.

Pingoud V, Sudina A, Geyer H, Bujnicki JM, Lurz R, Luder G, Morgan R, Kubareva E, Pingoud A. 2005c. Specificity changes in the evolution of type II restriction endonucleases: a biochemical and bioinformatic analysis of restriction enzymes that recognize unrelated sequences. *J Biol Chem* **280:** 4289–4298.

Pingoud A, Wilson GG, Wende W. 2014. Type II restriction endonucleases—a historical perspective and more. *Nucleic Acids Res* **42:** 7489–7527.

Powell LM, Dryden DT, Murray NE. 1998. Sequence-specific DNA binding by EcoKI, a type IA DNA restriction enzyme. *J Mol Biol* **283:** 963–976.

Powell LM, Lejeune E, Hussain FS, Cronshaw AD, Kelly SM, Price NC, Dryden DT. 2003. Assembly of EcoKI DNA methyltransferase requires the C-terminal region of the HsdM modification subunit. *Biophys Chem* **103:** 129–137.

Price C, Lingner J, Bickle TA, Firman K, Glover SW. 1989. Basis for changes in DNA recognition by the EcoR124 and EcoR124/3 type I DNA restriction and modification enzymes. *J Mol Biol* **205:** 115–125.

Prohaska SJ, Stadler PF, Krakauer DC. 2010. Innovation in gene regulation: the case of chromatin computation. *J Theor Biol* **265:** 27–44.

Protsenko A, Zakharova M, Nagornykh M, Solonin A, Severinov K. 2009. Transcription regulation of restriction-modification system Ecl18kI. *Nucleic Acids Res* **37:** 5322–5330.

Putnam CD, Tainer JA. 2005. Protein mimicry of DNA and pathway regulation. *DNA Repair (Amst)* **4:** 1410–1420.

Raghavendra NK, Rao DN. 2004. Unidirectional translocation from recognition site and a necessary interaction with DNA end for cleavage by Type III restriction enzyme. *Nucleic Acids Res* **32:** 5703–5711.

Raleigh EA. 1992. Organization and function of the *mcrBC* genes of *Escherichia coli* K-12. *Mol Microbiol* **6:** 1079–1086.

Raleigh EA, Wilson G. 1986. *Escherichia coli* K-12 restricts DNA containing 5-methylcytosine. *Proc Natl Acad Sci* **83:** 9070–9074.

Ramalingam S, Kandavelou K, Rajenderan R, Chandrasegaran S. 2011. Creating designed zinc-finger nucleases with minimal cytotoxicity. *J Mol Biol* **405:** 630–641.

Ramalingam S, London V, Kandavelou K, Cebotaru L, Guggino W, Civin C, Chandrasegaran S. 2013. Generation and genetic engineering of human induced pluripotent stem cells using designed zinc finger nucleases. *Stem Cells Dev* **22:** 595–610.

Ramalingam S, Annaluru N, Kandavelou K, Chandrasegaran S. 2014. TALEN-mediated generation and genetic correction of disease-specific human induced pluripotent stem cells. *Curr Gene Ther* **14:** 461–472.

Ramanathan A, Agarwal PK. 2011. Evolutionarily conserved linkage between enzyme fold, flexibility, and catalysis. *PLoS Biol* **9:** e1001193. doi:10.1371/journal.pbio.1001193.

Ramanathan SP, van Aelst K, Sears A, Peakman LJ, Diffin FM, Szczelkun MD, Seidel R. 2009. Type III restriction enzymes communicate in 1D without looping between their target sites. *Proc Natl Acad Sci* **106:** 1748–1753.

Rao DN, Dryden DT, Bheemanaik S. 2014. Type III restriction-modification enzymes: a historical perspective. *Nucleic Acids Res* **42:** 45–55.

Rasko T, Der A, Klement E, Slaska-Kiss K, Posfai E, Medzihradszky KF, Marshak DR, Roberts RJ, Kiss A. 2010. BspRI restriction endonuclease: cloning, expression in *Escherichia coli* and sequential cleavage mechanism. *Nucleic Acids Res* **38:** 7155–7166.

Ratner D. 1974. The interaction bacterial and phage proteins with immobilized *Escherichia coli* RNA polymerase. *J Mol Biol* **88:** 373–383.

Reich S, Gossl I, Reuter M, Rabe JP, Kruger DH. 2004. Scanning force microscopy of DNA translocation by the Type III restriction enzyme EcoP15I. *J Mol Biol* **341:** 337–343.

Reuter M, Mucke M, Kruger DH. 2004. Structure and function of Type IIE restriction endonucleases—or: from a plasmid that restricts phage replication to a new molecular DNA recognition mechanism. In *Restriction endonucleases* (ed. Pingoud A), pp. 261–295. Springer, Berlin.

Revel HR, Luria SE. 1970. DNA-glucosylation in T-even phage: genetic determination and role in phagehost interaction. *Ann Rev Genet* **4:** 177–192.

Rezulak M, Borsuk I, Mruk I. 2016. Natural C-independent expression of restriction endonuclease in a C protein-associated restriction-modification system. *Nucleic Acids Res* **44:** 2646–2660.

Rifat D, Wright NT, Varney KM, Weber DJ, Black LW. 2008. Restriction endonuclease inhibitor IPI* of bacteriophage T4: a novel structure for a dedicated target. *J Mol Biol* **375:** 720–734.

Rimseliene R, Vaisvila R, Janulaitis A. 1995. The eco72IC gene specifies a trans-acting factor which influences expression of both DNA methyltransferase and endonuclease from the Eco72I restriction-modification system. *Gene* **157:** 217–219.

Rimseliene R, Maneliene Z, Lubys A, Janulaitis A. 2003. Engineering of restriction endonucleases: using methylation activity of the bifunctional endonuclease Eco57I to select the mutant with a novel sequence specificity. *J Mol Biol* **327:** 383–391.

Roberts RJ, Cheng X. 1998. Base flipping. *Ann Rev Biochem* **67:** 181–198.

Roberts RJ, Halford SE. 1993. Type II restriction enzymes. In *Nucleases*, 2nd ed. (ed. Linn SM, Lloyd RS, Roberts RJ), pp. 35–88. Cold Spring Harbor Laboratory Press, Cold Spring Harbor, NY.

Roberts RJ, Belfort M, Bestor T, Bhagwat AS, Bickle TA, Bitinaite J, Blumenthal RM, Degtyarev S, Dryden DT, Dybvig K, et al. 2003. A nomenclature for restriction enzymes, DNA methyltransferases, homing endonucleases and their genes. *Nucleic Acids Res* **31:** 1805–1812.

Roberts RJ, Vincze T, Posfai J, Macelis D. 2010. REBASE—a database for DNA restriction and modification: enzymes, genes and genomes. *Nucleic Acids Res* **38:** D234–D236.

Roberts GA, Cooper LP, White JH, Su TJ, Zipprich JT, Geary P, Kennedy C, Dryden DT. 2011. An investigation of the structural requirements for ATP hydrolysis and DNA cleavage by the EcoKI Type I DNA restriction and modification enzyme. *Nucleic Acids Res* **39**: 7667–7676.

Roberts RJ, Vincze T, Pósfai J, Macelis D. 2015. REBASE—a database for DNA restriction and modification: enzymes, genes and genomes. *Nucleic Acids Res* **43**: D298–D299.

Robertson BD, Meyer TF. 1992. Genetic variation in pathogenic bacteria. *Trends Genet* **8**: 422–427.

Rogers JM, Barrera LA, Reyon D, Sander JD, Kellis M, Joung JK, Bulyk ML. 2015. Context influences on TALE-DNA binding revealed by quantitative profiling. *Nat Commun* **6**: 7440. doi:10.1038/ncomms8440.

Rohs R, Jin X, West SM, Joshi R, Honig B, Mann RS. 2010. Origins of specificity in protein-DNA recognition. *Ann Rev Biochem* **79**: 233–269.

Rusling DA, Laurens N, Pernstich C, Wuite GJ, Halford SE. 2012. DNA looping by FokI: the impact of synapse geometry on loop topology at varied site orientations. *Nucleic Acids Res* **40**: 4977–4987.

Rutkauskas D, Petkelyte M, Naujalis P, Sasnauskas G, Tamulaitis G, Zaremba M, Šikšnys V. 2014. Restriction enzyme Ecl18kI-induced DNA looping dynamics by single-molecule FRET. *J Phys Chem B* **118**: 8575–8582.

Ryan KA, Lo RY. 1999. Characterization of a CACAG pentanucleotide repeat in *Pasteurella haemolytica* and its possible role in modulation of a novel type III restriction-modification system. *Nucleic Acids Res* **27**: 1505–1511.

Sanchez-Romero MA, Cota I, Casadesus J. 2015. DNA methylation in bacteria: from the methyl group to the methylome. *Curr Opin Microbiol* **25**: 9–16.

Sanders KL, Catto LE, Bellamy SR, Halford SE. 2009. Targeting individual subunits of the FokI restriction endonuclease to specific DNA strands. *Nucleic Acids Res* **37**: 2105–2115.

Sapienza PJ, Dela Torre CA, McCoy WHt, Jana SV, Jen-Jacobson L. 2005. Thermodynamic and kinetic basis for the relaxed DNA sequence specificity of "promiscuous" mutant EcoRI endonucleases. *J Mol Biol* **348**: 307–324.

Sapienza PJ, Rosenberg JM, Jen-Jacobson L. 2007. Structural and thermodynamic basis for enhanced DNA binding by a promiscuous mutant EcoRI endonuclease. *Structure* **15**: 1368–1382.

Sapienza PJ, Niu T, Kurpiewski MR, Grigorescu A, Jen-Jacobson L. 2014. Thermodynamic and structural basis for relaxation of specificity in protein-DNA recognition. *J Mol Biol* **426**: 84–104.

Sapranauskas R, Sasnauskas G, Lagunavicius A, Vilkaitis G, Lubys A, Šikšnys V. 2000. Novel subtype of type IIs restriction enzymes. BfiI endonuclease exhibits similarities to the EDTA-resistant nuclease Nuc of *Salmonella typhimurium*. *J Biol Chem* **275**: 30878–30885.

Saravanan M, Bujnicki JM, Cymerman IA, Rao DN, Nagaraja V. 2004. Type II restriction endonuclease R.KpnI is a member of the HNH nuclease superfamily. *Nucleic Acids Res* **32**: 6129–6135.

Saravanan M, Vasu K, Ghosh S, Nagaraja V. 2007a. Dual role for Zn^{2+} in maintaining structural integrity and inducing DNA sequence specificity in a promiscuous endonuclease. *J Biol Chem* **282**: 32320–32326.

Saravanan M, Vasu K, Kanakaraj R, Rao DN, Nagaraja V. 2007b. R.KpnI, an HNH superfamily REase, exhibits differential discrimination at non-canonical sequences in the presence of Ca^{2+} and Mg^{2+}. *Nucleic Acids Res* **35**: 2777–2786.

Sarrade-Loucheur A, Xu SY, Chan SH. 2013. The role of the methyltransferase domain of bifunctional restriction enzyme RM.BpuSI in cleavage activity. *PloS One* **8**: e80967. doi:10.1371/journal.pone.0080967.

Sasnauskas G, Halford SE, Šikšnys V. 2003. How the BfiI restriction enzyme uses one active site to cut two DNA strands. *Proc Natl Acad Sci* **100:** 6410–6415.

Sasnauskas G, Connolly BA, Halford SE, Šikšnys V. 2007. Site-specific DNA transesterification catalyzed by a restriction enzyme. *Proc Natl Acad Sci* **104:** 2115–2120.

Sasnauskas G, Connolly BA, Halford SE, Šikšnys V. 2008. Template-directed addition of nucleosides to DNA by the BfiI restriction enzyme. *Nucleic Acids Res* **36:** 3969–3977.

Sasnauskas G, Zakrys L, Zaremba M, Cosstick R, Gaynor JW, Halford SE, Šikšnys V. 2010. A novel mechanism for the scission of double-stranded DNA: BfiI cuts both 3′–5′ and 5′–3′ strands by rotating a single active site. *Nucleic Acids Res* **38:** 2399–2410.

Sasnauskas G, Kostiuk G, Tamulaitis G, Šikšnys V. 2011. Target site cleavage by the monomeric restriction enzyme BcnI requires translocation to a random DNA sequence and a switch in enzyme orientation. *Nucleic Acids Res* **39:** 8844–8856.

Sasnauskas G, Tamulaitienė G, Tamulaitis G, Calyseva J, Laime M, Šikšnys V. 2015a. Biochemical and structural characterization of a monomeric Type IIE REase. In *7th NEB meeting on restriction and modification (Gdansk).* Short talk S1.

Sasnauskas G, Zagorskaite E, Kauneckaite K, Tamulaitienė G, Šikšnys V. 2015b. Structure-guided sequence specificity engineering of the modification-dependent restriction endonuclease LpnPI. *Nucleic Acids Res* **43:** 6144–6155.

Sasnauskas G, Tamulaitienė G, Tamulaitis G, Calyseva J, Laime M, Rimseliene R, Lubys A, Šikšnys V. 2017. UbaLAI is a monomeric Type IIE restriction enzyme. *Nucleic Acids Res* **45:** 9583–9594.

Sawaya MR, Zhu Z, Mersha F, Chan SH, Dabur R, Xu SY, Balendiran GK. 2005. Crystal structure of the restriction-modification system control element C.BclI and mapping of its binding site. *Structure* **13:** 1837–1847.

Schierling B, Noel AJ, Wende W, Hien le T, Volkov E, Kubareva E, Oretskaya T, Kokkinidis M, Rompp A, Spengler B, et al. 2010. Controlling the enzymatic activity of a restriction enzyme by light. *Proc Natl Acad Sci* **107:** 1361–1366.

Schierling B, Dannemann N, Gabsalilow L, Wende W, Cathomen T, Pingoud A. 2012. A novel zinc-finger nuclease platform with a sequence-specific cleavage module. *Nucleic Acids Res* **40:** 2623–2638.

Schwarz FW, Toth J, van Aelst K, Cui G, Clausing S, Szczelkun MD, Seidel R. 2013. The helicase-like domains of type III restriction enzymes trigger long-range diffusion along DNA. *Science* **340:** 353–356.

Schwarz FW, van Aelst K, Toth J, Seidel R, Szczelkun MD. 2011. DNA cleavage site selection by Type III restriction enzymes provides evidence for head-on protein collisions following 1D bidirectional motion. *Nucleic Acids Res* **39:** 8042–8051.

Sears LE, Zhou B, Aliotta JM, Morgan RD, Kong H. 1996. *Bae*I, another unusual *Bcg*I-like restriction endonuclease. *Nucleic Acids Res* **24:** 3590–3592.

Seib KL, Peak IR, Jennings MP. 2002. Phase variable restriction-modification systems in *Moraxella catarrhalis*. *FEMS Immunol Med Microbiol* **32:** 159–165.

Seidel R, Dekker C. 2007. Single-molecule studies of nucleic acid motors. *Curr Opin Struct Biol* **17:** 80–86.

Seidel R, van Noort J, van der Scheer C, Bloom JG, Dekker NH, Dutta CF, Blundell A, Robinson T, Firman K, Dekker C. 2004. Real-time observation of DNA translocation by the type I restriction modification enzyme EcoR124I. *Nat Struct Mol Biol* **11:** 838–843.

Seidel R, Bloom JG, van Noort J, Dutta CF, Dekker NH, Firman K, Szczelkun MD, Dekker C. 2005. Dynamics of initiation, termination and reinitiation of DNA translocation by the motor protein EcoR124I. *EMBO J* **24:** 4188–4197.

Seidel R, Bloom JG, Dekker C, Szczelkun MD. 2008. Motor step size and ATP coupling efficiency of the dsDNA translocase EcoR124I. *EMBO J* **27:** 1388–1398.

Semenova E, Minakhin L, Bogdanova E, Nagornykh M, Vasilov A, Heyduk T, Solonin A, Zakharova M, Severinov K. 2005. Transcription regulation of the EcoRV restriction-modification system. *Nucleic Acids Res* **33:** 6942–6951.

Serfiotis-Mitsa D, Roberts GA, Cooper LP, White JH, Nutley M, Cooper A, Blakely GW, Dryden DT. 2008. The Orf18 gene product from conjugative transposon Tn*916* is an ArdA antirestriction protein that inhibits type I DNA restriction-modification systems. *J Mol Biol* **383:** 970–981.

Serfiotis-Mitsa D, Herbert AP, Roberts GA, Soares DC, White JH, Blakely GW, Uhrin D, Dryden DT. 2010. The structure of the KlcA and ArdB proteins reveals a novel fold and antirestriction activity against Type I DNA restriction systems in vivo but not in vitro. *Nucleic Acids Res* **38:** 1723–1737.

Shao C, Wang C, Zang J. 2014. Structural basis for the substrate selectivity of PvuRts1I, a 5-hydroxymethylcytosine DNA restriction endonuclease. *Acta Crystallogr, Sect D: Biol Crystallogr* **70:** 2477–2486.

Shen BW, Heiter DF, Chan SH, Wang H, Xu SY, Morgan RD, Wilson GG, Stoddard BL. 2010. Unusual target site disruption by the rare-cutting HNH restriction endonuclease PacI. *Structure* **18:** 734–743.

Shen BW, Xu D, Chan SH, Zheng Y, Zhu Z, Xu SY, Stoddard BL. 2011. Characterization and crystal structure of the type IIG restriction endonuclease RM.BpuSI. *Nucleic Acids Res* **39:** 8223–8236.

Shen BW, Heiter DF, Lunnen KD, Wilson GG, Stoddard BL. 2015. Crystal structure of a 8-bp REase, SwaI. In *7th NEB meeting on restriction and modification (Gdansk): Poster P3.*

Shen BW, Heiter DF, Lunnen KD, Wilson GG, Stoddard BL. 2017. DNA recognition by the SwaI restriction endonuclease involves unusual distortion of an 8 base pair A:T-rich target. *Nucleic Acids Res* **45:** 1516–1528.

Shilov I, Tashlitsky V, Khodoun M, Vasil'ev S, Alekseev Y, Kuzubov A, Kubareva E, Karyagina A. 1998. DNA-methyltransferase SsoII interaction with own promoter region binding site. *Nucleic Acids Res* **26:** 2659–2664.

Shlyakhtenko LS, Gilmore J, Portillo A, Tamulaitis G, Šikšnys V, Lyubchenko YL. 2007. Direct visualization of the EcoRII-DNA triple synaptic complex by atomic force microscopy. *Biochemistry* **46:** 11128–11136.

Simoncsits A, Tjornhammar ML, Rasko T, Kiss A, Pongor S. 2001. Covalent joining of the subunits of a homodimeric type II restriction endonuclease: single-chain PvuII endonuclease. *J Mol Biol* **309:** 89–97.

Simons M, Szczelkun MD. 2011. Recycling of protein subunits during DNA translocation and cleavage by Type I restriction-modification enzymes. *Nucleic Acids Res* **39:** 7656–7666.

Simons M, Diffin FM, Szczelkun MD. 2014. ClpXP protease targets long-lived DNA translocation states of a helicase-like motor to cause restriction alleviation. *Nucleic Acids Res* **42:** 12082–12091.

Singleton MR, Dillingham MS, Wigley DB. 2007. Structure and mechanism of helicases and nucleic acid translocases. *Ann Rev Biochem* **76:** 23–50.

Sisakova E, Stanley LK, Weiserova M, Szczelkun MD. 2008a. A RecB-family nuclease motif in the Type I restriction endonuclease EcoR124I. *Nucleic Acids Res* **36:** 3939–3949.

Sisakova E, Weiserova M, Dekker C, Seidel R, Szczelkun MD. 2008b. The interrelationship of helicase and nuclease domains during DNA translocation by the molecular motor EcoR124I. *J Mol Biol* **384:** 1273–1286.

Sisakova E, van Aelst K, Diffin FM, Szczelkun MD. 2013. The Type ISP restriction-modification enzymes LlaBIII and LlaGI use a translocation-collision mechanism to cleave non-specific DNA distant from their recognition sites. *Nucleic Acids Res* **41**: 1071–1080.

Šikšnys V, Gražulis S, Huber R. 2004. Structure and function of the tetrameric restriction enzymes. In *Restriction endonucleases* (ed. Pingoud A), pp. 237–259. Springer, Berlin.

Šikšnys V, Skirgaila R, Sasnauskas G, Urbanke C, Cherny D, Grazulis S, Huber R. 1999. The *Cfr*10I restriction enzyme is functional as a tetramer. *J Mol Biol* **291**: 1105–1118.

Sitaraman R, Dybvig K. 1997. The hsd loci of *Mycoplasma pulmonis*: organization, rearrangements and expression of genes. *Mol Microbioly* **26**: 109–120.

Siwek W, Czapinska H, Bochtler M, Bujnicki JM, Skowronek K. 2012. Crystal structure and mechanism of action of the N6-methyladenine-dependent type IIM restriction endonuclease R.DpnI. *Nucleic Acids Res* **40**: 7563–7572.

Skirgaila R, Šikšnys V. 1998. Ca^{2+}-ions stimulate DNA binding specificity of Cfr10I restriction enzyme. *Biol Chem* **379**: 595–598.

Skirgaila R, Grazulis S, Bozic D, Huber R, Šikšnys V. 1998. Structure-based redesign of the catalytic/metal binding site of Cfr10I restriction endonuclease reveals importance of spatial rather than sequence conservation of active centre residues. *J Mol Biol* **279**: 473–481.

Skoglund A, Bjorkholm B, Nilsson C, Andersson AF, Jernberg C, Schirwitz K, Enroth C, Krabbe M, Engstrand L. 2007. Functional analysis of the M.HpyAIV DNA methyltransferase of *Helicobacter pylori*. *J Bacteriol* **189**: 8914–8921.

Skowron P, Kaczorowski T, Tucholski J, Podhajska AJ. 1993. Atypical DNA-binding properties of class-IIS restriction endonucleases: evidence for recognition of the cognate sequence by a *Fok*I monomer. *Gene* **125**: 1–10.

Skowron PM, Harasimowicz R, Rutkowska SM. 1996. GCN4 eukaryotic transcription factor/ *Fok*I endonuclease-mediated 'Achilles' heel cleavage': quantitative study of protein-DNA interaction. *Gene* **170**: 1–8.

Skowron PM, Majewski J, Zylicz-Stachula A, Rutkowska SM, Jaworowska I, Harasimowicz-Slowinska RI. 2003. A new *Thermus* sp. class-IIS enzyme sub-family: isolation of a 'twin' endonuclease TspDTI with a novel specificity 5'-ATGAA-3', related to TspGWI, TaqII and Tth111II. *Nucleic Acids Res* **31**: e74. PMC167652.

Smith RM, Diffin FM, Savery NJ, Josephsen J, Szczelkun MD. 2009a. DNA cleavage and methylation specificity of the single polypeptide restriction-modification enzyme LlaGI. *Nucleic Acids Res* **37**: 7206–7218.

Smith RM, Josephsen J, Szczelkun MD. 2009b. An Mrr-family nuclease motif in the single polypeptide restriction-modification enzyme LlaGI. *Nucleic Acids Res* **37**: 7231–7238.

Smith RM, Josephsen J, Szczelkun MD. 2009c. The single polypeptide restriction-modification enzyme LlaGI is a self-contained molecular motor that translocates DNA loops. *Nucleic Acids Res* **37**: 7219–7230.

Smith RM, Jacklin AJ, Marshall JJ, Sobott F, Halford SE. 2013a. Organization of the BcgI restriction-modification protein for the transfer of one methyl group to DNA. *Nucleic Acids Res* **41**: 405–417.

Smith RM, Marshall JJ, Jacklin AJ, Retter SE, Halford SE, Sobott F. 2013b. Organization of the BcgI restriction-modification protein for the cleavage of eight phosphodiester bonds in DNA. *Nucleic Acids Res* **41**: 391–404.

Smith RM, Pernstich C, Halford SE. 2014. TstI, a Type II restriction-modification protein with DNA recognition, cleavage and methylation functions in a single polypeptide. *Nucleic Acids Res* **42**: 5809–5822.

Sohail A, Ives CL, Brooks JE. 1995. Purification and characterization of C.BamHI, a regulator of the BamHI restriction-modification system. *Gene* **157:** 227–228.

Sokolowska M, Kaus-Drobek M, Czapinska H, Tamulaitis G, Šikšnys V, Bochtler M. 2007a. Restriction endonucleases that resemble a component of the bacterial DNA repair machinery. *Cell Mol Life Sci* **64:** 2351–2357.

Sokolowska M, Kaus-Drobek M, Czapinska H, Tamulaitis G, Szczepanowski RH, Urbanke C, Šikšnys V, Bochtler M. 2007b. Monomeric restriction endonuclease BcnI in the apo form and in an asymmetric complex with target DNA. *J Mol Biol* **369:** 722–734.

Sokolowska M, Czapinska H, Bochtler M. 2009. Crystal structure of the ββα-Me type II restriction endonuclease Hpy99I with target DNA. *Nucleic Acids Res* **37:** 3799–3810.

Sokolowska M, Czapinska H, Bochtler M. 2011. Hpy188I–DNA pre- and post-cleavage complexes—snapshots of the GIY-YIG nuclease mediated catalysis. *Nucleic Acids Res* **39:** 1554–1564.

Srikhanta YN, Maguire TL, Stacey KJ, Grimmond SM, Jennings MP. 2005. The phasevarion: a genetic system controlling coordinated, random switching of expression of multiple genes. *Proc Natl Acad Sci* **102:** 5547–5551.

Srikhanta YN, Fox KL, Jennings MP. 2010. The phasevarion: phase variation of type III DNA methyltransferases controls coordinated switching in multiple genes. *Nat Rev Microbiol* **8:** 196–206.

Stahl F, Wende W, Jeltsch A, Pingoud A. 1996. Introduction of asymmetry in the naturally symmetric restriction endonuclease EcoRV to investigate intersubunit communication in the homodimeric protein. *Proc Natl Acad Sci* **93:** 6175–6180.

Stanley LK, Seidel R, van der Scheer C, Dekker NH, Szczelkun MD, Dekker C. 2006. When a helicase is not a helicase: dsDNA tracking by the motor protein EcoR124I. *EMBO J* **25:** 2230–2239.

Steczkiewicz K, Muszewska A, Knizewski L, Rychlewski L, Ginalski K. 2012. Sequence, structure and functional diversity of PD-(D/E)XK phosphodiesterase superfamily. *Nucleic Acids Res* **40:** 7016–7045.

Stephanou AS, Roberts GA, Cooper LP, Clarke DJ, Thomson AR, MacKay CL, Nutley M, Cooper A, Dryden DT. 2009a. Dissection of the DNA mimicry of the bacteriophage T7 Ocr protein using chemical modification. *J Mol Biol* **391:** 565–576.

Stephanou AS, Roberts GA, Tock MR, Pritchard EH, Turkington R, Nutley M, Cooper A, Dryden DT. 2009b. A mutational analysis of DNA mimicry by ocr, the gene 0.3 antirestriction protein of bacteriophage T7. *Biochem Biophys Res Commun* **378:** 129–132.

Stephenson FH, Ballard BT, Boyer HW, Rosenberg JM, Greene PJ. 1989. Comparison of the nucleotide and amino acid sequences of the RsrI and EcoRI restriction endonucleases. *Gene* **85:** 1–13.

Stern A, Sorek R. 2011. The phage-host arms race: shaping the evolution of microbes. *Bioessays* **33:** 43–51.

Stewart FJ, Raleigh EA. 1998. Dependence of McrBC cleavage on distance between recognition elements. *Biol Chem* **379:** 611–616.

Stewart FJ, Panne D, Bickle TA, Raleigh EA. 2000. Methyl-specific DNA binding by McrBC, a modification-dependent restriction enzyme. *J Mol Biol* **298:** 611–622.

Stoddard BL. 2005. Homing endonuclease structure and function. *Q. Rev Biophys* **38:** 49–95.

Stower H. 2014. Epigenetics: reprogramming with TET. *Nat Rev Genet* **15:** 66. doi:10.1038/nrg3659.

Streeter SD, Papapanagiotou I, McGeehan JE, Kneale GG. 2004. DNA footprinting and bio-physical characterization of the controller protein C.AhdI suggests the basis of a genetic switch. *Nucleic Acids Res* **32:** 6445–6453.

Studier FW. 1975. Gene 0.3 of bacteriophage T7 acts to overcome the DNA restriction system of the host. *J Mol Biol* **94:** 283–295.

Studier FW. 2013. Phage T7 is neither modified nor restircted by EcoKI or EcoBI, which led to the discovery of the T7 *0.3* gene encoding Ocr, as well as the collision model for Type I enzymes. http://library.cshl.edu/Meetings/restriction-enzymes/v-Studier.php.

Studier FW, Movva NR. 1976. SAMase gene of bacteriophage T3 is responsible for overcoming host restriction. *J Virol* **19:** 136–145.

Su TJ, Tock MR, Egelhaaf SU, Poon WC, Dryden DT. 2005. DNA bending by M.EcoKI methyltransferase is coupled to nucleotide flipping. *Nucleic Acids Res* **33:** 3235–3244.

Sudina AE, Zatsepin TS, Pingoud V, Pingoud A, Oretskaya TS, Kubareva EA. 2005. Affinity modification of the restriction endonuclease *Sso*II by 2′-aldehyde-containing double stranded DNAs. *Biochem Biokhim* **70:** 941–947.

Sugisaki H, Kanazawa S. 1981. New restriction endonucleases from *Flavobacterium okeanokoites* (*Fok*I) and *Micrococcus luteus* (*Mlu*I). *Gene* **16:** 73–78.

Sugisaki H, Kita K, Takanami M. 1989. The *Fok*I restriction-modification system. II. Presence of two domains in *Fok*I methylase responsible for modification of different DNA strands. *J Biol Chem* **264:** 5757–5761.

Sukackaite R, Gražulis S, Tamulaitis G, Šikšnys V. 2012. The recognition domain of the methyl-specific endonuclease McrBC flips out 5-methylcytosine. *Nucleic Acids Res* **40:** 7552–7562.

Sutherland E, Coe L, Raleigh EA. 1992. McrBC: a multisubunit GTP-dependent restriction endonuclease. *J Mol Biol* **225:** 327–348.

Szczelkun MD. 2011. Translocation, switching and gating: potential roles for ATP in long-range communication on DNA by Type III restriction endonucleases. *Biochem Soc Trans* **39:** 589–594.

Szczelkun MD. 2013. Roles for helicases as ATP-dependent molecular switches. *Adv Exp Med Biol* **767:** 225–244.

Szczelkun MD, Friedhoff P, Seidel R. 2010. Maintaining a sense of direction during long-range communication on DNA. *Biochem Soc Trans* **38:** 404–409.

Szczepanowski RH, Carpenter MA, Czapinska H, Zaremba M, Tamulaitis G, Šikšnys V, Bhagwat AS, Bochtler M. 2008. Central base pair flipping and discrimination by PspGI. *Nucleic Acids Res* **36:** 6109–6117.

Szczepek M, Brondani V, Buchel J, Serrano L, Segal DJ, Cathomen T. 2007. Structure-based redesign of the dimerization interface reduces the toxicity of zinc-finger nucleases. *Nat Biotechnol* **25:** 786–793.

Szczepek M, Mackeldanz P, Moncke-Buchner E, Alves J, Kruger DH, Reuter M. 2009. Molecular analysis of restriction endonuclease EcoRII from *Escherichia coli* reveals precise regulation of its enzymatic activity by autoinhibition. *Mol Microbiol* **72:** 1011–1021.

Szwagierczak A, Brachmann A, Schmidt CS, Bultmann S, Leonhardt H, Spada F. 2011. Characterization of PvuRts1I endonuclease as a tool to investigate genomic 5-hydroxymethyl-cytosine. *Nucleic Acids Res* **39:** 5149–5156.

Szybalski W, Kim SC, Hasan N, Podhajska AJ. 1991. Class-IIS restriction enzymes—a review. *Gene* **100:** 13–26.

Tahiliani M, Koh KP, Shen Y, Pastor WA, Bandukwala H, Brudno Y, Agarwal S, Iyer LM, Liu DR, Aravind L, et al. 2009. Conversion of 5-methylcytosine to 5-hydroxymethylcytosine in mammalian DNA by MLL partner TET1. *Science* **324:** 930–935.

Tamulaitienė G, Silanskas A, Grazulis S, Zaremba M, Siksnys V. 2014. Crystal structure of the R-protein of the multisubunit ATP-dependent restriction endonuclease NgoAVII. *Nucleic Acids Res* **42:** 14022–14030.

Tamulaitienė G, Jovaisaite V, Tamulaitis G, Songailiene I, Manakova E, Zaremba M, Gražulis S, Xu SY, Šikšnys V. 2017. Restriction endonuclease AgeI is a monomer which dimerizes to cleave DNA. *Nucleic Acids Res* **45:** 3547–3558.

Tamulaitis G, Mucke M, Šikšnys V. 2006a. Biochemical and mutational analysis of EcoRII functional domains reveals evolutionary links between restriction enzymes. *FEBS Lett* **580:** 1665–1671.

Tamulaitis G, Sasnauskas G, Mucke M, Šikšnys V. 2006b. Simultaneous binding of three recognition sites is necessary for a concerted plasmid DNA cleavage by EcoRII restriction endonuclease. *J Mol Biol* **358:** 406–419.

Tamulaitis G, Zaremba M, Szczepanowski RH, Bochtler M, Šikšnys V. 2007. Nucleotide flipping by restriction enzymes analyzed by 2-aminopurine steady-state fluorescence. *Nucleic Acids Res* **35:** 4792–4799.

Tamulaitis G, Zaremba M, Szczepanowski RH, Bochtler M, Šikšnys V. 2008. How PspGI, catalytic domain of EcoRII and Ecl18kI acquire specificities for different DNA targets. *Nucleic Acids Res* **36:** 6101–6108.

Tamulaitis G, Rutkauskas M, Zaremba M, Gražulis S, Tamulaitiene G, Šikšnys V. 2015. Functional significance of protein assemblies predicted by the crystal structure of the restriction endonuclease BsaWI. *Nucleic Acids Res* **43:** 8100–8110.

Tan A, Atack JM, Jennings MP, Seib KL. 2016. The capricious nature of bacterial pathogens: phasevarions and vaccine development. *Front Immunol* **7:** 586. PMC5149525.

Tao T, Blumenthal RM. 1992. Sequence and characterization of pvuIIR, the PvuII endonuclease gene, and of pvuIIC, its regulatory gene. *J Bacteriol* **174:** 3395–3398.

Tao T, Bourne JC, Blumenthal RM. 1991. A family of regulatory genes associated with type II restriction-modification systems. *J Bacteriol* **173:** 1367–1375.

Tautz N, Kaluza K, Frey B, Jarsch M, Schmitz GG, Kessler C. 1990. SgrAI, a novel class-II restriction endonuclease from *Streptomyces griseus* recognizing the octanucleotide sequence 5′-CR/CCGGYG-3′[corrected]. *Nucleic Acids Res* **18:** 3087. PMCID: 330872.

Taylor IA, Davis KG, Watts D, Kneale GG. 1994. DNA-binding induces a major structural transition in a type I methyltransferase. *EMBO J* **13:** 5772–5778.

Thanisch K, Schneider K, Morbitzer R, Solovei I, Lahaye T, Bultmann S, Leonhardt H. 2014. Targeting and tracing of specific DNA sequences with dTALEs in living cells. *Nucleic Acids Res* **42:** e38. doi:10.1093/nar/gkt1348.

The 7th NEB Meeting on DNA Restriction and Modification, August 24–29, 2015. Uniwersytet Gdanski ulica, Kladki 24, Gdańsk, Poland.

Thomas CB, Gumport RI. 2006. Dimerization of the bacterial RsrI N6-adenine DNA methyltransferase. *Nucleic Acids Res* **34:** 806–815.

Thomas AT, Brammar WJ, Wilkins BM. 2003. Plasmid R16 ArdA protein preferentially targets restriction activity of the type I restriction-modification system EcoKI. *J Bacteriol* **185:** 2022–2025.

Tock MR, Dryden DT. 2005. The biology of restriction and anti-restriction. *Curr Opin Microbiol* **8:** 466–472.

Too PH, Zhu Z, Chan SH, Xu SY. 2010. Engineering Nt.BtsCI and Nb.BtsCI nicking enzymes and applications in generating long overhangs. *Nucleic Acids Res* **38:** 1294–1303.

Toth J, Bollins J, Szczelkun MD. 2015. Re-evaluating the kinetics of ATP hydrolysis during initiation of DNA sliding by Type III restriction enzymes. *Nucleic Acids Res* **43:** 10870–10881.

Townson SA, Samuelson JC, Vanamee ES, Edwards TA, Escalante CR, Xu SY, Aggarwal AK. 2004. Crystal structure of BstYI at 1.85A resolution: a thermophilic restriction endonuclease with overlapping specificities to BamHI and BglII. *J Mol Biol* **338:** 725–733.

Townson SA, Samuelson JC, Xu SY, Aggarwal AK. 2005. Implications for switching restriction enzyme specificities from the structure of BstYI bound to a BglII DNA sequence. *Structure* **13:** 791–801.

Tucholski J, Skowron PM, Podhajska AJ. 1995. *Mme*I, a class-IIS restriction endonuclease: purification and characterization. *Gene* **157:** 87–92.

Tucholski J, Zmijewski JW, Podhajska AJ. 1998. Two intertwined methylation activities of the MmeI restriction-modification class-IIS system from *Methylophilus methylotrophus. Gene* **223:** 293–302.

Tuteja N, Tuteja R. 2004. Prokaryotic and eukaryotic DNA helicases. Essential molecular motor proteins for cellular machinery. *Eur J Biochem/FEBS* **271:** 1835–1848.

Tyndall C, Meister J, Bickle TA. 1994. The *Escherichia coli* prr region encodes a functional type IC DNA restriction system closely integrated with an anticodon nuclease gene. *J Mol Biol* **237:** 266–274.

Umate P, Tuteja N, Tuteja R. 2011. Genome-wide comprehensive analysis of human helicases. *Commun Integr Biol* **4:** 118–137.

Urnov FD, Miller JC, Lee YL, Beausejour CM, Rock JM, Augustus S, Jamieson AC, Porteus MH, Gregory PD, Holmes MC. 2005. Highly efficient endogenous human gene correction using designed zinc-finger nucleases. *Nature* **435:** 646–651.

Urnov FD, Rebar EJ, Holmes MC, Zhang HS, Gregory PD. 2010. Genome editing with engineered zinc finger nucleases. *Nat Rev Genet* **11:** 636–646.

Uyen NT, Nishi K, Park SY, Choi JW, Lee HJ, Kim JS. 2008. Crystallization and preliminary X-ray diffraction analysis of the HsdR subunit of a putative type I restriction enzyme from *Vibrio vulnificus* YJ016. *Acta Crystallographica, Sect F: Struct Biol Cryst Commun* **64:** 926–928.

Uyen NT, Park SY, Choi JW, Lee HJ, Nishi K, Kim JS. 2009. The fragment structure of a putative HsdR subunit of a type I restriction enzyme from *Vibrio vulnificus* YJ016: implications for DNA restriction and translocation activity. *Nucleic Acids Res* **37:** 6960–6969.

van Aelst K, Toth J, Ramanathan SP, Schwarz FW, Seidel R, Szczelkun MD. 2010. Type III restriction enzymes cleave DNA by long-range interaction between sites in both head-to-head and tail-to-tail inverted repeat. *Proc Natl Acad Sci* **107:** 9123–9128.

van Aelst K, Saikrishnan K, Szczelkun MD. 2015. Mapping DNA cleavage by the Type ISP restriction-modification enzymes following long-range communication between DNA sites in different orientations. *Nucleic Acids Res* **43:** 10430–10443.

van Belkum A, Scherer S, van Alphen L, Verbrugh H. 1998. Short-sequence DNA repeats in prokaryotic genomes. *Microbiol Mol Biol Rev* **62:** 275–293.

van den Broek B, Vanzi F, Normanno D, Pavone FS, Wuite GJ. 2006. Real-time observation of DNA looping dynamics of Type IIE restriction enzymes NaeI and NarI. *Nucleic Acids Res* **34:** 167–174.

van der Woude MW. 2006. Re-examining the role and random nature of phase variation. *FEMS Microbiol Lett* **254:** 190–197.

van der Woude MW, Baumler AJ. 2004. Phase and antigenic variation in bacteria. *Clin Microbiol Rev* **17:** 581–611.

van Ham SM, van Alphen L, Mooi FR, van Putten JP. 1993. Phase variation of *H. influenzae* fimbriae: transcriptional control of two divergent genes through a variable combined promoter region. *Cell* **73:** 1187–1196.

van Noort J, van der Heijden T, Dutta CF, Firman K, Dekker C. 2004. Initiation of translocation by Type I restriction-modification enzymes is associated with a short DNA extrusion.
Nucleic Acids Res **32:** 6540–6547.

Vanamee ES, Santagata S, Aggarwal AK. 2001. FokI requires two specific DNA sites for cleavage.
J Mol Biol **309:** 69–78.

Vanamee ES, Hsieh P, Zhu Z, Yates D, Garman E, Xu S, Aggarwal AK. 2003. Glucocorticoid
receptor-like Zn(Cys)4 motifs in *Bsl*I restriction endonuclease. *J Mol Biol* **334:** 595–603.

Vanamee ES, Viadiu H, Kucera R, Dorner L, Picone S, Schildkraut I, Aggarwal AK. 2005. A view
of consecutive binding events from structures of tetrameric endonuclease *Sfi*I bound to DNA.
EMBO J **24:** 4198–4208.

Vanamee ES, Berriman J, Aggarwal AK. 2007. An EM view of the FokI synaptic complex by
single particle analysis. *J Mol Biol* **370:** 207–212.

Vanamee ES, Viadiu H, Chan SH, Ummat A, Hartline AM, Xu SY, Aggarwal AK. 2011. Asymmetric DNA recognition by the OkrAI endonuclease, an isoschizomer of BamHI. *Nucleic
Acids Res* **39:** 712–719.

VanderVeen LA, Harris TM, Jen-Jacobson L, Marnett LJ. 2008. Formation of DNA-protein
cross-links between γ-hydroxypropanodeoxyguanosine and EcoRI. *Chem Res Toxicol* **21:**
1733–1738.

Vasu K, Nagaraja V. 2013. Diverse functions of restriction-modification systems in addition to
cellular defense. *Microbiol Mol Biol Rev* **77:** 53–72.

Vasu K, Nagamalleswari E, Nagaraja V. 2012. Promiscuous restriction is a cellular defense strategy that confers fitness advantage to bacteria. *Proc Natl Acad Sci* **109:** E1287–E1293.

Vasu K, Nagamalleswari E, Zahran M, Imhof P, Xu SY, Zhu Z, Chan SH, Nagaraja V. 2013.
Increasing cleavage specificity and activity of restriction endonuclease KpnI. *Nucleic Acids
Res* **41:** 9812–9824.

Veiga H, Pinho MG. 2009. Inactivation of the SauI type I restriction-modification system is not
sufficient to generate *Staphylococcus aureus* strains capable of efficiently accepting foreign
DNA. *Appl Environ Microbiol* **75:** 3034–3038.

Venclovas C, Timinskas A, Šikšnys V. 1994. Five-stranded β-sheet sandwiched with two α-helices: a structural link between restriction endonucleases EcoRI and EcoRV. *Proteins* **20:** 279–
282.

Veron N, Peters AH. 2011. Epigenetics: Tet proteins in the limelight. *Nature* **473:** 293–294.

Viadiu H, Vanamee ES, Jacobson EM, Schildkraut I, Aggarwal AK. 2003. Crystallization of
restriction endonuclease *Sfi*I in complex with DNA. *Acta Crystallogr, Sect D: Biol Crystallogr*
59: 1493–1495.

Vitkute J, Stankevicius K, Tamulaitienė G, Maneliene Z, Timinskas A, Berg DE, Janulaitis A.
2001. Specificities of eleven different DNA methyltransferases of *Helicobacter pylori* strain
26695. *J Bacteriol* **183:** 443–450.

Wah DA, Hirsch JA, Dorner LF, Schildkraut I, Aggarwal AK. 1997. Structure of the multimodular endonuclease FokI bound to DNA. *Nature* **388:** 97–100.

Wah DA, Bitinaite J, Schildkraut I, Aggarwal AK. 1998. Structure of *Fok*I has implications for
DNA cleavage. *Proc Natl Acad Sci* **95:** 10564–10569.

Waite-Rees PA, Keating CJ, Moran LS, Slatko BE, Hornstra LJ, Benner JS. 1991. Characterization and expression of the *Escherichia coli* Mrr restriction system. *J Bacteriol* **173:**
5207–5219.

Waldron DE, Lindsay JA. 2006. SauI: a novel lineage-specific type I restriction-modification system that blocks horizontal gene transfer into *Staphylococcus aureus* and between *S. aureus* isolates of different lineages. *J Bacteriol* **188:** 5578–5585.

Walkinshaw MD, Taylor P, Sturrock SS, Atanasiu C, Berge T, Henderson RM, Edwardson JM, Dryden DT. 2002. Structure of Ocr from bacteriophage T7, a protein that mimics B-form DNA. *Mol Cell* **9:** 187–194.

Wang L, Chen S, Vergin KL, Giovannoni SJ, Chan SW, DeMott MS, Taghizadeh K, Cordero OX, Cutler M, Timberlake S, et al. 2011a. DNA phosphorothioation is widespread and quantized in bacterial genomes. *Proc Natl Acad Sci* **108:** 2963–2968.

Wang H, Guan S, Quimby A, Cohen-Karni D, Pradhan S, Wilson G, Roberts RJ, Zhu Z, Zheng Y. 2011b. Comparative characterization of the PvuRts1I family of restriction enzymes and their application in mapping genomic 5-hydroxymethylcytosine. *Nucleic Acids Res* **39:** 9294–9305.

Ward DF, Murray NE. 1979. Convergent transcription in bacteriophage λ: interference with gene expression. *J Mol Biol* **133:** 249–266.

Warren RA. 1980. Modified bases in bacteriophage DNAs. *Annu Rev Microbiol* **34:** 137–158.

Watanabe M, Yuzawa H, Handa N, Kobayashi I. 2006. Hyperthermophilic DNA methyltransferase M.PabI from the archaeon *Pyrococcus abyssi*. *Appl Environ Microbiol* **72:** 5367–5375.

Waugh DS, Sauer RT. 1993. Single amino acid substitutions uncouple the DNA binding and strand scission activities of *Fok* I endonuclease. *Proc Natl Acad Sci* **90:** 9596–9600.

Waugh DS, Sauer RT. 1994. A novel class of *Fok*I restriction endonuclease mutants that cleave hemi-methylated substrates. *J Biol Chem* **269:** 12298–12303.

Webb B, Sali A. 2016. Comparative protein structure modeling using MODELLER. In *Current protocols in protein science* Vol. 86 pp. 2.9.1–2.9.37. John Wiley & Sons, Hoboken, NJ.

Webb M, Taylor IA, Firman K, Kneale GG. 1995. Probing the domain structure of the type IC DNA methyltransferase M.EcoR124I by limited proteolysis. *J Mol Biol* **250:** 181–190.

Weigele P, Raleigh EA. 2016. Biosynthesis and function of modified bases in bacteria and their viruses. *Chem Rev.* doi:10.1021/acs.chemrev.1026b00114.

Weiser JN, Williams A, Moxon ER. 1990. Phase-variable lipopolysaccharide structures enhance the invasive capacity of *Haemophilus influenzae*. *Infect Immun* **58:** 3455–3457.

Welsh AJ, Halford SE, Scott DJ. 2004. Analysis of Type II restriction endonucleases that interact with two recognition sites. In *Restriction endonucleases* (ed. Pingoud A), pp. 297–317. Springer, Berlin.

Wende W, Stahl F, Pingoud A. 1996. The production and characterization of artificial heterodimers of the restriction endonuclease EcoRV. *Biol Chem* **377:** 625–632.

Wentzell LM, Nobbs TJ, Halford SE. 1995. The SfiI restriction endonuclease makes a four-strand DNA break at two copies of its recognition sequence. *J Mol Biol* **248:** 581–595.

Wilkins BM. 2002. Plasmid promiscuity: meeting the challenge of DNA immigration control. *Environ Microbiol* **4:** 495–500.

Williams K, Savageau MA, Blumenthal RM. 2013. A bistable hysteretic switch in an activator-repressor regulated restriction-modification system. *Nucleic Acids Res* **41:** 6045–6057.

Wilson GG, Murray NE. 1991. Restriction and modification systems. *Annu Rev Genet* **25:** 585–627.

Winkler FK, Banner DW, Oefner C, Tsernoglou D, Brown RS, Heathman SP, Bryan RK, Martin PD, Petratos K, Wilson KS. 1993. The crystal structure of EcoRV endonuclease and of its complexes with cognate and non-cognate DNA fragments. *EMBO J* **12:** 1781–1795.

Winter M. 1997. "Investigation of de novo methylation activity in mutants of the EcoKI methyltransferase." PhD thesis, University of Edinburgh, Edinburgh.

Wion D, Casadesus J. 2006. N6-methyl-adenine: an epigenetic signal for DNA-protein interactions. *Nat Rev Microbiol* **4:** 183–192.

Wolfes H, Alves J, Fliess A, Geiger R, Pingoud A. 1986. Site directed mutagenesis experiments suggest that Glu 111, Glu 144 and Arg 145 are essential for endonucleolytic activity of EcoRI. *Nucleic Acids Res* **14:** 9063–9080.

Wood KM, Daniels LE, Halford SE. 2005. Long-range communications between DNA sites by the dimeric restriction endonuclease SgrAI. *J Mol Biol* **350:** 240–253.

Wright GD. 2010. Antibiotic resistance in the environment: a link to the clinic? *Curr Opin Microbiol* **13:** 589–594.

Wu J, Kandavelou K, Chandrasegaran S. 2007. Custom-designed zinc finger nucleases: what is next? *Cell Mol Life Scis* **64:** 2933–2944.

Wyatt GR, Cohen SS. 1953. The bases of the nucleic acids of some bacterial and animal viruses: the occurrence of 5-hydroxymethylcytosine. *Biochem J* **55:** 774–782.

Wyszomirski KH, Curth U, Alves J, Mackeldanz P, Moncke-Buchner E, Schutkowski M, Kruger DH, Reuter M. 2012. Type III restriction endonuclease EcoP15I is a heterotrimeric complex containing one Res subunit with several DNA-binding regions and ATPase activity. *Nucleic Acids Res* **40:** 3610–3622.

Xiao Y, Jung C, Marx AD, Winkler I, Wyman C, Lebbink JH, Friedhoff P, Cristovao M. 2011. Generation of DNA nanocircles containing mismatched bases. *BioTechniques* **51:** p259–262, 264–255.

Xu Q, Morgan RD, Roberts RJ, Blaser MJ. 2000a. Identification of type II restriction and modification systems in *Helicobacter pylori* reveals their substantial diversity among strains. *Proc Natl Acad Sci* **97:** 9671–9676.

Xu Q, Stickel S, Roberts RJ, Blaser MJ, Morgan RD. 2000b. Purification of the novel endonuclease, Hpy188I, and cloning of its restriction-modification genes reveal evidence of its horizontal transfer to the *Helicobacter pylori* genome. *J Biol Chem* **275:** 17086–17093.

Xu SY, Zhu Z, Zhang P, Chan SH, Samuelson JC, Xiao J, Ingalls D, Wilson GG. 2007. Discovery of natural nicking endonucleases Nb.BsrDI and Nb.BtsI and engineering of top-strand nicking variants from BsrDI and BtsI. *Nucleic Acids Res* **35:** 4608–4618.

Xu T, Liang J, Chen S, Wang L, He X, You D, Wang Z, Li A, Xu Z, Zhou X, et al. 2009. DNA phosphorothioation in *Streptomyces lividans*: mutational analysis of the dnd locus. *BMC Microbiol* **9:** 41 doi:10.1186/1471-2180-1189-1141.

Xu SY, Corvaglia AR, Chan SH, Zheng Y, Linder P. 2011. A type IV modification-dependent restriction enzyme SauUSI from *Staphylococcus aureus* subsp. aureus USA300. *Nucleic Acids Res* **39:** 5597–5610.

Xu S-y, Gidwani S, Heiter D. 2015. Rearranging the subdomains in the BsaXI S subunit to generate a new specificity. In *7th NEB meeting on restriction and modification (Gdansk)*: Poster P34.

Xu SY, Klein P, Degtyarev S, Roberts RJ. 2016. Expression and purification of the modification-dependent restriction enzyme BisI and its homologous enzymes. *Sci Rep* **6:** 28579. doi:10.1038/srep28579.

Yang Z, Horton JR, Maunus R, Wilson GG, Roberts RJ, Cheng X. 2005. Structure of HinP1I endonuclease reveals a striking similarity to the monomeric restriction enzyme MspI. *Nucleic Acids Res* **33:** 1892–1901.

Yanik M, Alzubi J, Lahaye T, Cathomen T, Pingoud A, Wende W. 2013. TALE-PvuII fusion proteins—novel tools for gene targeting. *PloS One* **8:** e82539. doi:10.1371/journal.pone .0082539.

Yonezawa A, Sugiura Y. 1994. DNA binding mode of class-IIS restriction endonuclease *Fok*I revealed by DNA footprinting analysis. *Biochim Biophys Acta* **1219:** 369–379.

Yuan R, Hamilton DL, Burckhardt J. 1980. DNA translocation by the restriction enzyme from *E. coli* K. *Cell* **20**: 237–244.

Yunusova AK, Rogulin EA, Artyukh RI, Zheleznaya LA, Matvienko NI. 2006. Nickase and a protein encoded by an open reading frame downstream from the nickase BspD6I gene form a restriction endonuclease complex. *Biochem Biokhim* **71**: 815–820.

Zabeau M, Friedman S, Van Montagu M, Schell J. 1980. The *ral* gene of phage λ. I. Identification of a non-essential gene that modulates restriction and modification in *E. coli*. *Mol Gen Genet* **179**: 63–73.

Zagorskaite E, Sasnauskas G. 2014. Chemical display of pyrimidine bases flipped out by modification-dependent restriction endonucleases of MspJI and PvuRts1I families. *PloS One* **9**: e114580. doi:10.1371/journal.pone.0114580.

Zaleski P, Wojciechowski M, Piekarowicz A. 2005. The role of Dam methylation in phase variation of *Haemophilus influenzae* genes involved in defence against phage infection. *Microbiology* **151**: 3361–3369.

Zaremba M, Urbanke C, Halford SE, Šikšnys V. 2004. Generation of the BfiI restriction endonuclease from the fusion of a DNA recognition domain to a non-specific nuclease from the phospholipase D superfamily. *J Mol Biol* **336**: 81–92.

Zaremba M, Sasnauskas G, Urbanke C, Šikšnys V. 2005. Conversion of the tetrameric restriction endonuclease Bse634I into a dimer: oligomeric structure-stability-function correlations. *J Mol Biol* **348**: 459–478.

Zaremba M, Sasnauskas G, Urbanke C, Šikšnys V. 2006. Allosteric communication network in the tetrameric restriction endonuclease Bse634I. *J Mol Biol* **363**: 800–812.

Zaremba M, Owsicka A, Tamulaitis G, Sasnauskas G, Shlyakhtenko LS, Lushnikov AY, Lyubchenko YL, Laurens N, van den Broek B, Wuite GJ, et al. 2010. DNA synapsis through transient tetramerization triggers cleavage by Ecl18kI restriction enzyme. *Nucleic Acids Res* **38**: 7142–7154.

Zaremba M, Sasnauskas G, Šikšnys V. 2012. The link between restriction endonuclease fidelity and oligomeric state: a study with Bse634I. *FEBS Lett* **586**: 3324–3329.

Zaremba M, Toliusis P, Grigaitis R, Manakova E, Silanskas A, Tamulaitienė G, Szczelkun MD, Šikšnys V. 2014. DNA cleavage by CgII and NgoAVII requires interaction between N- and R-proteins and extensive nucleotide hydrolysis. *Nucleic Acids Res* **42**: 13887–13896.

Zaremba M, Toliusis P, Silanskas A, Manakova E, Szczelkun M, Šikšnys V. 2015. DNA cleavage by CgII requires assembly of a heterotetramer of R- and H proteins and extensive ATP hydrolysis. *7th NEB meeting on restriction and modification (Gdansk)*: Short talk S8.

Zavil'gel'skii GB. 2000. [Antirestriction]. *Mol Biol* **34**: 854–862.

Zavil'gel'skii GB, Rastorguev SM. 2009. [Antirestriction proteins ardA and Ocr as effective inhibitors of the type I restriction-modification enzymes]. *Mol Biol* **43**: 264–273.

Zavil'gel'skii GB, Kotova V, Rastorguev SM. 2009. [Antirestriction and antimodification activities of the T7 Ocr protein: effect of mutations in interface]. *Mol Biol* **43**: 103–110.

Zavil'gel'skii GV, Kotova V, Rastorguev SM. 2011. [Antimodification activity of the ArdA and Ocr proteins]. *Genetika* **47**: 159–167.

Zhang M, Huang J, Deng M, Weng X, Ma H, Zhou X. 2009. Sensitive and visual detection of adenosine by a rationally designed FokI-based biosensing strategy. *Chemistry Asian J* **4**: 1420–1423.

Zheleznaya LA, Kainov DE, Yunusova AK, Matvienko NI. 2003. Regulatory C protein of the EcoRV modification-restriction system. *Biochem Biokhim* **68**: 105–110.

Zheng Y, Cohen-Karni D, Xu D, Chin HG, Wilson G, Pradhan S, Roberts RJ. 2010. A unique family of Mrr-like modification-dependent restriction endonucleases. *Nucleic Acids Res* **38:** 5527–5534.

Zhou EX, Reuter M, Meehan EJ, Chen L. 2002. A new crystal form of restriction endonuclease EcoRII that diffracts to 2.8 Å resolution. *Acta Crystallogr, Sect D: Biol Crystallogr* **58:** 1343–1345.

Zhou XE, Wang Y, Reuter M, Mackeldanz P, Kruger DH, Meehan EJ, Chen L. 2003. A single mutation of restriction endonuclease EcoRII led to a new crystal form that diffracts to 2.1 Å resolution. *Acta Crystallogr, Sect D: Biol Crystallogr* **59:** 910–912.

Zhou XE, Wang Y, Reuter M, Mucke M, Kruger DH, Meehan EJ, Chen L. 2004. Crystal structure of type IIE restriction endonuclease EcoRII reveals an autoinhibition mechanism by a novel effector-binding fold. *J Mol Biol* **335:** 307–319.

Zhou X, He X, Liang J, Li A, Xu T, Kieser T, Helmann JD, Deng Z. 2005. A novel DNA modification by sulphur. *Mol Microbiol* **57:** 1428–1438.

Zylicz-Stachula A, Bujnicki JM, Skowron PM. 2009. Cloning and analysis of a bifunctional methyltransferase/restriction endonuclease TspGWI, the prototype of a *Thermus* sp. enzyme family. *BMC Mol Biol* **10:** 52. doi:10.1186/1471-2199-1110-1152.

Zylicz-Stachula A, Jezewska-Frackowiak J, Skowron PM. 2014. Cofactor analogue-induced chemical reactivation of endonuclease activity in a DNA cleavage/methylation deficient TspGWI N(4)(7)(3)A variant in the NPPY motif. *Mol Biol Rep* **41:** 2313–2323.

WWW RESOURCES

http://library.cshl.edu/Meetings/restriction-enzymes/v-GeoffWilson.php Wilson G. 2013. The cloning efforts at NEB.

http://library.cshl.edu/Meetings/restriction-enzymes/v-Janulaitis.php Janulaitis A. 2013. Science and politics: three phases of commercialization at Fermentas.

http://library.cshl.edu/Meetings/restriction-enzymes/v-Lubys.php Lubys A. 2013. The cloning efforts at Fermentas.

http://library.cshl.edu/Meetings/restriction-enzymes/v-Roberts.php Roberts R. 2013. Many more REases at CSHL, the start of REBASE and more recent work.

http://library.cshl.edu/Meetings/restriction-enzymes/v-Studier.php Studier FW. 2013. Phage T7 is neither modified nor restricted by EcoKI or EcoBI, which led to the discovery of the T7 *0.3* gene encoding Ocr, as well as to the collision model for Type I enzymes.

http://rebase.neb.com/cgi-bin/cryyearbar 2017. REBASE crystals. Restriction enzyme structures per year.

http://rebase.neb.com/rebase/rebase.html REBASE. The restriction enzyme database.

http://scop.mrc-lmb.cam.ac.uk SCOP. Structural classification of proteins.

https://www.knaw.nl/en/awards/laureates/dr-h-p-heinekenprijs-voor-biochemie-en-biofy sica/alec-j-jeffreys-1950-groot-brittannia Koninklijke Nederlandse Academie van Wetenschappen site (Royal Dutch Academy of "Arts and Sciences"), which is hosting an Alec J. Jeffreys biography.

https://www.ncbi.nlm.nih.gov/pmc/articles/PMC3831583/pdf/2041-2223-4-21.pdf

https://www.ncbi.nlm.nih.gov/pubmed/20011117 Gitschier J. 2009. DNA fingerprints: an interview with professor Sir Alec Jeffreys.

APPENDIX 1: TABLE OF SELECTED REases STUDIED IN RECENT YEARS

Type II name and restriction site	Sub-type[a]	Details[b]	Reference(s)[c]
AhdI GACNNN↓NNGTC	H P	Hybrid IIP with tetrameric M&S MTase, missing link Type I and II?	Marks et al. 2003; Pingoud et al. 2014
AspCNI GCSGC	P	PLD REase cleaves poorly at high concentrations.	Heiter et al. 2015
BbvCI CCTCAGC (−5/−2)	A T	Two catalytic sites from different subunits, each cleaving own strand; useful as nicking enzyme.	Bellamy et al. 2005; Heiter et al. 2005
BcgI (10/12) CGA(N6)TGC (12/10)	B G H S	Cleaves four bonds in concerted action.	Kong and Smith 1998; Marshall et al. 2007, 2011; Marshall and Halford 2010; Halford 2013; Smith et al. 2013a,b; Pingoud et al. 2014
BcnI CC↓SGG	P	Monomer localizes target site by 1D and 3D diffusion, nicks one DNA strand, turns 180°, and cleaves the second strand; the switch in orientation proceeds without dissociation into bulk solution (Sasnauskas et al. 2011); design of nicking endonucleases (Kostiuk et al. 2011).	Janulaitis et al. 1983; **Sokolowska et al. 2007a**; Kostiuk et al. 2011; Sasnauskas et al. 2011; Kostiuk et al. 2015
BfiI ACTGGG (5/4)	S	PLD. The first non-PD-(D/E)XK restriction enzyme identified (Sapranauskas et al. 2000). Well-studied; carboxy-terminal DBD resembles that of B3-like plant transcription factors; cleaves one strand at a time via covalent intermediate; catalyzes both DNA hydrolysis and transesterification reactions (Sasnauskas et al. 2007).	Sapranauskas et al. 2000; Lagunavicius et al. 2003; Sasnauskas et al. 2003, 2007, 2008, 2010; Zaremba et al. 2004; **Gražulis et al. 2005**; Marshall and Halford 2010; Golovenko et al. 2014; Pingoud et al. 2014

Name / Sequence	Type	Description	References
BisI Gm5CNGC	M	Gm5CNGC; relatives widespread requiring different numbers of m5C.	Xu et al. 2016
BpuSI GGGAC (10/14)	G S	RM-BpuSI has FokI-like carboxy-terminal domain; accompanied by two MTases.	Niv et al. 2007; **Shen et al. 2011**; Sarrade-Loucheur et al. 2013; Pingoud et al. 2014
Bse634I R↓CCGGY	F P	Tetramer. Dimeric mutant still has same fidelity as tetramer. Member of CCGG family.	**Gražulis et al. 2002**; Zaremba et al. 2005, 2006, 2012; Manakova et al. 2012
BspD6I GACTC (4/6)	S	N-BspD6I is a site-specific nickase in a heterodimeric complex with a catalytic subunit.	**Kachalova et al. 2008**
BspMI ACCTGC (4/8)	S	Detection relative orientation of two target sites.	Kingston et al. 2003
BspRI GG↓CC	P	Monomeric REase cuts dsDNA recognition site in two independent binding events.	Rasko et al. 2010
BstYI R↓GATCY	P	Thermophilic REase with overlapping specificity to BamHI and BglII.	Townson et al. 2004, 2005
BsaXI (9/12) ACNNNNNCTCC (10/7)	B	Domain swapping and circular permutation of TRD subdomains (or deletion) → active protein with altered specificity/poor protein yields or inactivity.	Xu et al. 2015
Cfr10I R↓CCGGY	F	Tetramer. Importance of spatial rather than sequence conservation of aa at active center. Member of CCGG family.	**Bozic et al. 1996**; Skirgaila and Šikšnys 1998; Skirgaila et al. 1998; Šikšnys et al. 1999
CglI GCSGC	P	Heterotetramer of R + H with extensive ATP hydrolysis. R protein is similar to BfiI.	Zaremba et al. 2014, 2015

Continued

APPENDIX 1. Table of selected REases studied in recent years (*continued*)

Type II name and restriction site	Sub-type[a]	Details[b]	Reference(s)[c]
DpnI Gm6A↓TC	M P	Methylation-dependent; cuts as monomer, one strand at a time; amino-terminal PD domain and carboxy-terminal winged helix (wH) allosteric activator domain; both domains bind methylated DNA with sequence specificity.	Lacks and Greenberg 1975; **Siwek et al. 2012;** Mierzejewska et al. 2014; Pingoud et al. 2014
Ecl18kl ↓CCNGG	P	The first REase which flips out nucleotides from its target site to accommodate interrupted CCGG sequences in the conserved active site. Transient tetramerization triggers cleavage. Member of CCGG family.	Denjmukhametov et al. 1998; Bochtler et al. 2006; Tamulaitis et al. 2008; Fedotova et al. 2009; Protsenko et al. 2009; Zaremba et al. 2010; Burenina et al. 2013; Rutkauskas et al. 2014
Eco29kl CCGC↓GG	P	GIY-YIG structure.	**Mak et al. 2010;** Pertzev et al. 1997; Ibryashkina et al. 2007; Orlowski and Bujnicki 2008; Mokrishcheva et al. 2011
Eco57I CTGAAG (16/14)	E G S	Accompanied by one MTase (cuts 1½ turn away). Sequence specificity was altered by the methylation activity–based selection technique.	Janulaitis et al. 1992a,b; Rimseliene et al. 2003; Pingoud et al. 2014
EcoRI G↓AATTC	P	The first structure of a restriction enzyme; one of the most extensively studied "classical" restriction enzyme. Studies on inhibition by Cu^{2+} ions; relaxed specificity and structure of mutants that cleave EcoRI star sites; role of flanking sequences; regulation *ecoRIRM* operon.	**McClarin et al. 1986; Kim et al. 1990;** Kurpiewski et al. 2004; Sapienza et al. 2005, 2007; Liu and Kobayashi 2007; Liu et al. 2007; VanderVeen et al. 2008; Ji et al. 2014; Sapienza et al. 2014;

Continued

EcoRII ↓CCWGG	E P	For the first time it was shown that REase can be activated to cleave refractory DNA recognition sites (Kruger et al. 1988). The only demonstrated restriction enzyme which interacts with three recognition sites to effectively cleave one DNA site (Tamulaitis et al. 2006b). The first case of autoinhibition, a mechanism described for many transcription factors and signal transducing proteins (Zhou et al. 2004). Member of CCGG family.	Kruger et al. 1988; Zhou et al. 2002, 2003, 2004; Reuter et al. 2004; Kruger and Reuter 2005; Tamulaitis et al. 2006a,b, 2008; Shlyakhtenko et al. 2007; Gilmore et al. 2009; Golovenko et al. 2009; Szczepek et al. 2009
EcoRV GAT↓ATC	P	One of the most extensively studied "classical" restriction enzymes. Single molecule studies; tracking of single quantum dot labeled EcoRV sliding along DNA manipulated by double optical tweezers indicated that during sliding, EcoRV stays in close contact with the DNA.	**Winkler et al. 1993; Kostrewa and Winkler 1995;** Bonnet et al. 2008; Biebricher et al. 2009
FokI GGATG (9/13)	S	Early, best-known IIS; two MTases fused in single protein; crystals indicate catalytic domain hidden behind DNA-binding domain; DNA-cleavage domain used for engineering purposes.	Sugisaki and Kanazawa 1981; Miller et al. 1985; Nwankwo and Wilson 1987; Mandecki and Bolling 1988; Kaczorowski et al. 1989; Kita et al. 1989a,b; Landry et al. 1989; Looney et al. 1989; Sugisaki et al. 1989; Goszczynski and McGhee 1991; Szybalski et al. 1991; Li et al. 1992, 1993; Li and Chandrasegaran 1993; Skowron et al. 1993, 1996; Waugh and Sauer 1993; Kim et al. 1994, 1996a,b, 1997, 1998; Waugh and Sauer 1994;

APPENDIX 1. Table of selected REases studied in recent years (*continued*)

Type II name and restriction site	Sub-type[a]	Details[b]	Reference(s)[c]
			Yonezawa and Sugiura 1994; **Hirsch et al. 1997**; **Wah et al. 1997**, 1998; Bitinaite et al. 1998; Leismann et al. 1998; Chandrasegaran and Smith 1999; Friedrich et al. 2000; Vanamee et al. 2001; Bibikova et al. 2002; Urnov et al. 2005; Catto et al. 2006, 2008; Gemmen et al. 2006; Bellamy et al. 2007; Miller et al. 2007; Szczepek et al. 2007; Vanamee et al. 2007; Mino et al. 2009, 2014; Mori et al. 2009; Sanders et al. 2009; Zhang et al. 2009; Guo et al. 2010; Imanishi et al. 2010; Klug 2010a,b; Carroll 2011a,b; Gabriel et al. 2011; Halford et al. 2011; Handel and Cathomen 2011; Li et al. 2011; Pattanayak et al. 2011; Ramalingam et al. 2011; Handel et al. 2012; Laurens et al. 2012; Pernstich and Halford 2012; Rusling et al. 2012; Bhakta et al. 2013; Ramalingam et al. 2013; Guilinger et al. 2014a; Pingoud et al. 2014
HinP1I G↓CGC	P	Monomeric REase with structural (but no sequence) similarity to MspI; back-to-back dimer with two active sites and two DNA duplexes bound on the outer surfaces of the dimer facing away from each other. The function of the base flipping is unclear and it seems to be part of great DNA distortions.	Yang et al. 2005; Horton et al. 2006

Enzyme / Recognition sequence	Type	Comments	References
HphI GGTGA (8/7)	S	HNH REase	Cymerman et al. 2006
Hpy188I TCN↓GA	P	GIY-YIG structure with highly specific recognition structure, in contrast with HEases or DNA repair enzymes.	Xu et al. 2000b; Kaminska et al. 2008; Orlowski and Bujnicki 2008; Sokolowska et al. 2011
KpnI GGTAC↓C	P	The first member of a HNH family REase-mediated death benefits population; computer model, but no crystal structure available.	Chandrashekaran et al. 2004; Saravanan et al. 2004; 2007a,b; Gupta et al. 2010; Vasu et al. 2012, 2013; Vasu and Nagaraja 2013
MnlI CCTC (7/6)	S	Member of Type IIS with the HNH-type active site within carboxy-terminal domain, DNA recognition amino-terminal domain.	Kriukiene et al. 2005; Kriukiene 2006
NaeI GCC↓GGC	E	Needs to interact with two copies of the recognition sequence for efficient cleavage of one. Real-time observation of DNA looping dynamics of NaeI and NarI compared.	**Huai et al. 2000**; van den Broek et al. 2006
NarI GG↓CGCC	E	Real-time observation of DNA looping dynamics of NaeI and NarI compared.	van den Broek et al. 2006
Mva1269I GAATGC (1/−1)	S	Monomeric REase with two EcoRI-like and FokI-like catalytic domains; design of nicking endonucleases (Armalyte et al. 2005).	Armalyte et al. 2005; Pingoud et al. 2014
MmeI TCCRAC (20/18)	G S	No separate MTase (cuts two turns away, crystal); changes in the S domain alter recognition site for R and M (like Type I enzymes); altered specificities could be predicted.	Boyd et al. 1986; Tucholski et al. 1995, 1998; Nakonieczna et al. 2007, 2009; Morgan et al. 2008, 2009; Morgan and Luyten 2009; Callahan et al. 2011, 2016; Pingoud et al. 2014

Continued

APPENDIX 1. Table of selected REases studied in recent years (*continued*)

Type II name and restriction site	Sub-type[a]	Details[b]	Reference(s)[c]
NotI GC↓GGCCGC	P	Well-known 8-bp cutter; structure reveals unique metal binding fold (also in other putative endonucleases) with an iron atom within Cys4 motif	Lambert et al. 2008
PabI GTA↓C	P	"half-pipe"; cuts GTA/C; no true REase, as it flips all four purines out of the helix; DNA adenine glycosylase excises adenines.	**Miyazono et al. 2007**, 2014; Ishikawa et al. 2005; Watanabe et al. 2006; Pingoud et al. 2014; Kojima and Kobayashi 2015
PacI TTAAT↓TAA	P	HNH REase; 8–base pair rare-cutting homodimer, each subunit with two Zn^{2+}-bound motifs surrounding a beta-beta-alpha-metal catalytic site.	**Shen et al. 2010**
PspGI ↓CCWGG	P	Use nucleotide flipping as a part of its DNA recognition mechanism. Member of CCGG family.	Pingoud et al. 2003; Szczepanowski et al. 2008; Tamulaitis et al. 2008
SfiI GGCCNNNN↓NGGCC	F P	The first characterized homotetrameric enzyme: Structures reveal two different binding states of SfiI: one with both DNA-binding sites fully occupied and the other with fully and partially occupied sites.	**Wentzell et al. 1995**; Viadiu et al. 2003; Embleton et al. 2004; Vanamee et al. 2005; Bellamy et al. 2007, 2008, 2009; Laurens et al. 2009

Enzyme / Sequence		Description	References
SgrAI CR↓CCGGYG	F P	Filaments by cryoEM; Member of CCGG family; preferentially cleaves concertedly at two sites; assembles into homotetramers, then other molecules join to generate helical structures with one DNA-bound homodimer after another.	Laue et al. 1990; Tautz et al. 1990; Capoluongo et al. 2000; Bitinaite and Schildkraut 2002; Daniels et al. 2003; Hingorani-Varma and Bitinaite 2003; Šikšnys et al. 2004; Wood et al. 2005; Dunten et al. 2008, 2009; Park et al. 2010; Little et al. 2011; **Lyumkis et al. 2013**; Ma et al. 2013b; Horton 2015
SsoII ↓CCNGG	P	Member of CCGG family; tested for applications such as genome surgery.	Kubareva et al. 1992, 2000; Pingoud et al. 2005, 2014; Sudina et al. 2005; Bochtler et al. 2006; Pingoud and Silva 2007; Fedotova et al. 2009; Schierling et al. 2010; Hien le et al. 2011; Abrosimova et al. 2013
SwaI ATTT↓AAAT	P	8–base pair cutter of AT-rich site.	Dedkov and Degtyarev 1998; Shen et al. 2015
TseI G↓CWGC	P	Cuts A:A and T:T mismatch in CAG and CTG repeats (useful for typing in Huntington's disease).	Ma et al. 2013a
TspGWI ACGGA (11/9)	G S	Monomeric bifunctional R-M with independent REase and MTase activities that can be uncoupled; resembles Type I, but lacks translocation domain ("half" Type I).	Skowron et al. 2003; Zylicz-Stachula et al. 2009, 2014
TstI (8/13) CACNNNNNNTCC (12/7)	B	DNA recognition, cleavage, and methylation in one polypeptide.	Smith et al. 2014

Continued

APPENDIX 1. Table of selected REases studied in recent years (continued)

Type II name and restriction site	Sub-type[a]	Details[b]	Reference(s)[c]
Other Types [d]		Structure.	
McrB-N Rm5C(N30-35/)-(N30-3000)-Rm5C	IV	Structure recognition domain McrBC; flips out the modified cytosine.	**Sukackaite et al. 2012**
PvuRts1I mC(N11-13/N9-10)G	IV	Amino-terminal, atypical PD-(D/E)XK REase domain and carboxy-terminal SRA domain with potential pocket for a flipped hm5C or ghm5C. Epigenome studies (Szwagierczak et al. 2011; Wang et al. 2011). This one and other family members could also be classified as Type IIM (like DpnI). Also a Type IIS. Unlike Type IV (McrBC), PvuRts1I family enzymes cut at a fixed distance from the recognition site.	Janosi et al. 1994; Szwagierczak et al. 2011; Wang et al. 2011b; **Kazrani et al. 2014**; **Shao et al. 2014**; Zagorskaite and Sasnauskas 2014
EcoKI (AACNNNNNNTGC) &	I	Structure based on EM analysis.	**Kennaway et al. 2012**
EcoR124I (GAANNNNNNRTCG)	I	Structure HsdR.	**Csefalvay et al. 2015**

Name	Type	Comments	References
EcoP15I (CAGCAG)	III	Structure.	Gupta et al. 2015
MspJI (and other family members, e.g., LpnPI, AspBHI, ...)	IIM, IIS (methylation-dependent)	This is a large family of methylation-dependent (Type IIM and IIS) restriction enzymes. Structures of MspJI (apo- [Horton et al. 2012] and DNA bound, with a flipped base in the SRA domain (Horton et al. 2014c); structures of apo-LpnPI (Sasnauskas et al. 2015b), apo-AspBHI (Horton et al. 2014b) are also available. Base flipping in solution (Zagorskaite and Sasnauskas 2014). Applications in genome methylation studies (Zheng et al. 2010; Cohen-Karni et al. 2011).	Zheng et al. 2010; Cohen-Karni et al. 2011; **Horton et al. 2012**, 2014b,c; Zagorskaite and Sasnauskas 2014; Sasnauskas et al. 2015b

Update September 2017; courtesy of Vilnius group.

a Subtypes as listed on REBASE, with some adaptations by Pingoud et al. (2014) to indicate the overlap between different subtypes.

b The REases that were the first of their (sub)type (adapted from Horton 2015) are indicated in bold: For example, BcnI was the first Type II enzyme to be identified as a monomer (Sokolowska et al. 2007).

c References include older papers to the respective REases, where suitable: For example, SgrAI was the first REase to be shown to form filaments by cryoEM a few years ago (Lyumkis et al. 2013), but the enzyme was already first reported in 1990.

d See Parts B, C, and D within the chapter (on Type I, III, and IV, respectively) for details and other references.

Summary and Conclusions

This book started with experiments in the early 1950s on a barrier to phage infection and "host-controlled variation," which led to the discovery of DNA restriction and modification. Importantly, this modification was reversible and did not lead to mutations. This would herald the end of the distinction between genotype and phenotype and is reminiscent of the current nature versus nurture debate. The next breakthrough came in the shape of the *Escherichia coli* (EcoKI and EcoBI) restriction enzymes, which required *S*-adenosylmethionine (SAM) and ATP for restriction, followed by the discovery of the first enzymes (HindII and EcoRI) that did not require these cofactors. When EcoRI was found to produce staggered DNA ends, which could "restick" in vivo and/or in vitro, genetic engineering was born. Further studies resulted into the division in three types of enzymes: *E. coli*–like (Type I), EcoRI-like (Type II), and phage P1–like (Type III). The 1970s saw a revolution in recombinant DNA technology, while Alec Jeffreys started his analysis of eukaryotic DNA repeats, which would result in the development of the invaluable DNA finger printing technique allowing the solution of paternity cases, the identification of criminals and their victims, and the exoneration of the falsely accused. However, it was also the decade of cloning trouble due to modification-dependent (later named Type IV) restriction enzymes in *E. coli* that destroy cloned methylated DNA from other organisms.

The list of achievements of the 1980s includes DNA sequences of restriction genes (e.g., EcoRI and EcoKI) and many new restriction enzymes derived from different strains. For extensive biochemical and structural analysis, EcoRI and EcoRV became the enzymes of choice, because large amounts ("bathtubs" full) had to be produced for this work. Plasmids and phages such as lambda with mutant restriction enzyme sites helped to reveal the incredible specificity and fidelity of the enzymes for their DNA recognition site. Regular updates by Rich Roberts from 1976 onward of lists of enzymes and their properties eventually led to the REBASE website in the early 1990s. By 1993, nearly 1000 Type II enzymes with 200 specificities had been identified in many bacterial species, which would lead to a subdivision into 11 subtypes in 2003. In 2004 Alfred Pingoud edited a specialized book with 16 chapters dedicated to Type II restriction enzymes (including the first crystal structures), with a single chapter on the ATP-dependent Type I and III "molecular motors"

Chapter doi:10.1101/restrictionenzymes_Summ-Conc

of the SF2 superfamily. Mutant enzymes with longer recognition sequences were high on the wish list with the goal of preparing tools for gene therapy. This proved very difficult, although FokI looked promising with its separate recognition and catalytic cleavage domains, allowing the construction of chimeric fusion proteins. Different constructs of Type II catalytic domains with zinc fingers and improvements using TALE proteins met with varying success, and the off-target activity remained a worry. The new RNA-based method of genome editing using the CRISPR–Cas9 system may present a better alternative but also has its drawbacks, as reported in recent publications. The diverse studies described in this book provide clear evidence for tight control of potential genome alterations, and bacterial populations are not alone in this. This control allows a certain carefully contained level of changes in order to generate heterogeneity within populations, and large eukaryotic genomes have developed similar mechanisms to stabilize the genome. Taken together it is the proper balance between what is best for an individual cell versus what benefits the population and/or organism as a whole. Clearly, microorganisms in bacterial populations, and cells in tissues and individuals, do not allow their DNA to be so easily manipulated as genome engineers wish, and gene targeting will require further investigations and refinements before being of true benefit to human welfare.

The Type II enzymes prove to be incredibly versatile and diverse: Recognition sites can be palindromic, asymmetric, with ambiguities or indifferent internal bases, and/or differentially sensitive to methylated bases; the enzymes might have one or two catalytic sites, cleave DNA in one or two steps, with or without sliding and detaching from the DNA, and with or without looping. Crystal structures in combination with database searches have been useful to build evolutionary trees, which questioned the view held until the mid-1990s that the baffling lack of common features suggested independent convergence and not divergence from a common ancestor. Despite the lack of sequence similarity, the majority of the Type II enzymes have a common catalytic core, mainly with the PD...(D/E)XK motif, but also the HNH and GIY-YIG structural domains, and some the PLD domain. These catalytic domains seem to have been "mixed and matched" during the course of evolution. The structures of the current 50-odd Type II enzymes (and the first Type I and Type III enzymes) are revealing new and unexpected details about the mechanisms employed to prevent indiscriminate restriction: The catalytic cleavage domain may be hidden behind the recognition domain requiring a large conformational change dictated by the recognition domain or the methyltransferase to access the DNA; it may require dimerization (or multimerization) and/or two unmodified recognition sites; and it may require distortions, bending, or contortions of the DNA to properly position the two nucleolytic sites. A variety of other types of control of restriction emerged over the years and involve transcription regulation, control by "C" proteins, or the cognate methyltransferase (in the case of Type II systems). Various plasmids and phages employ DNA mimics and other proteins to inhibit Type I enzymes. Rather spectacular was the finding that the restriction subunit of EcoKI was degraded by the bacterial host's ClpXP protease *during* translocation of the EcoKI complex (but not before), when modification was impaired. Such extraordinary control of restriction was in sharp contrast to that of Type II enzymes, which destroy their own host DNA under similar circumstances. This led to the debate on recognition of "self" versus "non-self." The "pro-self" camp stated that the restriction enzyme would not only destroy incoming DNA, but it would also enhance the frequency of horizontal transfer of restriction-modification systems by generating recombinogenic free DNA ends in the cell. These ideas are not mutually exclusive.

By 1993, only two dozen Type I and III enzymes were known, but this changed with the advent of whole-genome sequencing projects, which would lead to the identification of many more (putative) Type II restriction enzymes and to the realization that Type I and III enzymes are quite common in bacteria and archaea, like Type II enzymes. The evolution of the Type I DNA specificity genes became a hot topic in the 1980s, as new specificities could be generated via homologous recombination, unequal crossing-over, and transposition. Decades later this finding has become a finding of great importance to understand life-threatening bacterial infections in humans and other organisms. In 1988 Bill Studier proposed the collision model for Type I enzymes based on his work with phage T7, which proved to be correct in the following years and is apparently not limited to Type I enzymes. Extensive modeling of Type I enzymes led to the tentative conclusion for a common ancestor with one monomeric recognition domain and a separate catalytic domain for methylation, whereas the ATP-dependent molecular motor domains of Type I and III enzymes were assigned to the SF2 helicase superfamily, but proved to be translocases that do not open up the double helix. The breakthrough in 2012 concerned the structures of the Type I EcoKI and EcoR124I enzymes, which in the absence of crystals relied on single-molecule studies, and computer-assisted EM single-particle reconstructions. In 2015, the first structure of a Type III enzyme was published, that of EcoP15I, which indicated a division of labor of the two modification subunits: one for DNA recognition, the other for methylation. This threw light on the differential usage of ATP by Type III enzymes, which involves a large conformational change. The first long-awaited structures led to a comparison of the Type I enzymes with the Type IIB and IIG enzymes: A Type IIB is a motorless Type I system, whereas a Type IIG would be a half of a motorless Type I system.

Of interest also are the Type IIG enzymes that are combined Type I–like restriction-modification enzymes that are either SAM-dependent or at least stimulated by SAM (like Type III enzymes). The major conformational change to DNA or protein, or both, to reposition the catalytic cleavage site of many Type II enzymes is apparently not limited to Type II enzymes and appears to be dictated in Type I and III enzymes by the MTase and also by cofactor SAM. The necessity for large conformational changes before cleavage indicates that perhaps the Type I, II, and III enzymes are not so different after all.

The new class of Type IV enzymes, defined in 2003 as modification-dependent restriction enzymes, recognizes a wide variety of DNA modifications at cytosine or adenine residues, and these enzymes have become important tools for research into DNA modifications in all kingdoms. The Type IV enzymes are highly diverse—a mix and match of various cleavage and DNA recognition domains. The Type IV enzymes present models to study eukaryotic modifications, and their role in the study of epigenetic phenomena will be obvious. Identifying new Type IV enzymes is of great importance for these studies, but unfortunately their genes are not easy to detect in whole-genome sequences.

The role of restriction and modification enzymes in bacterial pathogens and also in bacteria and archaea in our gut (the "microbiome") has become a topic of great interest to the medical field. The presence of different Type I enzymes allows typing of different "non-typeable" pathogenic strains and should also be useful to analyze diet- or disease-induced changes in the microbiome. Restriction enzymes are linked to virulence via phase variation, which may be a common strategy to create phenotypic heterogeneity, in order to hide from the host immune system, or survive environmental changes as a population. Phase variation can occur via hypermutation at simple repeat sequences, or homopolymeric tracts located within the reading frame or promoter region in a subset of genes. This switches promoters "on" or "off" and causes frameshifts and/or alternative usage of translation initiation codons in different reading frames. And, obviously, if this switch affects a methyltransferase, this will have major effects on the methylome.

What will the future bring? More crystal structures may bring more surprises and remain a model for the much more complex eukaryotic DNA-recognition-cum-restriction-and/or-modification complexes such as that of, for example, the Xeroderma pigmentosum disease, or other complexes involved in DNA repair, recombination, or replication. The studies on phase variation will continue, with emphasis on the impact on microorganisms in the human world outside the laboratory, especially those of pathogens, and those in the gut. Of interest are some reports that need to be followed up: Type I enzymes may be linked to stress responses *via* associated anticodon nucleases or become phosphorylated, whereas both Type I and IV enzymes have been reported to be able to cut a replication fork. An important issue that still needs to be resolved is that of maintenance versus de novo methylation of EcoKI. Although EcoKI has a preference for methylation of hemimethylated DNA, in the presence of the small lambda Ral protein, the enzyme efficiently methylates recognition sites with either one or no methyl groups (i.e., it changes the enzyme from a maintenance to a de novo methyltransferase). Noreen Murray managed to generate EcoKI* mutants with Ral-independent de novo methylation. Interestingly, such mutants have a single high-affinity SAM-binding site in contrast to the wild-type enzyme, which has two high-affinity sites. How the ability to bind only one SAM molecule properly results in a de novo enzyme, without apparently affecting the methylation reaction as such, requires further investigation and remains a subject close to the author's heart.

The History of Restriction Enzymes
October 19–21, 2013 Meeting Program

The following pages show the program for the October 19–21, 2013 meeting on The History of Restriction Enzymes held at Cold Spring Harbor Laboratory.[1] Videos and slides from the meeting are available at http://library.cshl .edu/meetings/restriction-enzymes/program.php.

Meeting photos: (*top left*) Stu Linn, Matt Meselson; (*top right*) Rich Roberts, Thomas Kelly; (*bottom left*) Bruce Stillman, Ham Smith; (*bottom right*) Mila Pollock, John Rosenberg, Herb Boyer.

---

[1]Courtesy Cold Spring Harbor Laboratory Archives.

Chapter doi:10.1101/restrictionenzymes_AppA

Cold Spring Harbor Laboratory
Genentech Center for the History of
Molecular Biology and Biotechnology

The History of Restriction Enzymes
October 19–21, 2013

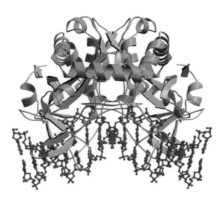

Structure of the BglI dimer bound to DNA containing its recognition sequence, with a crystallographic two-fold running vertically through the centre of the complex.

Grace Auditorium
Cold Spring Harbor Laboratory
Cold Spring Harbor, NY

The History of Restriction Enzymes
October 19–21, 2013

Organizers: Herb Boyer, Stu Linn, Mila Pollock and Rich Roberts

Saturday, October 19

7:30pm Session 1: The Beginnings of the Field

Chairperson: Stu Linn

Tom Bickle: Variations on a Theme, the families of restriction/modification enzymes

Werner Arber: Microbial Genetics is at the origin of molecular genetics

Ham Smith: Discovery of the first Type II restriction enzyme and its aftermath

Ludmila Pollock: Illuminating the history of science, the importance of scientific archives

Bruno Strasser: Restriction enzymes: between nature, culture and politics

Sunday, October 20

9:00am Session 2: The Restriction Enzyme field begins to grow
Chairperson: David Dryden

Matt Meselson: The discovery of EcoKI

Herb Boyer: The discovery of EcoRI and its uses in recombinant DNA

Clyde Hutchison: Restriction Enzymes and DNA sequencing
Ken Horiuchi: Work in Norton Zinder's lab and in Japan

11:00am Session 3: Restriction Enzymes become commercial reagents
Chairperson: Srinivasan Chandrasegaran
Rich Roberts: Many more REs at CSHL, the start of REBASE and more recent work

11:30am Conversations

Ira Schildkraut: First commercial sales and the start of NEB

Jack Chirikjian: Starting Bethesda Research Laboratories

Bill Linton: The early days of Promega

Arvydas Janulaitis: Science and politics: three phases of commercialization at Fermentas

Sunday, October 20, continued

2:00pm   **Session 4: Cloning and sequencing of RM systems**
    Chairperson: Ham Smith
    Geoff Wilson: The cloning efforts at NEB
    Arvydas Lubys: the cloning efforts at Fermentas
    Rick Morgan: The MmeI family and engineering opportunities

4:00pm   **Session 5: The biochemistry of restriction enzymes**
    Chairperson: V. Nagaraja
    Steve Halford: Type II restriction enzymes: searching for one site and then two
    Alfred Pingoud: Sequence specific recognition and engineering
    Andrzej Piekarowicz: H.influenzae and N.gonorrhoeae RM systems and their biological implications

Monday, October 21

9:00am   **Session 6: Structural Studies**
    Chairperson: Herb Boyer
    John Rosenberg: EcoRI structure
    Aneel Aggarwal: The current state of structural studies
    Xiaodong Cheng: Methylase and RE strucures

11:00am   **Session 7: Short Contributions and Summary**
    Chairperson: Rich Roberts

    Short solicited contributions
    Robert Yuan
    Lise Raleigh
    David Dryden
    Ichizo Kobayashi

    Other contributors
    Bill Studier
    Gary Wilson
    Paul Roy

    Stu Linn: Summarization and Close of Meeting

The History of Restriction Enzymes

As enzymes that cleave DNA at specific recognition sites, the Type II restriction enzymes are among the workhorses of molecular biology, genetics, and biotechnology. Following their discovery in 1970, a flurry of activity led to their use on many fronts including mapping Restriction Fragment Length Polymorphisms, and laid the basis for studies of molecular evolution. More than 3,000 restriction enzymes are known today of which more than 625 are commercially available.

Although the phenomenon of restriction was already well known from the work of Luria, Bertani, and others in the 1950s, it was not until the late 1960s that the existence of restriction enzymes was shown genetically and biochemically. The work of Arber, Dussoix, Linn, Meselson, and Yuan throughout the 1960s led to the identification of the enzymes they called "restriction endonuclease," later classified as Type I restriction enzymes.

In 1970, two papers from Hamilton Smith's lab at Johns Hopkins University described a new type of restriction enzyme found in *Haemophilus influenza,* strain RD, now know as Type II restriction enzymes. Kathleen Danna and Dan Nathans first showed the usefulness of this enzyme in 1972, by making a physical map of SV40 DNA. This encouraged many groups to look for more restriction enzymes with different specialties and within a few years more than 100 such enzymes had been described. A database of these enzymes, now called REBASE, was started in 1975, and it was in that year that the first restriction enzymes were made commercially available by New England BioLabs.

The current meeting on the History of Restriction Enzymes, is the first meeting to bring together the scientists who were involved with the discoveries and research on restriction enzymes dating back to the 1950s to the present time. The speakers from around the world include scientists who made key discoveries in the field and their students and collaborators, as well as historians of science and archivists from the institutions where the groundbreaking work on restriction enzymes was carried out.

 Cold Spring Harbor Laboratory is a world-renowned research and education institution with research programs focusing on cancer, neurobiology, plant genetics, genomics and bioinformatics.

Speaker	Affiliation
Aneel Aggarwal	Mount Sinai School of Medicine, New York, NY, USA
Werner Arber	Emeritus Professor, Biozentrum Basel, University of Basel, Switzerland
Tom Bickle	Emeritus Professor, Biozentrum Basel, University of Basel, Switzerland
Herb Boyer	Formerly at University of California, San Francisco, CA, USA
Srinivasan Chandrasegaran	Johns Hopkins University, Baltimore, MD, USA
Xiaodong Cheng	Emory University School of Medicine, Atlanta, GA, USA; University of Texas MD Anderson Cancer Center, Houston, TX, USA
Jack Chirikjian[1]	Georgetown University, Washington, DC, USA
David Dryden	Edinburgh University until 2013; currently at Durham University, Department of Biosciences, Durham, UK
Stephen Halford	Emeritus Professor, Bristol University, UK
Ken Horiuchi	Emeritus Professor, National Institute of Genetics, Japan; Senior Research Associate, Rockefeller University, New York, NY, USA
Clyde Hutchison	Distinguished Professor Synthetic Biology Group, J. Craig Venter Institute, San Diego, CA, USA; Kenan Professor Emeritus, University of North Carolina, Chapel Hill, NC, USA
Arvydas Janulaitis	Institute of Biotechnology, Vilnius, Lithuania
Ichizo Kobayashi	University of Tokyo, Minato-ku, Tokyo, Japan
Stu Linn	Emeritus Professor, University of California, Berkeley, CA, USA
Bill Linton	Promega, Madison, WI, USA
Arvydas Lubys	Institute of Biotechnology, Vilnius, Lithuania
Matt Meselson	Harvard University, Cambridge, MA, USA
Rick Morgan	New England Biolabs, Ipswich, MA, USA
Andrzej Piekarowicz	University of Warsaw, Warsaw, Poland
Alfred Pingoud	Justus-Liebig-University, Giessen, Germany
Mila Pollock	Executive Director CSHL Library and Archives, CSHL, Cold Spring Harbor, NY, USA
Lise Raleigh	New England Biolabs, Ipswich, MA, USA
Rich Roberts	Chief Scientific Officer, New England Biolabs, Ipswich, MA, USA
John Rosenberg	University of Pittsburgh, Pittsburgh, PA, USA
Ira Schildkraut	Scientist Emeritus, New England Biolabs, Ipswich, MA, USA
Ham Smith	Head, Synthetic Biology Group, J. Craig Venter Institute, San Diego, CA, USA
Bruno Strasser	Université de Genève, Geneva, Suisse
Bill Studier	Brookhaven National Laboratory, Upton, NY, USA
Geoff Wilson	New England Biolabs, Ipswich, MA, USA
Robert Yuan	Faculty, Osher Lifelong Learning Institute, Lewes, DE, USA

[1] Sadly, Jack passed away in September 2018.

Modern-Day Applications
of Restriction Enzymes

In this book, the focus has been on the history of the development of the four types of REases with respect to their genetics, structure, and function, both in vivo and in vitro. Initially Type II REases such as EcoRI, HindIII, BamHI, and PstI have been used for genetic engineering of phage and plasmid vectors because of the presence or absence of their recognition sites in these phages and plasmids. These vectors have been extensively used for cloning, subcloning, DNA mapping, synthesis of large genetic scaffolds, and the study of chromatin structures and dynamics. In this appendix, S. Hong Chan discusses modern day applications of REases and nicking endonucleases (NEases) in molecular biology. Some of these REases (e.g., EcoP15I, FokI, MmeI, and NotI) have been discussed in detail in this book and are listed in Appendix 1 of Chapter 8. Many other REases can be found on the REBASE website (http://rebase.neb.com/rebase/rebase.html).

APPLICATIONS OF RESTRICTION ENDONUCLEASES: MOLECULAR CLONING AND BEYOND

Siu-Hong Chan

New England Biolabs, Inc.

From Molecular Cloning to Gene Assembly

Molecular Cloning

Restriction endonucleases and DNA ligases together facilitate a robust "cut and paste" workflow in which a defined DNA fragment (derived from cDNA or a cloned fragment) can be moved from one organism to another. The vehicles for cloning, plasmid vectors, were also created using this simple "cut and paste" methodology; original vectors, such as pSC101 and pBR322 (Cohen 2013), have gone through numerous generations of cutting and pasting with modules

---

Chapter doi:10.1101/restrictionenzymes_AppB

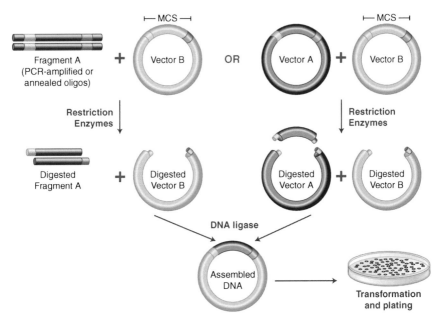

FIGURE 1. Traditional cloning workflow. Using PCR, restriction sites are added to both ends of a dsDNA, which is then digested by the corresponding REases. The cleaved DNA can then be ligated to a plasmid vector cleaved by the same or compatible REases with T4 DNA ligase. DNA fragments can also be moved from one vector into another by digesting with REases and ligating to compatible ends of the target vector. (Courtesy of NEB.)

to become the backbone of many present-day vectors. For example, inserting promoters and origins of replication of eukaryotic viruses into these bacteriophage-derived plasmids has generated shuttle vectors, which are functional in both prokaryotic and eukaryotic cells. This cloning workflow, joined by DNA amplification technologies, such as PCR and RT-PCR, has facilitated the study of the molecular mechanisms of life (see Fig. 1).

In vitro DNA Assembly Technologies

Synthetic biology is a rapidly growing field in which defined components are used to create biological systems with precise control over the processes involved for the study of biological processes and the creation of useful biological devices (Ellis et al. 2011). Novel technologies such as BioBrick and USER Enzyme emerged to facilitate the building of such biological systems. Recently, more robust approaches such as Golden Gate Assembly, NEBuilder HiFi DNA Assembly, and Gibson Assembly have been widely adopted by the synthetic biology community. These approaches allow for parallel and seamless assembly of multiple DNA fragments without resorting to nonstandard bases.

BioBrick

The BioBricks community sought to create thousands of standardized parts for quick gene assembly (Stephanopoulos 2012). The BioBricks framework, together with the annual International Genetically Engineered Machines (iGEM) competition (www.igem.org), has elicited great interest from university and high school students around the world and helped inspire a whole new generation of synthetic biology scientists. Based on the traditional REase-ligation methodology, however, BioBrick and its derivative methodologies (such as BglBricks; Anderson et al. 2010) introduce scar sequences at the junctions and require multiple cloning cycles to create a working biological system.

USER Enzyme

USER (Uracil-Specific Excision Reagent) Enzyme is one of the first scar-less cloning technologies. It exploits the action of uracil DNA glycosylase and a pyrimidine lyase at a uracil incorporated into the PCR products through the primers (Nour-Eldin et al. 2010). USER Enzyme can therefore generate $3'$ overhangs of custom sequences. Annealing of complementary overhangs allows multiple pieces of DNA to join together simultaneously and in order.

Golden Gate Assembly

Golden Gate Assembly and its derivative methods (Engler et al. 2008; Sarrion-Perdigones et al. 2011) exploit the ability of Type IIS REases to cleave DNA outside of the recognition sequence. The inserts and the cloning vectors are designed to place the Type IIS recognition site distal to the cleavage site, such that the Type IIS REase can remove the recognition sequence from the assembly. The advantages of such arrangement are threefold: (1) The overhang sequence created is not dictated by the REase and therefore no scar sequence is introduced; (2) the fragment-specific sequence of the overhangs allows orderly assembly of multiple fragments simultaneously; and (3) as the restriction site is eliminated from the ligated product, digestion and ligation can be carried out simultaneously. The net result is the ordered and seamless assembly of DNA fragments in one reaction. The accuracy of the assembly is dependent on the length of the overhang sequences. Therefore, Type IIS REases that create four-base overhangs (such as BsaI/BsaI-HF v2, BsmBI, Esp3I, BbsI/BbsI-HF, and EarI) are preferred. The downside of these Type IIS REase-based methods is

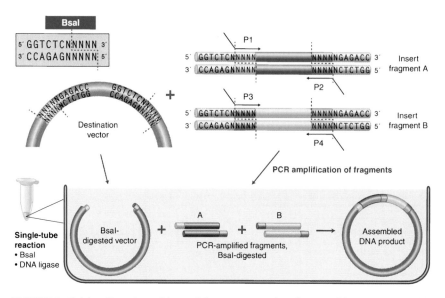

FIGURE 2. Golden Gate Assembly workflow. In its simplest form, Golden Gate Assembly requires a BsaI recognition site (GGTCTC) added to both ends of a dsDNA fragment distal to the cleavage site, such that the BsaI site is eliminated by digestion with BsaI. Upon cleavage, the overhanging sequences of the adjoining fragments anneal to each other. DNA ligase then seals the nicks to create a new covalently linked DNA molecule. Multiple pieces of DNA can be cleaved and ligated simultaneously. (Courtesy of NEB.)

that the small number of overhanging bases can lead to misligation of fragments with similar overhang sequences (Engler et al. 2009). Research has identified ligase bias on mismatch ligation sites to help guide the design of ligation junctions for high-fidelity assembly (Potapov et al. 2018a,b). As in REase-based cloning, it is also necessary to verify that the Type IIS REase sites used are not present in the fragments. Nonetheless, Golden Gate Assembly is a robust technology that generates multiple site-directed mutations (Yan et al. 2012) and assembles multiple DNA fragments into large contigs (Scior et al. 2011; Werner et al. 2012). As open source methods and reagents have become increasingly available (see www.addgene.org), Golden Gate Assembly has been widely used in the construction of genetic circuits (Halleran et al. 2018; Kong et al. 2017) among other applications. (See Fig. 2.)

Gibson Assembly, NEBuilder HiFi DNA Assembly, and Exonuclease-Based DNA Assembly Methods

Named after Daniel G. Gibson, Gibson Assembly is a robust exonuclease-based method to assembly DNA seamlessly and in sequence under isothermal conditions. It takes advantage of three complementary enzymatic activities to achieve a one-pot assembly of multiple pieces of DNA into a large contig: a 5' exonuclease generates long 3' overhangs, a polymerase fills in the gaps of the annealed single-stranded regions, and a DNA ligase seals the nicks of the annealed sequences (Gibson et al. 2009).

In addition to gene assembly, this and other commercially available assembly technologies (NEBuilder HiFi DNA Assembly from New England Biolabs and In-Fusion from Takara) can also be used for cloning; the assembly of a DNA insert with a linearized vector, followed by transformation, can be completed within a few hours. Other applications of these gene assembly methods include introduction of multiple mutations, assembly of plasmid vectors from chemically synthesized oligonucleotides, and creation of combinatorial libraries of genes and pathways. Reviews of DNA assembly methods are available in the literature (Merryman and Gibson 2012; Casini et al. 2015). (See Fig. 3.)

From DNA Mapping to Chromatin Structural Dynamics

With only a handful of REases available in the early 1970s, Kathleen Danna in Daniel Nathan's group mapped the functional units of (simian virus) SV40 DNA (Danna and Nathans 1972) and thus commenced the era of eukaryotic gene mapping and comparative genomes. It has since evolved into sophisticated methodologies that allow the detection of single-nucleotide polymorphisms (SNPs) and insertions/deletions (indels) (Kudva et al. 2004), driving technologies that have facilitated genome-wide studies such as the mapping of epigenetic marks and chromatin structural dynamics and population-wide research such as population genomics of traits and genetic disorders.

Construction of DNA Libraries

Serial analysis of gene expression (SAGE) has been widely used to identify mutations in cancer research and study gene expression in transcriptome research. REases are key to creating ditags and concatamers in SAGE-type analyses. NlaIII is instrumental as an anchoring enzyme because of its unique property of recognizing a 4-bp sequence CATG and creating a 4-nt overhang of the same sequence. The use of Type IIS enzymes that cleave even further away from the recognition sequence as tagging enzyme allows the creation of longer ditags for higher information content of SAGE analyses: FokI and BsmFI in SAGE (Velculescu et al. 1995), MmeI in LongSAGE (Høgh and Nielsen 2008), and EcoP15I in SuperSAGE (Matsumura et al. 2012) and DeepSAGE (Nielsen 2008).

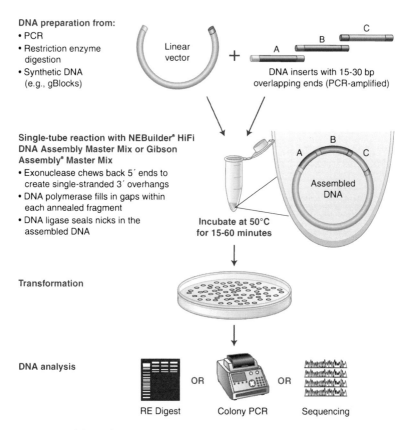

FIGURE 3. NEBuilder and Gibson Assembly workflow. NEBuilder and Gibson Assembly both employ three enzymatic activities in a single-tube reaction: 5′ exonuclease, the 3′-extension activity of a DNA polymerase, and DNA ligase activity. The 5′-exonuclease activity chews back the 5′-end sequences and exposes the complementary sequence for annealing. The polymerase activity then fills in the gaps on the annealed regions. A DNA ligase then seals the nick and covalently links the DNA fragments together. The overlapping sequence of adjoining fragments is much longer than those used in Golden Gate Assembly and therefore results in a higher percentage of correct assemblies. (Courtesy of NEB.)

Although REases do not allow for the random fragmentation of DNA that most deep sequencing technologies require, they are used in target enrichment methodologies: hairpin adaptor ligation (Singh et al. 2011) and HaloPlex enrichment (Agilent). The long-reach REase AcuI and USER Enzyme were used to insert tags into sample DNA, which was then amplified using rolling circle amplification to form long single-stranded DNA "nanoballs" that served as template in a high-density ChIP-based sequencing by ligation methodology developed (Drmanac et al. 2010). ApeKI is used to generate DNA library for a genotyping-by-sequencing technology for the study of sequence diversity of maize (Elshire et al. 2011).

Mapping Epigenetic Modifications

REases have an extraordinary ability to discriminate the methylation status of the target bases. This property has been exploited to map modified bases within a genomic context. Before the advent of deep sequencing technologies, a two-dimensional gel electrophoresis–based mapping technique called restriction landmark genome scanning (RLGS) used NotI (GC^GGCCGC), AscI (GG^CGCGCC), EagI (C^GGCCG), or BssHII (G^CGCGC) to interrogate changes in the methylation patterns of the genome during development of normal and cancer cells. Methylation-sensitive amplification polymorphism (MSAP) takes advantage of the differential sensitivity of MspI and HpaII toward the methylation status of the second C in the sequence CCGG to map m5C, hm5C, and 5-glucosyl hydroxymethylcytosine (Reyna-López et al. 1997; Davis and Vaisvila 2011; Mastan et al. 2012). The REases that recognize and cleave DNA at 5-mC or 5-hmC sites, such as MspJI, FspEI, and LpnPI, are also potential tools for high-throughput mapping of the cytosine epigenetic markers in complex genomes (Cohen-Karni et al. 2011; Wang et al. 2011).

Point-of-Care DNA Amplification and Detection Using Nicking Endonucleases

By generating sequence- and strand-specific nicks on dsDNA, nicking endonucleases (NEases) open the door to applications that cannot be achieved by REases. In the presence of a strand-displacing DNA polymerase such as Bst DNA polymerase, the 3'-hydroxyl end of the nicked site can be extended for hundreds of nucleosides. Because the NEase site is regenerated, repeated nicking–extension cycles result in amplification of specific single-strand segments of the sample DNA without the need for thermocycling. Nicking enzyme–based isothermal DNA amplification technologies such as rolling circle amplification, NESA, EXPAR, and related amplification schemes have been shown to be capable of detecting very low levels of DNA (Dawson et al. 2009; Murakami et al. 2009). Similar schemes have been incorporated into molecular beacon technologies to amplify the signal (Li et al. 2008). The implementation of these sample and/or signal amplification schemes can lead to simple but sensitive and specific methods for the detection of target DNA molecules at point of care (e.g., NEAR, SDA, and EXPAR). This procedure is amenable to multiplexing and can potentially achieve higher fidelity than PCR. The combined activity of NEases and Bst DNA polymerase have also been used to introduce site-specific fluorescent labels into long/chromosomal DNA in vitro for visualization (nano-coding) (Kounovsky-Shafer et al. 2017). A general review of NEases and their applications has been published (Chan et al. 2011), and an excellent review of NEase-based DNA amplification and detection technologies and their application in molecular diagnostics is also available (Niemz et al. 2011).

Study of the Spatial Structure of Chromatin

Chromosome conformation capture (3C)-based technologies, such as 3C, 4C, 5C, and Hi-C, have revealed the important role of spatial proximity of genomic regions in genome rearrangements in cancer cells and the regulation of gene expression in general (Grob and Cavalli 2018). The core of the technologies—namely, 3C—involves the reversible formaldehyde crosslinking of DNA to proteins in the vicinity. After crosslinking, a six-base cutting restriction enzyme, normally HindIII or EcoRI, is used to cleave the contiguous genomic DNA into smaller units cross-linked to proteins. These DNA fragments, putatively spatial neighbors organized by the cross-linked proteins, are then ligated in such a way that intramolecular ligations are favored. For Hi-C-based technologies, the HindIII-cleaved DNA is filled in with biotiny-lated dA before ligation to allow for enrichment of the neighboring DNA downstream of the process. After the protein cross-links are reversed, the ligated DNA can be subjected to arrays of manipulations for loci-focused analysis (3C, 4C, or 5C) or deep-sequencing (Hi-C). Higher resolution of interacting regions can be achieved by using more frequent cutting restriction enzymes such as DpnII, MboI, and Sau3AI (four-base cutters) (Belaghzal et al. 2017).

Mapping of Open Chromatin Regions

The mammalian genome is largely packaged into chromatin consisting primarily of DNA and histones. Chromatin undergoes remodeling events that include switching between closed and open conformations to provide access to regulatory factors such as transcription factors. Hence, open chromatin profiling can provide information on the active regions of the genome under specific conditions. Tn5 transposase and DNase I–based sequencing methods (ATAC-seq and DNase-Seq, respectively) have been applied to map opened chromatin regions by virtue of the accessibility of the opened regions (Song and Crawford 2010; Buenrostro et al. 2013). Recently, a frequent nicking enzyme Nt.CviPII (recognition sequence = CCD, D being A, G, or T) has been used to establish the NicE-seq (nicking enzyme–assisted sequencing) method for open chromatin profiling at single-nucleotide resolution (Ponnaluri et al. 2017). In addition to being applicable to both native and formaldehyde-fixed cells,

NicE-Seq has been shown to require a lower sequencing burden than DNase hypersensitive and ATAC-seq sites.

Genome Editing

At the infancy of genome editing, REases and homing endonucleases were the only available tools for creating double-strand breaks at nonspecific locations of the genome of higher organisms for transgenesis (Ishibashi et al. 2012a,b). In the 2000Ts, engineered enzymes such as zinc-finger nucleases (ZFNs) and transcription activator–like effector nucleases (TALENs) were developed and allowed genome editing operations, such as gene knockout and knock-in. The challenge, however, is the selection and screening of appropriate cleavage sites and the engineering of the enzyme for specific target sites.

In 2012, two seminal papers describing the adaptation of the bacterial CRISPR/CRISPR-associated systems (Cas) as RNA-guided DNA endonucleases (Gasiunas et al. 2012; Jinek et al. 2012) set off a new era of genome editing. These CRISPR/Cas systems can be adapted to use a single-guide RNA to direct the endonuclease activity to a target site within a complex genomic context both in vitro and in vivo. It is relatively simple to design and screen for appropriate guide RNA sequences and there is no need to engineer the enzyme. The simplicity and elegance of the CRIPSR–Cas systems has democratized genome editing, gene expression manipulation (by using catalytically inactive CRISPR–Cas proteins), and even RNA targeting (e.g., the Cas13 systems [Abudayyeh et al. 2017] and the prokaryotic Argonaute proteins [Dayeh et al. 2018]). The application of the CRISPR–Cas systems has been extensively reviewed (Nuñez et al. 2016; Huang et al. 2018; Knott and Doudna 2018). Comparative reviews of the three major genome editing methods are also available in the literature (Guha and Edgell 2017; Jaganathan et al. 2018; Yang and Wu 2018).

Looking Forward

Although restriction endonucleases have been one of the major forces that transformed molecular biology in the past decades, novel technologies such as Gibson Assembly, NEBuilder HiFi DNA Assembly, and Golden Gate Assembly continue to extend our ability to create new DNA molecules in vitro and CRISPR–Cas to edit and manipulate genomes in vivo. Restriction endonucleases and nicking endonucleases, meanwhile, have found new applications beyond molecular cloning in the age of deep sequencing and molecular diagnostics; genome-wide mapping of epigenetic marks, chromatin structural dynamics, and isothermal amplification and detection of genetic markers are all exciting and invaluable tools in the study of the molecular biology of life.

As new technologies and new tools emerge, these highly sequence-specific endonucleases may find even more unique and exciting applications that help us understand the molecular mechanisms of life.

REFERENCES

Abudayyeh OO, Gootenberg JS, Essletzbichler P, Han S, Joung J, Belanto JJ, Verdine V, Cox DBT, Kellner MJ, Regev A, et al. 2017. RNA targeting with CRISPR-Cas13. *Nature* **550:** 280–284.

Anderson JC, Dueber JE, Leguia M, Wu GC, Goler JA, Arkin AP, Keasling JD. 2010. BglBricks: a flexible standard for biological part assembly. *J Biol Eng* **4:** 1.

Belaghzal H, Dekker J, Gibcus JH. 2017. Hi-C 2.0: An optimized Hi-C procedure for high-resolution genome-wide mapping of chromosome conformation. *Methods* **123:** 56–65.

Buenrostro JD, Giresi PG, Zaba LC, Chang HY, Greenleaf WJ. 2013. Transposition of native chromatin for fast and sensitive epigenomic profiling of open chromatin, DNA-binding proteins and nucleosome position. *Nat Methods* **10:** 1213–1218.

Casini A, Storch M, Baldwin GS, Ellis T. 2015. Bricks and blueprints: methods and standards for DNA assembly. *Nat Rev Mol Cell Biol* **16:** 568–576.

Chan S-H, Stoddard BL, Xu S-Y. 2011. Natural and engineered nicking endonucleases—from cleavage mechanism to engineering of strand-specificity. *Nucleic Acids Res* **39:** 1–18.

Cohen-Karni D, Xu D, Apone L, Fomenkov A, Sun Z, Davis PJ, Morey Kinney SR, Yamada-Mabuchi M, Xu S-y, Davis T, et al. 2011. The MspJI family of modification-dependent restriction endonucleases for epigenetic studies. *Proc Natl Acad Sci* **108:** 11040–11045.

Cohen SN. 2013. DNA cloning: a personal view after 40 years. *Proc Natl Acad Sci* **110:** 15521–15529.

Danna KJ, Nathans D. 1972. Bidirectional replication of Simian Virus 40 DNA. *Proc Natl Acad Sci* **69:** 3097–3100.

Davis T, Vaisvila R. 2011. High sensitivity 5-hydroxymethylcytosine detection in Balb/C brain tissue. *J Vis Exp* **48:** e2661.

Dawson ED, Taylor AW, Smagala JA, Rowlen KL. 2009. Molecular detection of *Streptococcus pyogenes* and *Streptococcus dysgalactiae* subsp. equisimilis. *Mol Biotechnol* **42:** 117–127.

Dayeh DM, Cantara WA, Kitzrow JP, Musier-Forsyth K, Nakanishi K. 2018. Argonaute-based programmable RNase as a tool for cleavage of highly-structured RNA. *Nucleic Acids Res* **46:** e98.

Drmanac R, Sparks AB, Callow MJ, Halpern AL, Burns NL, Kermani BG, Carnevali P, Nazarenko I, Nilsen GB, Yeung G, et al. 2010. Human genome sequencing using unchained base reads on self-assembling DNA nanoarrays. *Science* **327:** 78–81.

Ellis T, Adie T, Baldwin GS. 2011. DNA assembly for synthetic biology: from parts to pathways and beyond. *Integr Biol (Camb)* **3:** 109–118.

Elshire RJ, Glaubitz JC, Sun Q, Poland JA, Kawamoto K, Buckler ES, Mitchell SE. 2011. A Robust, simple genotyping-by-sequencing (GBS) approach for high diversity species. *PLoS One* **6:** e19379.

Engler C, Gruetzner R, Kandzia R, Marillonnet S. 2009. Golden Gate shuffling: a one-Pot DNA shuffling method based on Type IIs restriction enzymes. *PLoS One* **4:** e5553.

Engler C, Kandzia R, Marillonnet S. 2008. A one pot, one step, precision cloning method with high throughput capability. *PLoS One* **3:** e3647.

Gasiunas G, Barrangou R, Horvath P, Siksnys V. 2012. Cas9-crRNA ribonucleoprotein complex mediates specific DNA cleavage for adaptive immunity in bacteria. *Proc Natl Acad Sci* **109:** E2579–E2586.

Gibson DG, Yong L, Chuang R, Venter JC, Hutchison CA 3rd. 2009. Enzymatic assembly of DNA molecules up to several hundred kilobases. *Nat Methods* **44:** 343–345.

Grob S, Cavalli G. 2018. Technical review: a hitchhiker's guide to chromosome conformation capture. *Methods Mol Biol* **1675:** 233–246.

Guha TK, Edgell DR. 2017. Applications of alternative nucleases in the age of CRISPR/Cas9. *Int J Mol Sci* **18:** 2565.

Halleran AD, Swaminathan A, Murray RM. 2018. Single day construction of multigene circuits with 3G assembly. *ACS Synth Biol* **7:** 1477–1480.

Høgh AL, Nielsen KL. 2008. SAGE and LongSAGE. *Methods Mol Biol* **387:** 3–24.

Huang C-H, Lee K-C, Doudna JA. 2018. Applications of CRISPR–Cas enzymes in cancer therapeutics and detection. *Trends Cancer* **4:** 499–512.

Ishibashi S, Kroll KL, Amaya E. 2012a. Generating transgenic frog embryos by restriction enzyme mediated integration (REMI). In *Methods in molecular biology (Clifton, N.J.)*, Vol. 917, pp. 185–203. Springer, New York.

Ishibashi S, Love NR, Amaya E. 2012b. A simple method of transgenesis using I-Sce I meganuclease in *Xenopus*. In *Methods in molecular biology (Clifton, N.J.)*, Vol. 917, pp. 205–218. Springer, New York.

Jaganathan D, Ramasamy K, Sellamuthu G, Jayabalan S, Venkataraman G. 2018. CRISPR for crop improvement: an update review. *Front Plant Sci* **9:** 985.

Jinek M, Chylinski K, Fonfara I, Hauer M, Doudna JA, Charpentier E. 2012. A programmable dual-RNA-guided DNA endonuclease in adaptive bacterial immunity. *Science* **337:** 816–821.

Knott GJ, Doudna JA. 2018. CRISPR–Cas guides the future of genetic engineering. *Science* **361:** 866–869.

Kong DS, Thorsen TA, Babb J, Wick ST, Gam JJ, Weiss R, Carr PA. 2017. Open-source, community-driven microfluidics with Metafluidics. *Nat Biotechnol* **35:** 523–529.

Kounovsky-Shafer KL, Hernandez-Ortiz JP, Potamousis K, Tsvid G, Place M, Ravindran P, Jo K, Zhou S, Odijk T, de Pablo JJ, et al. 2017. Electrostatic confinement and manipulation of DNA molecules for genome analysis. *Proc Natl Acad Sci* **114:** 13400–13405.

Kudva IT, Griffin RW, Murray M, John M, Perna NT, Barrett TJ, Calderwood SB. 2004. Insertions, deletions, and single-nucleotide polymorphisms at rare restriction enzyme sites enhance discriminatory power of polymorphic amplified typing sequences, a novel strain typing system for *Escherichia coli* O157:H7. *J Clin Microbiol* **42:** 2388–2397.

Li JJ, Chu Y, Lee BY-H, Xie XS. 2008. Enzymatic signal amplification of molecular beacons for sensitive DNA detection. *Nucleic Acids Res* **36:** e36.

Mastan SG, Rathore MS, Bhatt VD, Yadav P, Chikara J. 2012. Assessment of changes in DNA methylation by methylation-sensitive amplification polymorphism in *Jatropha curcas* L. subjected to salinity stress. *Gene* **508:** 125–129.

Matsumura H, Urasaki N, Yoshida K, Krüger DH, Kahl G, Terauchi R. 2012. SuperSAGE: powerful serial analysis of gene expression. In *Methods in molecular biology (Clifton, N.J.)*, Vol. 883, pp. 1–17. Springer, New York.

Merryman C, Gibson DG. 2012. Methods and applications for assembling large DNA constructs. *Metab Eng* **14:** 196–204.

Murakami T, Sumaoka J, Komiyama M. 2009. Sensitive isothermal detection of nucleic-acid sequence by primer generation-rolling circle amplification. *Nucleic Acids Res* **37:** e19.

Nielsen KL. 2008. DeepSAGE: higher sensitivity and multiplexing of samples using a simpler experimental protocol. *Methods Mol Biol* **387:** 81–94.

Niemz A, Ferguson TM, Boyle DS. 2011. Point-of-care nucleic acid testing for infectious diseases. *Trends Biotechnol* **29:** 240–250.

Nour-Eldin HH, Geu-Flores F, Halkier BA. 2010. USER cloning and USER fusion: the ideal cloning techniques for small and big laboratories. *Methods Mol Biol* **643:** 185–200.

Nuñez JK, Harrington LB, Doudna JA. 2016. Chemical and piophysical modulation of Cas9 for tunable genome engineering. *ACS Chem Biol* **11:** 681–688.

Ponnaluri VKC, Zhang G, Estève P-O, Spracklin G, Sian S, Xu S-Y, Benoukraf T, Pradhan S. 2017. NicE-seq: high resolution open chromatin profiling. *Genome Biol* **18:** 122.

Potapov V, Ong JL, Kucera RB, Langhorst BW, Bilotti K, Pryor JM, Cantor EJ, Canton B, Knight TF, Evans TC, et al. 2018a. Comprehensive profiling of four base overhang ligation fidelity by T4 DNA Ligase and application to DNA assembly. *ACS Synth Biol* **7:** 2665–2674.

Potapov V, Ong JL, Langhorst BW, Bilotti K, Cahoon D, Canton B, Knight TF, Evans TC, Lohman GJS. 2018b. A single-molecule sequencing assay for the comprehensive profiling of T4 DNA ligase fidelity and bias during DNA end-joining. *Nucleic Acids Res* **46:** e79.

Reyna-López GE, Simpson J, Ruiz-Herrera J. 1997. Differences in DNA methylation patterns are detectable during the dimorphic transition of fungi by amplification of restriction polymorphisms. *Mol Gen Genet* **253:** 703–710.

Sarrion-Perdigones A, Falconi EE, Zandalinas SI, Juárez P, Fernández-del-Carmen A, Granell A, Orzaez D. 2011. GoldenBraid: an iterative cloning system for standardized assembly of reusable genetic modules. *PLoS One* **6:** e21622.

Scior A, Preissler S, Koch M, Deuerling E. 2011. Directed PCR-free engineering of highly repetitive DNA sequences. *BMC Biotechnol* **11:** 87.

Singh P, Nayak R, Kwon YM. 2011. Target-enrichment through amplification of hairpin-ligated universal targets for next-generation sequencing analysis. *Methods Mol Biol* **733:** 267–278.

Song L, Crawford GE. 2010. DNase-seq: a high-resolution technique for mapping active gene regulatory elements across the genome from mammalian cells. *Cold Spring Harb Protoc* **2010:** pdb.prot5384.

Stephanopoulos G. 2012. Synthetic biology and metabolic engineering. *ACS Synth Biol* **1:** 514–525.

Velculescu VE, Zhang L, Vogelstein B, Kinzler KW. 1995. Serial analysis of gene expression. *Science* **270:** 484–487.

Wang H, Guan S, Quimby A, Cohen-Karni D, Pradhan S, Wilson G, Roberts RJ, Zhu Z, Zheng Y. 2011. Comparative characterization of the PvuRts1I family of restriction enzymes and their application in mapping genomic 5-hydroxymethylcytosine. *Nucleic Acids Res* **39:** 9294–9305.

Werner S, Engler C, Weber E, Gruetzner R, Marillonnet S. 2012. Fast track assembly of multigene constructs using Golden Gate cloning and the MoClo system. *Bioeng Bugs* **3:** 38–43.

Yan P, Gao X, Shen W, Zhou P, Duan J. 2012. Parallel assembly for multiple site-directed mutagenesis of plasmids. *Anal Biochem* **430:** 65–67.

Yang H, Wu Z. 2018. Genome editing of pigs for agriculture and biomedicine. *Front Genet* **9:** 360.

Index

Page references followed by f denote figures, those followed by t denote tables.

About the Author

Reading Biology and Chemistry (BSc) and Molecular Genetics (MSc) at the University of Leiden led Wil A.M. Loenen to a lifelong passion for DNA. Aided by British Council Scholarships and an EEC grant, she obtained a PhD in the Faculty of Science at the University of Leicester. Under supervision of Bill Brammar, she made a widely used cloning vector (SCI 389), lambda L47.1, and cloned PstI. After various postdocs working with microorganisms and bacterial viruses in the United States, United Kingdom, and The Netherlands, including an important period with Noreen Murray at the University of Edinburgh, Loenen switched to immunology at the Netherlands Cancer Institute. Among others she was involved in the cloning and analysis of the human CD27 cDNA, cloned the human CD27 gene, and murine CD27. This led to a model on the role of CD27 in lymphomagenesis in a second PhD in the Faculty of Medicine at Leiden University Medical Center (with Kees Melief and Bob Lowenberg as advisors), a guest-editorship of Seminars in Immunology, and contributions to the online cytokine database of Academic Press. For a while she worked with Cathrien Bruggeman at Maastricht University and wrote a review on cytomegalovirus.

The seeds for this book were planted in Edinburgh at the farewell party ending the long professorship of Noreen Murray. With Murray's and David Dryden's help, Loenen wrote her first historical perspective on restriction enzymes, in this case celebrating 50 years of EcoKI, the *E. coli* K12 enzyme. This led to further reviews on restriction enzymes, as well as the role of *S*-adenosylmethionine (SAM). SAM proves to be not only an essential cofactor for EcoKI and relatives, but also an ancient, highly versatile cofactor to a myriad of other compounds, with a pivotal role in health, disease, and aging. Leiden University Medical Center has hosted Loenen for many years, allowing her access to e-mail and library facilities while working from home in nearby Leiderdorp. She is active in Women's Science Networks, keeping up with long-term friends around the world, enjoys reading, BBB (bee-, bird-, and butterfly-friendly) gardening, and last, but not least, beach walking with dog Sam.

Selected Publications

Restriction

Loenen WAM, Brammar WJB. 1980. A bacteriophage lambda vector for cloning large DNA fragments made with several restriction enzymes. *Gene* **10:** 249–259.

Loenen WAM, Murray NE. 1986. Modification enhancement by the restriction alleviation protein (Ral) of bacteriophage lambda. *J Mol Biol* **190:** 11–22.

Loenen WAM, Daniel AS, Braymer HD, Murray NE. 1987. Organization and sequence of the *hsd* genes of *Escherichia coli* K-12. *J Mol Biol* **198:** 159–170.

Loenen WAM. 2003. Tracking EcoKI and DNA fifty years on: a golden story full of surprises. *Nucleic Acids Res* **31:** 7059–7069.

Loenen WAM, Raleigh EA. 2014. The other face of restriction: modification-dependent enzymes. *Nucleic Acids Res* **42:** 56–69.

Loenen WAM, Dryden DTF, Raleigh EA, Wilson GG. 2014. Type I restriction enzymes and their relatives. *Nucleic Acids Res* **42:** 20–44.

Loenen WAM, Dryden DTF, Raleigh EA, Wilson GG, Murray NE. 2014. Highlights of the DNA cutters: a short history of the restriction enzymes. *Nucleic Acids Res* **42:** 3–19.

CD27

Camerini D, Walz G, Loenen WAM, Borst J, Seed B. 1991. The T cell activation antigen CD27 is a member of the nerve growth factor/tumor necrosis factor receptor gene family. *J Immunol* **147:** 3165–3169.

Loenen WAM, Gravestein LA, Beumer S, Melief CJ, Hagemeijer A, Borst J. 1992. Genomic organization and chromosomal localization of the human CD27 gene. *J Immunol* **149:** 3937–3943.

Gravestein LA, Blom B, Nolten LA, de Vries E, van der Horst G, Ossendorp F, Borst J, Loenen WAM. 1993. Cloning and expression of murine CD27: comparison with 4-1BB, another lymphocyte-specific member of the nerve growth factor receptor family. *Eur J Immunol* **23:** 943–950.

Loenen WAM. 1997. "Molecular analysis of CD27 to elucidate the role in lymphocyte development." PhD thesis, Faculty of Medicine, RUL NL.

Loenen WAM. 1998. CD27 and (TNFR) relatives in the immune system: their role in health and disease. *Sem Immunol* **10:** 417–422. Guest Editor.

Loenen WAM, Bruggeman CA, Wiertz EJ. 2001. Immune evasion by human cytomegalovirus: lessons in immunology and cell biology. *Sem Immunol* **13:** 41–49.

SAM

Loenen WAM. 2006. *S*-adenosylmethionine: jack of all trades and master of everything? *Biochem Soc Trans* **34:** 330–333.

Loenen WAM. 2010. *S*-adenosylmethionine: simple agent of methylation and secret to aging and metabolism? In *Epigenetics of aging* (ed. Tollefsbol TO), pp. 107–131. Springer, Berlin.

Loenen WAM. 2017. *S*-adenosylmethionine: a promising avenue in anti-aging medicine? In *Anti-aging drugs: from basic research to clinical practice* (ed. Vaiserman AM), Chap. 18. Royal Society of Chemistry, London.

Loenen WAM. 2018. *S*-adenosylmethionine metabolism and aging. In *Epigenetics of aging and longevity* (ed. Moskalev A, Vaiserman A), pp. 59–93. Elsevier, New York.